So Simple a Beginning

So Simple a Beginning

HOW FOUR PHYSICAL PRINCIPLES SHAPE OUR LIVING WORLD

Raghuveer Parthasarathy

PRINCETON UNIVERSITY PRESS
Princeton and Oxford

Copyright © 2022 by Princeton University Press

Princeton University Press is committed to the protection of copyright and the intellectual property our authors entrust to us. Copyright promotes the progress and integrity of knowledge. Thank you for supporting free speech and the global exchange of ideas by purchasing an authorized edition of this book. If you wish to reproduce or distribute any part of it in any form, please obtain permission.

Requests for permission to reproduce material from this work should be sent to permissions@press.princeton.edu

Published by Princeton University Press
41 William Street, Princeton, New Jersey 08540
6 Oxford Street, Woodstock, Oxfordshire OX20 1TR

press.princeton.edu

All Rights Reserved

Library of Congress Cataloging-in-Publication Data

Names: Parthasarathy, Raghuveer, 1976– author.
Title: So simple a beginning : how four physical principles shape our living world / Raghuveer Parthasarathy.
Description: Princeton : Princeton University Press, [2022] | Includes bibliographical references and index.
Identifiers: LCCN 2021026503 (print) | LCCN 2021026504 (ebook) | ISBN 9780691200408 (hardback) | ISBN 9780691231617 (ebook)
Subjects: LCSH: Biophysics. | BISAC: SCIENCE / Life Sciences / Biophysics | SCIENCE / Life Sciences / Anatomy & Physiology (see also Life Sciences / Human Anatomy & Physiology)
Classification: LCC QH505 .P37 2022 (print) | LCC QH505 (ebook) | DDC 571.4—dc23
LC record available at https://lccn.loc.gov/2021026503
LC ebook record available at https://lccn.loc.gov/2021026504

British Library Cataloging-in-Publication Data is available

Editorial: Jessica Yao, Ingrid Gnerlich, and Maria Garcia
Production Editorial: Natalie Baan
Text Design: Carmina Alvarez
Jacket Design: Jessica Massabrook
Production: Jacquie Poirier
Publicity: Kate Farquhar-Thomson and Sara Henning-Stout
Copyeditor: Jennifer McClain

This book has been composed in Charis

Printed on acid-free paper. ∞

Printed in the United States of America

1 3 5 7 9 10 8 6 4 2

In memory of my parents,
Kalyani and Sampath Parthasarathy

Contents

Introduction   1

Part I: The Ingredients of Life

1	DNA: A Code and a Cord	15
2	Proteins: Molecular Origami	29
3	Genes and the Mechanics of DNA	44
4	The Choreography of Genes	60
5	Membranes: A Liquid Skin	75
6	Predictable Randomness	90

Part II: Living Large

7	Assembling Embryos	105
8	Organs by Design	124
9	The Ecosystem inside You	134
10	A Sense of Scale	155

| 11 | Life at the Surface | 173 |
| 12 | Mysteries of Size and Shape | 183 |

Part III: Organisms by Design

13	How We Read DNA	199
14	Genetic Combinations	220
15	How We Write DNA	239
16	Designing the Future	260

	Acknowledgments	277
	References	279
	Index	317

So Simple a Beginning

Introduction

How does life work? This question may seem overwhelming, or even preposterous. How could any answer do justice to both a sprinting cheetah and a stationary tree, to the unique *you* along with the trillions of bacteria that live inside you? The experiences of even a single organism are breathtakingly varied: consider a chick's emergence from its egg, the first flap of its wings, the racing of its heart at the sight of a fox, and its transformation of food and water into eggs of its own. Could any intellectual framework encompass all of this?

The search for an answer—for some kind of unity amid the diversity of life—is reflected in our ancient urge to categorize living things based on similarities of appearance or behavior. Aristotle partitioned animals into groups using attributes such as laying eggs or bearing live young. Ancient Indian texts applied a variety of classifiers, including, similarly, manner of origin: "those born from an egg, those born from an embryonic sac, those born from moisture, and those born from sprout." Modern taxonomy emerged from the eighteenth-century work of Carl Linnaeus, who systematized the naming of organisms and developed a hierarchical classification scheme based on shared characteristics that we continue to find useful. Classification in itself, however, is not very satisfying. We want to know the *why*, not just the *what*, of the commonalities that unify living things.

In this book, we look for that *why* through the lens of physics, revealing a surprising elegance and order in biology. Of course, this isn't the only perspective that offers deep insights into life. There is the viewpoint of biochemistry, with which we understand how atoms join together to form the molecular components of organic matter, how energy is deposited in and extracted from chemical bonds, and how the incessant flux of matter and energy in chemical reactions constitutes the metabolism of living things. But it is difficult to use chemistry alone to zoom out from the scale of molecules to the scales of the animals and plants around us, or even the scale of single cells, and make sense of shape and form.

Another all-encompassing perspective is that of evolution. Since the mid-nineteenth-century epiphanies of Charles Darwin and Alfred Russel Wallace, we can see the traits of living creatures as manifestations of deeper historical processes. Similarities, whether of visible characteristics of anatomy or more hidden patterns in DNA sequences, can reflect shared ancestry with which we can deduce a tree of relationships linking all of life together. Differences emerge due to random chance and the varied pressures on survival imposed by creatures' environments; again, present forms reflect past history. Evolution provides a powerful framework for understanding life. It is not, however, one that we focus on in this book. In part, this is because there is already a large popular literature on the subject. More importantly, however, evolutionary principles alone don't illuminate the *why* as much as the *how*.

To illustrate what I mean by "why," consider the swim bladder, a pair of gas-filled sacs possessed by many, but not all, species of fish. Comparing creatures both extant and extinct reveals this organ's evolutionary history, with connections to the emergence of lungs in air-breathing animals that Darwin himself remarked upon. Understanding the function of a swim bladder, however, requires a bit of physics: the low density of the enclosed gas offsets the high density of bone in bony fishes, allowing the animal to maintain the same average density as its watery surroundings and thereby easily position itself at whatever depth it likes. A swim bladder is just one solution to the challenge of matching density. The fish might instead contain large amounts of low-density oil, or a skeleton composed of cartilage rather than bone, both of which are strategies adopted by sharks, which lack a swim bladder. The last common ancestor of cartilaginous and bony fish lived over 400 million years ago. Since then, the distinct evolutionary paths of the two groups have led to different solutions to the shared physical challenges of aquatic locomotion. We can state, with a point of view echoed throughout this book, that understanding the why of these anatomical features, related to control of density, highlights a hidden unity that fish share that transcends their evolutionary divergence. We should keep in mind, however, that the machinery of variation and natural selection—the enhanced odds of survival that accrued over generations to those creatures better able to navigate their aquatic world—provides the paths by which the forms we see arise.

There are other vantage points besides those of biochemistry and evolution from which to survey the breadth of life. Rather than list all the approaches we won't be exploring, however, let's turn to the one we will.

I've already hinted at the view of nature the rest of this book expands upon, which I identify as *biophysical*. The term implies a unification of biology and physics. It encapsulates the notion that the substances, shapes, and actions that constitute life are governed and constrained by the universal laws of physics, and that illuminating the connections between physical rules and biological manifestations reveals a framework upon which the dazzling variety of life is built. The notion of universality is central to the utility of physics, and to its appeal. The same principles of gravity apply to an apple falling from a tree and to planets orbiting the sun, and current work aims to further expand this framework to encompass the strange behavior of the quantum world. Biophysics extends to the living world the quest for unity that lies at the heart of physics.

To say that living things obey the laws of physics may seem trivial. After all, organisms are made up of the same fundamental particles that make up everything else and are therefore governed by the same rules. But one might expect the explicit role of physics to be over after physical forces set up the formation of atoms and molecules, with complex chemistry giving shape to further molecular rearrangements and the idiosyncratic predilections of cells and organisms being responsible for larger features. This is, however, incorrect. Just as physical forces direct the intricate branching of frost on a winter window and the rhythmic curves of vast desert dunes, and do so in ways that don't require subatomic particles for their explanation, physical mechanisms shape life at all scales. One of the great triumphs of physics, especially over the last half century, has been an understanding of how broad rules arise in all sorts of natural phenomena, clearing the underbrush of complexity to reveal deep principles. Magnets, for example, become nonmagnetic if heated above some specific "critical" temperature; though magnets can be made of many different elements and alloys, each with their own unique atomic-scale structure, the magnetic field of every magnet decays with exactly the same form as it approaches its critical temperature. Being a three-dimensional arrangement of interacting atoms, it turns out, suffices to determine the consequences of these interactions, regardless of atomic details. As another example, consider a shaken container of mixed nuts. One typically finds that the larger nuts rise to the top, giving this well-known phenomenon its name: the Brazil nut effect. The effect isn't particular to nuts, of course, and occurs in mixtures of cereal grains, rocks on riverbeds, and any collection of agitated, disordered objects. Its explanation involves general notions of what are called granular flows, and the ways in which any ensemble of colliding particles must create and fill in interstitial spaces in order to move.

Biophysics applies this quest for broadly applicable physical rules to the world of living things. This endeavor, though still incomplete, has already been far more successful than we might have dreamed even a few decades ago. Using physics, we can understand the bursting of DNA from viruses, fundamental limits on the speed of thought, and the regular spacing of our vertebrae. We can apply our insights to grow organs on slabs of plastic and read genomes using pulses of light. We uncover a simplicity and an elegance in the living world that is otherwise hidden. Simplicity emerges because a handful of principles rather than a morass of detail suffices for many explanations; elegance because of the unity shared by the living and nonliving world. This is an unusual point of view; I hope the pages to come will convince you of it.

Every quest for unity amid complexity risks the pitfall of hubris, however. There is the temptation to ignore the lessons that variety provides, or to force motley data into unreasonably simple frameworks. A physical perspective is especially prone to these missteps, perhaps because of the elegance of its theories and perhaps because of their historical successes. Despite being a physicist myself, I note that the caricature of physicists as blithely trampling, elephant-like, through adjacent fields of inquiry without adequately appreciating the treasures underfoot is not wholly inaccurate. Though this book is a celebration of biophysics, I'll describe some of its stumbles as well; chapter 12 in particular examines contentious issues of metabolism against which a biophysical approach may have failed.

. . .

What are the physical principles that govern living things? We could refer to laws related to fundamental forces, thermodynamics, probability, and so on, amenable to precise mathematical formulation. While rigorous, this would be rather dry, and would moreover obscure the overarching lessons that biophysicists have drawn from nature. Instead, I direct our attention to four concepts or motifs that arise repeatedly in biophysical explorations.

The first is *self-assembly*, the idea that the instructions for building with biological components—whether molecules, cells, or tissues—are encoded in the physical characteristics of the components themselves. It may seem obvious that an organism contains its own instructions. After all, one doesn't need to carve a tree into a tree shape or paste five arms on a starfish; the creatures organize their own forms. Their internal instructions, however, need not take the form of a task list written into one set of components and executed by another. Rather, the physical characteristics of biological materials often *are* the instructions. Features like size and shape can guide the arrangement of pieces into a larger whole, as can less visible attributes such as electrical charge, harnessing the laws of physics.

I'll illustrate with an example. If you've ever blown soap bubbles and watched them come together, you may have noticed that there's never a junction at which more than three bubbles meet. Four adjoining bubbles may look like the drawing on the left of the figure below (page 6), with boundaries like a bent letter H, but never like the drawing on the right, with boundaries like an X. Physical forces drive soap films to minimize their surface area, leading to incontrovertible rules for

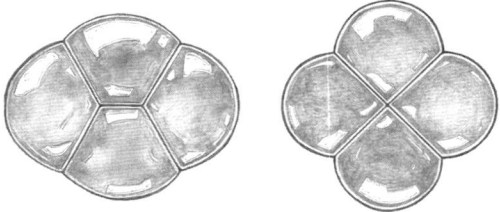

sets of bubbles that have been appreciated since their nineteenth-century elaboration by Belgian physicist Joseph Plateau. These rules prohibit any junction of four bubbles, as such a junction could never be part of a minimal-area surface. The arrangements of bubbles aren't haphazard. No external hand, however, is needed to guide them into their stereotyped pattern; the rules for their organization are embedded in their physical nature. For well over a century, scientists have noticed that arrangements of adjoining cells in all sorts of tissues resemble the arrangements of soap bubbles, and have investigated whether this is coincidence or a reflection of similar underlying mechanisms. In 2004, for example, Takashi Hayashi at the University of Tokyo and Richard Carthew at Northwestern University looked at the cluster of photoreceptor cells situated in each of a fruit fly's compound eyes. Normally, there are four, with exactly the same arrangement as four soap bubbles. Using mutant flies that developed 1, 2, 3, 5, and 6 photoreceptor cells per group, they found the same arrangements that one finds in assemblies of 1, 2, 3, 5, and 6 adjoining soap bubbles. The fly, it seems, relies on general physical mechanisms of surface area minimization to organize these crucial cells of its retina. Rather than painstakingly positioning cells, the fly makes the cells and lets them sort out their contacts, minimize their areas, and pattern themselves on their own. The cells, like the soap bubbles, assemble themselves. In countless other contexts as well, we similarly find that structure isn't drawn explicitly into the blueprints of an organism; rather, nature places the raw materials at the site and trusts that the laws of physics will put them together properly. Thankfully, the laws of physics are reliable workers.

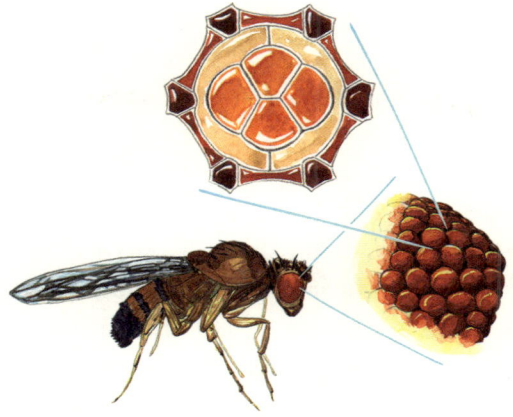

The second recurring motif is that of a *regulatory circuit*. The ubiquity of computers makes us familiar with the idea that machines can use rules of logic to transform inputs into outputs, making decisions based on signals from sensors or controllers. We're also comfortable with the idea that living creatures, ourselves included, make behavioral choices based on the stimuli in their environment, though the details of the computations are more mysterious. We'll see that decision-making circuitry is not just a feature of the large-scale world but is manifested in the microscopic activities of life's molecules, built in to their very structure and modes of interaction. The wet, squishy building blocks of life assemble into machines that can sense their environment, perform calculations, and make logical decisions.

A migrating cell in a developing embryo, for example, must stop its wandering when it reaches the appropriate destination, a decision determined in part by assessing the mechanical stiffness of the neighboring tissue. Cells adhere using proteins that jut out from their surfaces, and through these proteins they can tug on their surroundings. Some adhesion proteins can serve as sensors as well as anchors, with these two roles inexorably linked: for stiff surroundings, the protein molecules are stretched, as your arm would be if tugging on a thick tree branch from a few feet away; for soft surroundings, the proteins are bent, as your arm would be if pulling a towel on a clothesline, easily dragged toward you. The cell contains other components that can bind to sites on the adhesion protein only if those sites are exposed, which occurs only if the molecule is stretched—imagine the inside of your elbow, accessible as you tug on the tree but not the towel. This binding triggers events that culminate in the cell's decision to stop its wandering. The physical conformation of the protein, therefore, underpins a cell-scale machine that senses, calculates, and decides.

Our third concept is that of *predictable randomness*. The physical processes underlying the machinery of life are fundamentally random but, paradoxically, their average outcomes are reliably predictable. In the nonliving world, randomness is central to activities as diverse as the shuffling of cards and the collisions of gas molecules. Physics has long tackled the question of how robust features emerge from underlying chaos. We know, for example, why steady, consistently colored light shines from stars despite their churning interiors, and how energy can be extracted from the violent combustion of gasoline. The microscopic world is subject to incessant, vigorous, and fundamentally random motion that DNA and other cellular components must deal with, and even exploit. We can deduce the probable outcomes of random processes, which in many cases provide simple explanations of superficially complex phenomena. A virus reaching a cell that it may infect, for example, doesn't need to think (even if it were capable of thought) about how to find the specific surface proteins to which it can bind; it is buffeted by random forces that drag it everywhere, ensuring that its chaotic trajectory will intersect its target. Your immune system also makes use of randomness, generating an enormous variety of receptor proteins that might, by chance, recognize invaders that have never before been encountered. We devote all of chapter 6 to the randomness of microscopic motion, which finds echoes in discussions of genes and traits where randomness is also built into the way life works.

Our final recurring biophysical motif is that of *scaling*, the idea that physical forces depend on size and shape in ways that determine the forms accessible to living, growing, and evolving organisms. That size, shape, and physics are related is well appreciated for artificial structures. It's hard to build big buildings, for example. Before the advent of steel frames and other modern inventions, to attempt great heights or large interior spaces was to tempt collapse, as the weight of a structure could overwhelm the support its walls could provide. Simply scaling up a small building, maintaining the proportionality of its dimensions, fails. In modern language that we elaborate in chapter 10, gravity and other forces *scale* with size in different ways that we need to account for when designing buildings. Scaling concepts are similarly reflected in the sizes and shapes of animals but extend to much more than mechanical concerns. Scaling illuminates aspects of living forms, from the existence of lungs to (perhaps) the rate of our metabolism.

These four themes don't exist in isolation but can interact with and even depend on each other, as we'll see in the chapters to come. The precision of biological circuits often depends on the statistics of random motion. Random motion nudges the positions of biological components to facilitate their self-assembly. Self-assembly into larger structures is subject to the dictates of scaling laws. All these processes and principles together create the explanatory framework of biophysics.

· · ·

Understanding life brings with it the ability to influence life. This isn't in itself a new insight. Our knowledge of the immune system and the behavior of microorganisms, among other topics, has enabled us to triumph over a multitude of diseases that ravaged humanity in the past. In the twentieth century alone, for example, more than 300 million people died of smallpox, a disease that has now vanished thanks to the invention of vaccines. Our knowledge of genetics, biochemistry, and many other subjects lets us coax plants and animals to produce enough food for over seven billion people, four times as many as inhabited the planet just a hundred years ago. In recent years, we've learned how to alter organisms at their core, directly reading the information carried in genomes and rewriting it to modify form and function. As we'll see, these contemporary advances required taking seriously a biophysical view of life, acknowledging the tangible, physical character of DNA and other molecules to design tools that quite literally push, pull, cut, and connect life's pieces.

A biophysical perspective also helps us make sense of the implications of these new biotechnologies and the difficult choices they bring. We'll encounter, for example, methods to engineer the extinction of the mosquitoes that spread malaria, dengue fever, and other diseases, bringing to mind both the dismal legacy of human-induced extinctions and the uplifting histories of past eradications of disease. The decision whether to deploy such methods requires understanding how they work and how they differ from past tools. At a more personal level, our ability to read our own genetic code brings with it the prediction of likelihoods of various illnesses in ourselves or in our children; our nascent ability to edit genomes offers the chance to alter these likelihoods. What would it mean to alter the genome of an unborn child to try to avoid cystic fibrosis, or cancer, or depression? Whether to take such an action is both a deeply personal decision and one with serious ethical and societal implications. Making such decisions can, and should, be aided by an understanding of what genes, genomes, cells, and organisms actually are, and the processes that shape the relationships among them. As we'll see, the physical nature of life's materials, as well as fundamental issues related to randomness and uncertainty, influence what we can and cannot do with our new technologies.

・・・

Our exploration of biophysical themes includes examples spanning the variety of life. We consider the normal workings of organisms, including ourselves, as well as the pitfalls of disease and the intersections of biology and technology. In part I ("The Ingredients of Life"), our journey begins inside cells. We delineate the pieces that make up living things, materials like DNA and proteins that also exemplify a sort of universality, as they make up every living thing ever discovered. The molecular characters in this first part of the story will likely be familiar from high school biology, but we focus on the physical traits that guide their functions. We find stiff strands of DNA, two-dimensional liquids that define cell boundaries, and three-dimensional sculptures made of single molecules. In part II ("Living Large"), we expand our horizons to look at communities of cells, including embryos, organs, and the consortia of bacteria that live inside each of us. We also explore scaling relationships that govern the shapes of animals and plants, revealing why an elephant can never be as athletic as an antelope. In part III ("Organisms by Design"), we return to the microscopic world of DNA, but now, having developed deeper connections between molecules and organisms, we tackle the genome. We learn what it means to read, write, and edit DNA, learn how nature itself pointed us toward the tools that make these feats possible, and examine the opportunities and challenges these technologies present for our future.

As interesting as these topics and examples may be, their cumulative effect is greater than the sum of their parts. Biophysics transforms the way we look at the world. At the end of *On the Origin of Species*, Darwin writes:

> There is grandeur in this view of life, with its several powers, having been originally breathed into a few forms or into one; and that, whilst this planet has gone cycling on according to the fixed law of gravity, from so simple a beginning endless forms most beautiful and most wonderful have been, and are being, evolved.

I hope to convince you that Nature has a grandeur even deeper than what Darwin discerned. Rather than a contrast between the fixed, clockwork laws of physics and the generation of endless and beautiful forms, the two are inextricably linked. We can identify the crucial "simple beginning" not as the origin of life, nor the formation of our planet, but as the primeval emergence of the physical laws that characterize our universe. The influence of these laws on life didn't end billions of years ago, but rather shaped and continues to shape all the wonderful forms around us and within us. To discern simplicity amid complexity and to draw connections between life's diverse phenomena and universal physical concepts gives us a deeper appreciation of ourselves, our fellow living creatures, and the natural world that we inhabit. I hope you'll agree.

PART I

The Ingredients of Life

1
DNA: A Code and a Cord

A beige gelatinous slab speckled with bacterial colonies hangs in the National Portrait Gallery in London. The bacteria contain copies of DNA from the artwork's subject, Nobel laureate John Sulston. Though Sulston's friends probably wouldn't recognize the likeness, the artist, Marc Quinn, notes that the piece "is the most realistic portrait in the Portrait Gallery" because "it carries the actual instructions that led to the creation of John."

These days even small children are taught that DNA somehow makes you "you," setting the color of your eyes, the shape of your nose, your fondness for cilantro, and more. We've grown accustomed to the idea that DNA encodes instructions that govern us, but what does "encodes" actually mean?

We begin our biophysical exploration of the machineries and mechanisms of life with DNA; it is iconic and familiar, yet abstract in many descriptions we encounter. Understanding how instructions are embedded in DNA occupies us for a few chapters, as we introduce proteins, genes, and networks of interactions among these constituents. Scientists' conceptions of how these pieces form both a message and the means of reading this message have evolved over the past few decades, and DNA's intricacies are still being unraveled. Recent years have seen the development of breathtaking ways to manipulate the code contained in DNA, for ends not yet fully imagined, a subject we examine in part III. In this chapter, we focus on DNA alone, which already allows us to introduce connections between biology and physics, and between science and technology.

FOUR VIEWS OF DNA

More than just a set of abstract instructions, DNA is a substance with shape and structure whose physical properties are intimately tied to its functions. What *is* this substance? Is it solid or liquid, stiff or floppy, compressed or relaxed? DNA is multifaceted, and we can focus on different aspects of it depending on what we care about. Here are four views of DNA.

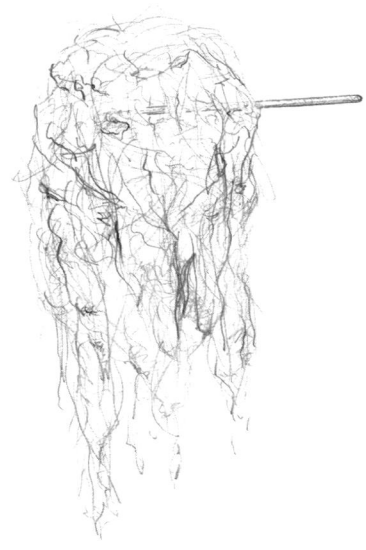

DNA is a colorless goo. We can hold DNA in our hands and see it with our unaided eyes. This isn't hard to do: a blender and some kitchen chemicals enable the extraction of DNA from a bowl of strawberries or a cup of peas. The recipe is roughly this: puree the fruits or vegetables in a blender, ripping their cells apart from each other. Add detergent to disintegrate cellular membranes. Sprinkle a dash of meat tenderizer or pineapple juice, supplying enzymes that digest proteins. DNA is now the only cellular component left intact. Add rubbing alcohol, which dissolves the protein pieces but not the DNA. The DNA clumps into long strands you can draw out with a toothpick, collecting a cloudy, stringy, white blob. That's DNA. It's not an awe-inspiring sight. I was once brought to tears extracting DNA during a classroom demonstration, but that was because I made the terrible decision to use onions as my source material—conveniently colorless, but a painful puree.

DNA is a code. At the other extreme of tangibility, we can think of DNA as an abstract code composed of four symbols. These symbols are often denoted by letters—A, T, C, G—but four colored squares work just as well. The particular sequence of symbols conveys information about how your cells should build what they need to build and do what they need to do. How much information? Let's compare it to the amount of digital information stored in a portable music player. These days we're used to thinking about "bits" and "bytes"—units of information. A bit (**binary digit**) is anything that can have just two values: yes or no, 0 or 1, a magnet pointing north or south. A one-gigabyte thumb drive has about eight billion bits of information, *giga* meaning "billion" and a byte being eight bits. Its contents can be expressed as a particular string of eight billion ones and zeros (. . . 01110100110010100011 01110100001101110001101011 . . .). How many bits is each person's DNA sequence? Three billion symbols—that is, three billion As, Ts, Cs, and Gs—make up your DNA. We could make a dictionary like this, for example, to translate each symbol into a binary code:

00 = A
01 = T
10 = C
11 = G

A sequence like ATTGC would be equivalent to 0001011110. Our three-billion-letter genome, therefore, carries six billion bits of information—less than a gigabyte, and probably a small fraction of the storage capacity of the phone in your pocket. This presents a puzzle: I seem much more complex than my phone, despite my apparent paucity of information! We grapple with the concept of complexity through much of this book. For now, there's a more immediate question: How does this abstract picture of codes and information relate to the stringy blob illustrated previously?

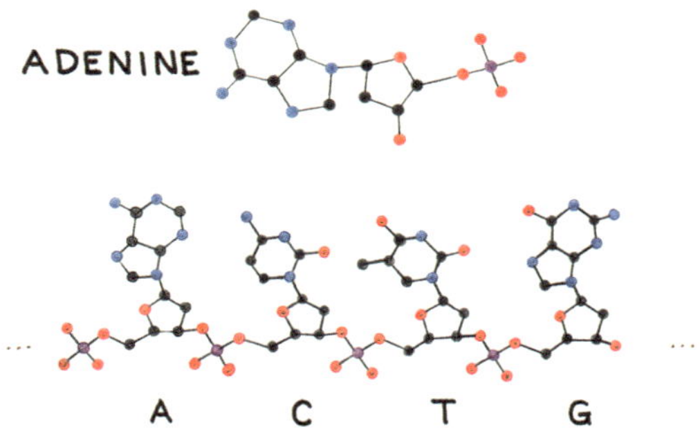

DNA is a molecule. Like all molecules, DNA consists of atoms, in its case atoms of carbon, hydrogen, oxygen, nitrogen, and phosphorus, held together by chemical bonds. The four symbols of the code mentioned above are really four assemblies of atoms, called *nucleotides*, stitched together to form a long chain. The upper illustration depicts all the atoms in an adenine nucleotide (A), with carbon atoms as black, nitrogen as blue, oxygen as red, phosphorus as purple, and black lines denoting chemical bonds. (For clarity, I omitted the many hydrogen atoms.) The lower illustration shows the atoms in the four-nucleotide sequence ACTG, with the ellipses (. . .) indicating where this segment would connect to adjacent nucleotides if it were part of a longer strand. Specifying the sequence of nucleotides in the chain suffices to identify a DNA molecule—the shorthand of ACTG is exactly equivalent to the array of atoms I've drawn, and it can't refer to any other set of atoms than this.

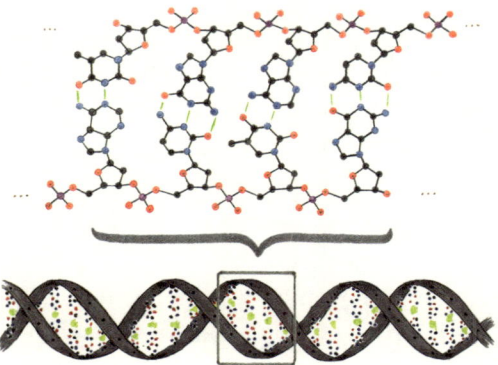

DNA is a double helix. The interactions between atoms determine the structure of a molecule, and the structure of the molecule governs its function. Any of the four nucleotides, A, T, C, G, can be linked to any other to form a single strand of DNA. But the nucleotides also interact *between* strands, though more weakly, in a specific way: A and T bind together, as do C and G. (We say that A and T are complementary, as are C and G.) A single strand of DNA, for example, AGCCTATGA, binds its complementary strand, TCGGATACT. The illustration shows the atoms in the double-stranded DNA formed from ACTG and its complement, TGAC, with the green lines indicating the interstrand bonds. Interactions among the atoms drive the two DNA strands to wind around each other like twisting ivy, forming a double helix. The illustration echoes the cartoon double helix we've all seen countless times; the smooth ribbons and well-ordered dots are a schematic of the more complicated arrangements of atoms and bonds in three-dimensional space.

The iconic double-helical form of DNA is functional as well as elegant. The two complementary strands convey redundant information: if I tell you the sequence of one strand, you know the other, since each nucleotide is the complement of its partner. This redundancy reveals how information can be transmitted from one cell to its two daughter cells as it divides: the DNA is unzipped and the complement of each original strand is synthesized, giving two DNA double helices from the original one.

James Watson and Francis Crick figured out the structure of double-stranded DNA in 1953, based on exquisite X-ray measurements performed by Rosalind Franklin and graduate student Raymond Gosling. (This story is a fascinating one, filled with cleverness and insight as well as ethical lapses and tragedy, and it is well told elsewhere; see the references.) Before this, no one knew what DNA molecules might look like. The most prominent hypothesis was from Linus Pauling, the chemist who first formulated the modern notion of a chemical bond, who suspected that it formed a three-stranded twisted fiber (a triple helix). The revelation of the double helix made it clear how DNA's structure might enable the transfer of genetic information by the copying of complementary strands. Other consequences of DNA's structure are less obvious, though, and we're still unraveling the mysteries of DNA today.

In living cells as well as in sterile test tube solutions, single strands of DNA will spontaneously wrap themselves into a double helix if their nucleotides complement each other. No external scaffolding or microscopic ropes and pulleys are required: the DNA contains within itself the mechanism of its organization, highlighting a theme of self-assembly that surfaces repeatedly in our explorations of life.

Each of the four views of DNA depicted above is useful, emphasizing attributes relevant to various roles. The fibrous goo extracted from pureed cells may be homely, but all of the more glamorous uses of DNA—plucking it from cancer cells to map the genes they carry, harvesting it from crime scenes to track down suspects, and more—must acknowledge the material, corporeal character of DNA if they are to work. As an abstract code, the sequence of symbols specifies the information carried by the molecule. When we say that your DNA is unique, it means that your sequence of nucleotides or colored squares is different from mine. (It's only slightly different; over 99% of our squares would match.) When we say we know the genome of some organism, we mean that we know the full sequence of symbols. This tells us a lot, but it also leaves a great deal undetermined, as we'll see. We might care about the atomic level of detail—the exact architecture of atoms making up DNA rather than just the symbolic code of constituents—if, for example, we're designing tools that cut DNA strands or splice them together, which we encounter in the context of genome editing in part III. More often, however, the arrangement of the units, the As, Cs, Gs, and Ts, suffices. The double-stranded helix describes how DNA is situated in space; the size, shape, stiffness, and electrical charge of double-stranded DNA govern its packaging in cells and the readout of information it contains. As our first illustration of the importance of the physical character of double-stranded DNA, let's look at a process that has transformed biotechnology: the polymerase chain reaction.

DOES DNA MELT?

Imagine you'd like to make an exact copy of some DNA. You could begin by separating the two strands of the double helix and then creating a new complementary strand for each of the resulting halves. That is, in fact, what your cells do every time they divide, with a particular protein machinery performing the initial unzipping of the double-stranded DNA. Outside of cells, we've developed another approach that allows us to replicate DNA at will, taking tiny amounts of DNA and transforming them into innumerable identical copies to yield enough material to run tests on or to transport to new targets. The tiny amounts of nucleotide strands could, for example, be swabbed from a crime scene to assess similarity to DNA from suspects; sampled from the amniotic fluid surrounding a fetus to search for genetic abnormalities or the presence of bacterial or viral DNA; excised from a tumor to map mutations in the nucleotide code indicative of cancer; or extracted from a Nobel Prize winner to be replicated and reinserted, in fragments, into bacteria that make up an artwork. Just as in natural replication, the artificial replication of DNA relies on drawing the double helix apart, which in turn relies on the physical phenomenon of phase transitions.

Rather than asking how we might separate the two strands of a DNA double helix, imagine asking how we could separate the tightly connected water molecules that make up an ice cube. We know the answer: we'd heat it. Above 0°C (32°F) ice melts into liquid water, in which each water molecule meanders around, only transiently bound to any other molecule. In general, temperature is the nemesis of attraction and order, a recurring theme in physics. For water, the transition between solid and liquid is sharp, occurring precisely at the "melting temperature," which is 0°C at standard atmospheric pressure. Even a few degrees below 0°C, water is solid; even slightly above, it's liquid. This transition isn't sharp for all substances, however. If you heat honey, it gets progressively less viscous (flowing more easily) as the temperature rises, rather than suddenly changing at one particular temperature.

Recalling that the bonds between the two strands of a DNA double helix are weaker than the bonds within a strand, we might expect that we could use heat to separate DNA strands without destroying them, and this is, in fact, the case. But is the transformation sharp or smooth? In other words, does DNA have a melting transition? The answer is important if we want separation to enable replication. If there *is* a melting transition, we'd be assured that by raising the temperature, even to just a few degrees above the transition, we'd have complete separation (upper illustration). If there isn't a melting transition, we'd likely have some unseparated DNA that couldn't be copied (lower illustration).

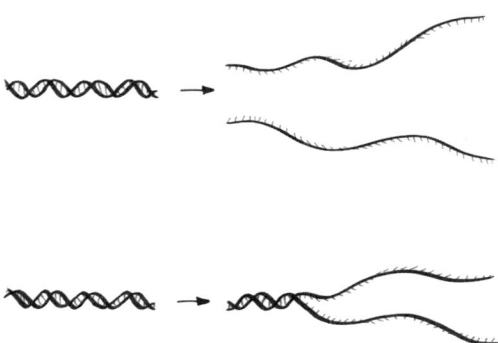

In the latter case, though we could keep heating until every last segment is completely separated, this would in practice likely require such high temperatures that the DNA itself, and any other biological molecules that are present, would be damaged.

It turns out that, for double-stranded DNA, the transition into single strands is a sharp one. DNA *does* melt. If we take a beaker containing DNA and heat it, the molecules are double-stranded up to a specific melting temperature, and just above this they separate into single strands. Not only can we measure this in the lab, but we understand its origins. Making sense of phase transitions—transformations between liquid, solid, and gas, or between magnetic and nonmagnetic forms, or between any of the different structures that materials might take—was one of the great triumphs of twentieth-century physics. The melting of DNA is a transition from order to disorder that maps onto other transitions, though with subtleties of its own.

All phase transitions reflect a conflict between order and disorder. Order is typically driven by the energy associated with attraction or alignment, while disorder is often driven by geometry, the ways in which constituents can arrange themselves in space. Temperature amplifies the impact of disorder. At low temperatures, the tendency to be ordered wins; at high temperatures, disorder is dominant. Water molecules, for example, are marshaled into the crystalline arrangement of ice when cold; if warmer, they succumb to the randomness of being liquid. Our statement that the melting transition is sharp means that there's a specific temperature separating these two outcomes and therefore a clear demarcation of the ordered and disordered phases.

The energy of ordering and the varieties of disorder both depend on the dimensions the substance can explore. The consequences for phase transitions are dramatic: in general, theory predicts that one-dimensional materials shouldn't exhibit sharp transformations between phases. Water molecules arranged in a row, for example, wouldn't melt at a specific point; disorder would emerge at even the coldest accessible temperature and would grow steadily as the temperature increases.

To a good approximation, the long chain of double-stranded DNA is one-dimensional—rungs on a ladder, one after the other. Its sharp

melting, readily observed in the lab, therefore seems to confound expectations. As strands come unbound, however, the liberated, flexible, single-stranded DNA bends and twists through three dimensions, subject to random forces common to all molecules whose character we explore further throughout this book. Though the motion is random, its consequences are robust and predictable—another recurring theme—as the resulting configurational freedom grants the overall separation a precise transition temperature, which is generally the case for three-dimensional materials. With experimental data and theoretical understanding in hand, we can even predict, for a given DNA sequence, the temperature at which it will come apart. These transition temperatures are typically around 95°C (200°F), a bit below the boiling point of water, with the exact value depending on the particular nucleotide sequence.

We can therefore separate DNA in a test tube just by heating it. The next task for our goal of replicating DNA is to create the complements of each strand. We could borrow the machinery that your cells use, called *DNA polymerase*, but these proteins would form a dysfunctional, rubbery glob, rather like a boiled egg white, at the temperatures necessary to melt the DNA. (Egg whites are, in fact, mostly protein.) There is, however, a clever way around this: We use DNA polymerase from bacteria that live in hot springs, creatures whose proteins have evolved to run smoothly when hot. Like all DNA polymerases, these require a little bit of double-stranded DNA to get started with their replication of a single strand. These ingredients were all in hand by the early 1980s; bacteriologists discovered and purified the hot-springs polymerase, for example, in 1976. In 1983, the recipe that combines nucleotides, polymerases, temperature, and DNA came to scientist Kary Mullis

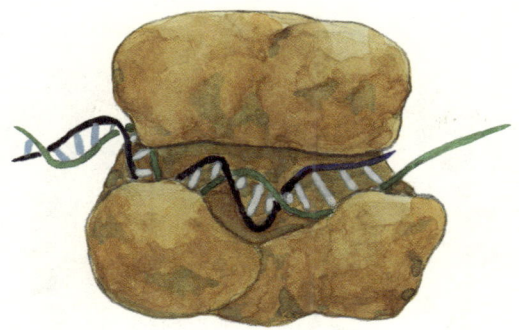

during a late-night drive through California's Coast Range, along with the realization that this would enable simple and almost limitless replication of DNA. The process is now known as the *polymerase chain reaction*, or PCR.

To perform PCR, we first gather the ingredients, dissolved in a saltwater solution: the little bit of DNA we wish to replicate, DNA polymerase proteins, an abundance of individual nucleotides (As, Cs, Gs, and Ts), and a lot of *primers*—segments of DNA just a few nucleotides long that are complements of the ends of the DNA to be copied.

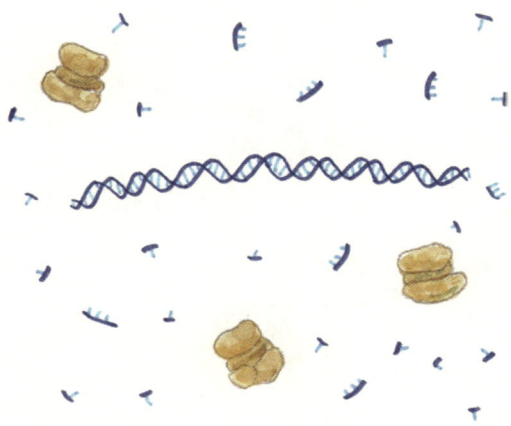

Next, we raise the temperature to around 95°C (200°F) so that the DNA melts, giving two single strands from the original double helix.

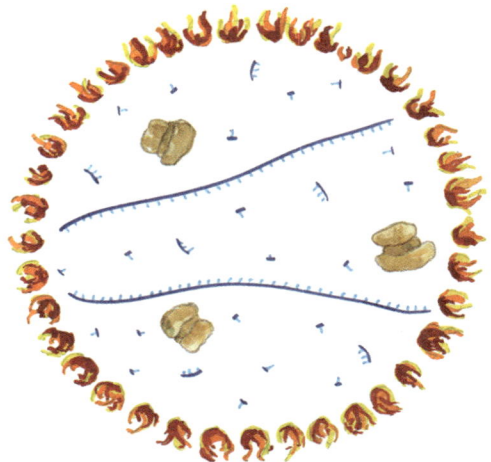

We next lower the temperature, so that the primers will bind to the ends of the single-stranded DNA. There are a lot of primers, so it's much more likely that the single-stranded DNA encounters a primer than its former partner strand; we put predictable randomness to work for us. The polymerase will then bind and craft the complementary DNA strand. When it finishes, we have two double-stranded DNA helices, from our original one.

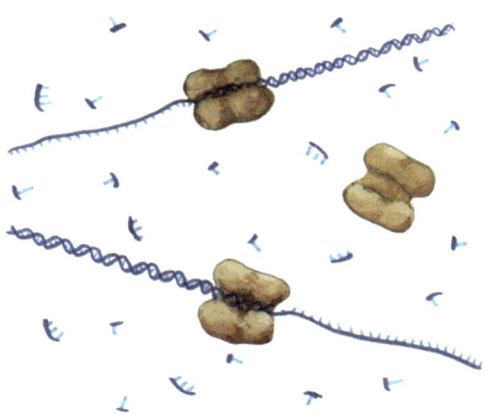

We then repeat. With one more cycle of warming and cooling, we end up with 4 DNA fragments; at the next round 8; then 16, 32, 64, ...

Ten doublings give 1024 pieces of DNA; twenty give over a million; thirty repetitions (not difficult with an automated heating and cooling machine) give over a billion!

We can therefore take a little bit of DNA and turn it into a *lot* of identical DNA, amplifying our original whisper into a chorus of replicas to be used in whatever diagnostic, forensic, or therapeutic application we like. In living things, copying DNA requires the elaborate dance of reproduction, creating an entire new cell or even a new organism. With the polymerase chain reaction, we can copy DNA at will. The recipe is beautifully simple. As Mullis himself wrote, the "almost universal first response" of molecular biologists upon learning about his invention was, "Why didn't I think of that?" Mullis continues, "And nobody really knows why; surely I don't. I just ran into it one night."

By replicating DNA, we can create enough copies to "sequence" it—in other words, to determine its particular pattern of nucleotides—using techniques that we describe in part III. With this approach, we've mapped the genomes of humans and many other organisms.

You may wonder, since PCR needs primers that bind to the DNA of interest, creating a short double-stranded span, don't we have to know the sequence we're amplifying before we even begin? Not quite. First of all, for many applications we do know some part of the genome of the organism whose DNA we're working with, and can design primers that bind to this part. More generally, though, we can cut the unknown DNA into fragments and link them to DNA that we know well, for example, the genomes of easily grown bacteria, using naturally occurring proteins that sew DNA strands together. Then primers to the known part of the DNA can seed the progression of DNA polymerase onto the unknown parts. It's with this approach, for example, that the genome of the woolly mammoth, an animal that's been extinct for several thousand years, was sequenced from ancient remains.

The polymerase chain reaction is essential to nearly all of what we do with DNA, and it is made possible by bringing together quirks of biology (like heat-loving microorganisms) and generalities of physics (melting phase transitions), each of which on their own may have seemed tangential to practical aims. PCR also highlights a message central to many modern technologies as well as natural processes: DNA is more than a code or an abstraction, but is a tangible, physical object. Especially with notions of self-assembly and predictable randomness in hand, we can understand and even invent methods to work with this crucial molecule.

There's a lot more we can, and will, ask about DNA: How much do you have? How is it stored, organized, and deciphered by your cells? How could you alter the code written into your (or your unborn child's) genome? These questions tie together physical properties and biological functions. To answer them, though, we need to introduce another key player on the cellular stage: proteins. In the next chapter, we see what a protein is and examine the interplay between proteins and DNA.

2

Proteins: Molecular Origami

At the core of nearly every action, every task, and every event in your body is a protein. Proteins in red blood cells soak up oxygen from the air you breathe. Proteins tug on other proteins to contract your muscles. Proteins extend and retract protrusions with which immune cells squeeze through your tissues. Proteins in your eyes capture light and trigger electrical impulses, while other proteins open and close gates that send these impulses to your brain. Many kinds of proteins are inside every cell, and many are outside as well, making up, for example, the elastic matrix of your flesh. So: What *are* proteins?

Like DNA, a protein is a molecule composed of a chain of simple units. In DNA, these units are any of four nucleotides. In proteins, these units are any of 20 amino acids. Double-stranded DNA, regardless of its nucleotide sequence, adopts a double-helical structure. In contrast, proteins have structures that are determined by their particular amino acid sequences. Every distinct protein has a different pattern of amino acids and therefore a different three-dimensional shape. The blueprints for the structure and the tools for its construction are encoded within the protein itself. Proteins provide perhaps the most striking manifestation of the theme of self-assembly, nature's encoding of instructions for organizing matter within the matter itself, to be activated and realized by universal physical forces. Though self-assembly isn't unique to living things—sand piles, for example, arrange themselves into cones tilted at specific angles and soap bubbles form themselves into spheres—it is ubiquitous in biology. Considering proteins, we'll see how forces generate forms, how the process usually succeeds but sometimes catastrophically fails, and how computers struggle with geometric calculations that molecules perform in microseconds.

PROTEINS IN THREE DIMENSIONS

An amino acid chain in water bends, twists, and folds into a particular shape. Two very common motifs found in proteins are helices and sheets.

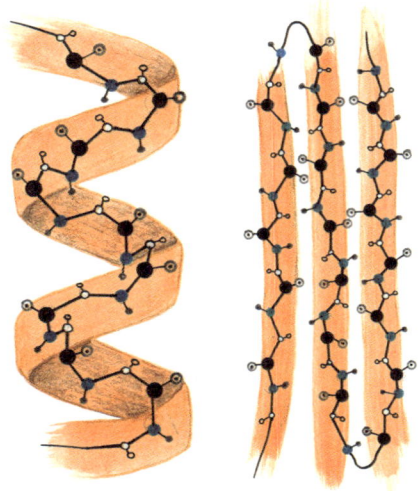

I haven't drawn all the atoms in these structures—just a few representative dots and the bonds between them. Helices and sheets are so common in protein structures that we often depict only stylized shapes—a smooth helix about one nanometer in diameter (one billionth of a meter), and a sheet or ribbon about a third of a nanometer wide—rather than showing every component atom.

The first protein to have its three-dimensional structure revealed was myoglobin, a protein that carries oxygen in muscle, in 1958. As with DNA and many other molecules, this was made possible by illumination with X-rays together with mathematical analysis of the resulting intensity patterns, in this case performed by a team led by John Kendrew at the University of Cambridge. X-ray imaging requires proteins to solidify into crystals, similar to the sugar crystals you might make in your kitchen, and even now there is an art to coaxing proteins to crystallize. Kendrew's team struggled with myoglobin from porpoises,

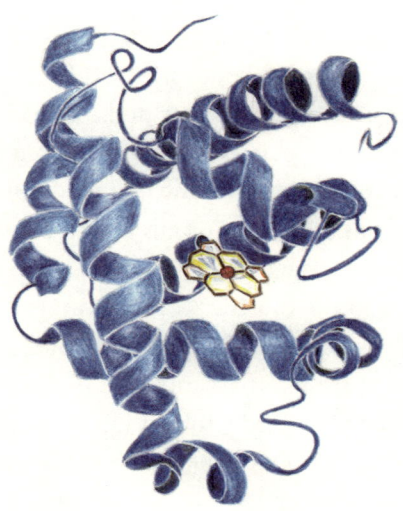

penguins, seals, and other creatures before stumbling on sperm whale meat, conveniently on hand in the freezer of Cambridge's Low Temperature Research Station. (Deep-diving, air-breathing sea creatures have high concentrations of myoglobin in their muscles, allowing them to

store more oxygen and surface less often, hence the focus on these animals.) The sperm whale protein formed "the most marvelous... gigantic crystals." From these, Kendrew and his team were able to determine that the 153-amino-acid chain of myoglobin folds into a structure composed of eight helices and some nonhelical spans that together hold onto a flat compound containing a single iron atom that binds oxygen (upper illustration, page 31).

We can again turn to the marine world for an example of a protein composed mainly of sheets. Green fluorescent protein, commonly called GFP, is a light-emitting protein first found in bioluminescent jellyfish. GFP is a chain of 238 amino acids, folded into a barrel of sheets about 3 nanometers wide that surrounds the part of the molecule responsible for color (lower illustration, page 31). The protein has become much more than an oceanic curiosity. GFP has been engineered into bacteria, fungi, plants, and even animals as diverse as fruit flies and zebrafish, in which it serves as a beacon allowing researchers to visualize particular types of cells and see how they grow, move, and divide. GFP can also be fused to other proteins, creating reporters that reveal where in cells these proteins are, how they behave as the cells perform various tasks, and how they associate with other proteins to form more complex architectures. A rich palette of fluorescent proteins now exists, emitting light in a rainbow of colors, derived from GFP or other proteins discovered in corals, with names ranging from the dull ("red fluorescent protein") to the more evocative ("tangerine," "cherry," "plum"—a whole series of fruit names). The ensemble has made possible multicolor imaging of the machinery of living things, with applications far removed from these proteins' marine origins.

PROTEIN PORTRAITS

The three-dimensional structure of a protein matters: it's intimately tied to the protein's chemical or physical tasks. In GFP, for example, the barrel shields the light-generating unit from water and dissolved oxygen that would quench its glow. A few more examples make the relationship between structure and function even more evident.

Thin membranes separate spaces within a cell and also demarcate the cell's inside and outside. Specific proteins embedded in membranes, often joined together into barrels or rings, transport atoms and molecules across. Ion channels are one class of these transporters, allowing specific charged atoms (ions), such as potassium, sodium, and chlorine, in and out of cells through a central pore that can be open or closed. Controlling flows of ions is a crucial task. The motion of your eyes

scanning this page and the thoughts racing through your brain are both manifestations of the electrical voltages generated by the redistribution of ions across membranes. Many toxins produced by animals such as snakes and scorpions work by interfering with ion channel proteins, shutting down the nervous system of their victims. The illustrated example on page 33 is a potassium channel, drawn end-on with the membrane, not shown, in the plane of the page. The red dot is a potassium ion, traveling toward or away from us and therefore entering or exiting a cell. The channel is actually composed of four identical protein molecules that loosely bind together to construct a membrane-spanning pore.

Channels can open and close, but other proteins can perform more elaborate gymnastics. I've depicted an assembly of two molecules of a protein called kinesin (lower illustration, page 33). As the name implies (think "kinetic"), it's involved in motion. Each kinesin protein takes the shape of a long stalk with a bulbous end, connected by a flexible amino acid joint. The two stalks bind together and can adhere to cargo that needs to be ferried from one place to another, for example, packets of chemicals synthesized deep in the interior of a neuron and stored for release near its edge. The whole complex can then walk along tracks within a cell. "Walking" isn't metaphorical: the two feet alternately bind and unbind from the tracks, ambling foot-over-foot to reach their destination. (Traditionally, these feet are called *heads* and the foot-over-foot motion is called *hand-over-hand* motion. Yes, the nomenclature is puzzling.) The tracks themselves are also made of proteins, arranged into rigid filaments; again, their three-dimensional shape allows them to play their role.

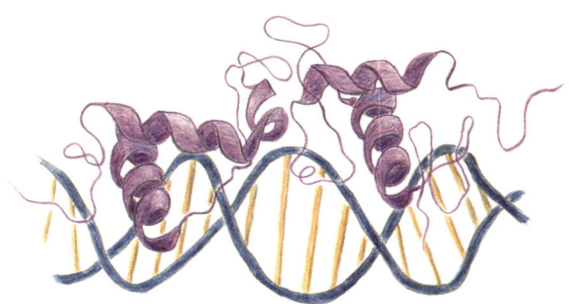

The structure of proteins influences their interactions with each other and with other molecules they encounter, such as DNA. Many proteins bind to DNA to guide the readout of its genetic information, as we explore in more detail in the next two chapters. These DNA-binding proteins must adopt a shape that conforms to the curves of the DNA double helix. Amino acid helices can nestle in the DNA grooves and are a common motif in these proteins' structures, illustrated here for a hormone-sensing molecule known as the glucocorticoid receptor. (A pair of these proteins works together; I've drawn only the DNA-adjacent regions of such a pair.) When the receptor encounters and latches onto a hormone called cortisol, its structure changes, and only then is it capable of binding to DNA, setting in motion a series of events that, among other things, inhibit the organism's inflammatory immune response. You've probably encountered cortisol in ointments, where it's often called hydrocortisone, and made use of its activation of receptor proteins to reduce the redness, itch, and swelling of your body's reaction to poison ivy, insect bites, and other irritants.

PROTEIN FOLDING

As we've seen, protein structure is intimately connected to protein function. A protein doesn't begin its life fully formed, however. Every protein is made by cellular machines that attach one amino acid to the next, sequentially, like paper clips linked together into a chain. There isn't any scaffold that gives structure to the string-like molecule, arranging it into stacked sheets, tangles of helices, or any of a near-infinite variety of possible forms. Rather, the protein *sculpts itself* into its proper shape. The amino acid sequence of the protein encodes the determinants of its structure; the protein self-assembles.

Each of the 20 amino acids has particular physical properties. Some have positive electrical charge; some are negative; some are neutral. Some are large; some are small. Some are greasy ("hydrophobic") and prefer to separate from water; some are "hydrophilic" and mix well with water. Imagine a protein with several positive amino acids in a row (red circles in the illustration), followed by a string of neutral hydrophilic amino acids (black circles), and then several negative amino acids (blue circles). Opposite charges attract; so left to its own devices, the protein folds to bring together the contrasting ends.

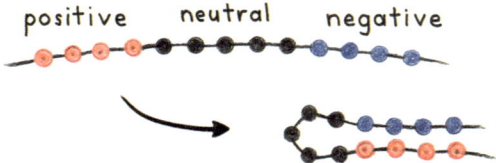

Or imagine a protein with amino acids that are hydrophobic (orange squares) and hydrophilic (black circles). The protein is surrounded by water—water makes up the majority of the cellular interior—and will fold to bury the hydrophobic bits in the center to be surrounded by their water-loving colleagues.

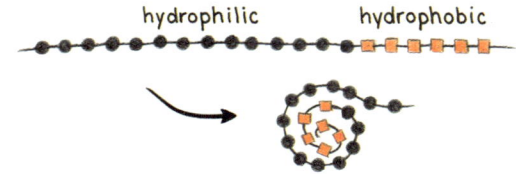

I've drawn these schematic illustrations in two dimensions for clarity. Really, you should imagine a roughly spherical core of hydrophobic amino acids surrounded by a shell of hydrophilic amino acids.

In any real protein, many such interactions between amino acids and each other and between amino acids and the surrounding water take place, giving rise to forces that pull the protein toward a particular conformation. Every protein is synthesized in the cell as a chain of amino acids, and that chain folds itself into its optimal three-dimensional shape. The technical term is, in fact, *protein folding*.

As with nearly everything in biology, this blunt picture isn't quite true. Some proteins, especially large ones that tend to aggregate, need a bit of help to fold, and a group of other proteins called *chaperones* comes to their aid. Assemblies of chaperone proteins contain chambers that protect the nascent protein from the complexities of the crowded cellular environment, facilitating the proper folding of the amino acid chain. Chaperones notwithstanding, the general idea of proteins possessing the plans for their own architecture is powerful and prevalent throughout the living world.

Every protein we've described above, and tens of thousands of others, folds within a fraction of a second into a three-dimensional form, bypassing the innumerable pitfalls and dead ends of shapes that don't quite satisfy the interactions their component parts prefer. This is an amazing feat—like a piece of paper spontaneously folding itself into a perfect origami sculpture. What's more, for the vast majority of proteins the sculpture is uniquely determined by the amino acid sequence. In other words, a given sequence always folds into the same shape. Every green fluorescent protein forms a barrel; every myoglobin folds into the same collection of curls.

A few illustrations might help convey how remarkable this self-organization is. Imagine a sequence of amino acids that, as usual, has positively charged, negatively charged, neutral hydrophilic, and hydrophobic amino acids. (Charged amino acids are always hydrophilic, by the way.) The chain could fold into the arrangement on the left, which is pretty good—the hydrophobic pieces are buried in the interior and opposite charges lie next to each other. But the arrangement on the right, of exactly the same sequence, is also pretty good, for the same reasons.

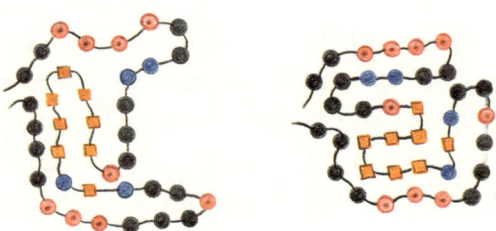

The two conformations are certainly different. We might imagine that if this protein needs to bind to some small molecule—a hormone, for example—the presence of a "pocket" in the first form but not the second would make the first functional and the second useless.

Understanding how an amino acid chain adopts a single, optimal shape turns out to be remarkably puzzling. For a random amino acid

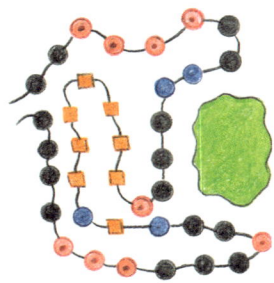

sequence—imagine blindly picking amino acids out of a hat and stringing them together—analysis of forces and energies reveals that we should expect a huge number of "pretty good" configurations, far too many to expect the chain to routinely find a unique endpoint to its folding. Nature sidesteps this multiplicity of possible forms; the proteins that actually exist in the real world aren't random, but rather are those that have been selected by four billion years of evolution. Organisms that encode amino acid sequences that don't fold into unique shapes are plagued by dysfunctional and perhaps even harmful proteins, and hence are less likely to live and reproduce. Those that persist are those that encode amino acid sequences with a clear, unique path to three-dimensional structure.

The result of all this, as we've seen, is the general principle of a one-to-one correspondence between amino acid sequence and protein structure for the proteins actually present in ourselves and other organisms. If we know the structure of one kinesin molecule, we know the structure of every kinesin molecule. If we know the structure of one cortisol receptor, we know the structure of every cortisol receptor. As with all rules of thumb, however, there are exceptions, and the exceptions to this one are exceptionally important.

One set of rule-breakers are the "intrinsically disordered proteins," which don't have a particular form at all. Examples include some of the proteins that make up the pores of the membrane surrounding cell nuclei. It's thought that the spaghetti of disordered proteins occupying the pore provides flexibility for the transport of different-sized objects into and out of the nucleus.

More interesting, in my opinion, are the proteins that have *a few* stable configurations—not one unique form and not an amorphous blur, but rather a couple of architectures that they might toggle between, like a light switch toggles between stable on and off positions. In the past few decades, we've discovered not only that proteins like this exist but that they contribute to some perplexing diseases. They also provide a warning, in case you needed it, not to indulge in cannibalism.

KURU AND CANNIBALISM

In Papua New Guinea in the 1950s, an epidemic of a strange disease characterized by tremors and uncontrollable fits of laughter struck villages of the Fore people, killing up to 200 people a year among a total tribal population of about 11,000. (For scale, imagine 150,000 New Yorkers dying eerie deaths every year.) From patterns of illness and contagion, anthropologists and medical researchers deduced that the disease, named *kuru* after the Fore "to shake," was spread by ritual cannibalism among the Fore: when a person died, family members consumed the body, an act of love and respect that helped liberate the spirit of the deceased. The Australian government that was ruling Papua New Guinea banned cannibalism around this time, which led to a steady decline in the prevalence of kuru. It took decades, however, to figure out the actual cause of the disease. The culprit was not a bacterium, a virus, or a parasite but rather a protein—an unusual protein that doesn't have one unique structure but can adopt one of two forms. In the "normal" form, the protein performs its usual functions. In the "misfolded" form, it does not. Even more perniciously, however, the misfolded protein induces others to adopt its aberrant shape and join with it to form fibrous aggregates. In this way, the deviant protein is infectious: ingesting proteins of the misfolded form, some of which make their way to the brain, induces structural changes in molecules with an otherwise benign amino acid sequence. This change amplifies itself through the victim's nervous system and is propagated further still if the victim dies and is eaten by another villager. The chain of events is reminiscent of Kurt Vonnegut's novel *Cat's Cradle*, in which the fictional "ice-nine" form of water is solid at room temperature and, upon contact with the normal form of water, induces its crystallization and transformation into more ice-nine. The resulting chain reaction is even more deadly than kuru. Unlike ice-nine, however, kuru is real.

Proteins that can fold into multiple forms and that act as infectious agents are called *prions*, and we now realize that they drive several diseases of humans and other animals. One of these is bovine spongiform encephalopathy, more evocatively known as mad cow disease. Like kuru, it is neurodegenerative, inducing tremors, excitability, and poor motor coordination, but in cattle rather than humans. An outbreak in the United Kingdom in the late 1980s infected about 200,000 cows, and over 4 million animals were killed to halt the spread of the epidemic. Before it was stopped, however, it spread to humans. Well over a hundred people died from the human analog, called variant Creutzfeldt-Jakob disease, almost certainly from eating diseased animals. How did the animals get infected? Cannibalism! Meat-and-bone meal, believed to enhance the animals' growth and productivity as well as providing a use for leftover parts, was commonly fed to farm animals. This livestock cannibalism was banned following these outbreaks, in the United Kingdom in 1989 and by now in most of the world, at least for ruminants like cows and sheep. (Meat-and-bone meal is, in general, still allowed as feed for other farm animals, such as chickens and pigs.)

The very existence of prions was contentious for quite a while. In the 1980s, researchers led by Stanley Prusiner at the University of California at San Francisco, following a decade of work, isolated the infectious agent of scrapie, the sheep analog of bovine spongiform encephalopathy, and identified it as a protein. Their announcement was met with intense skepticism—there's a sense of agency to the actions of bacteria, viruses, and parasites that a simple amino acid chain lacks, and it's understandably hard to imagine it propagating, amplifying itself, and causing disease. Nonetheless, painstaking analyses and elimination of other possibilities established the reality of the prion hypothesis.

Kuru and mad cows aside, prion or prion-like proteins pop up in other major diseases. Most notably, Alzheimer's disease is often accompanied by aggregates of misfolded proteins that resemble those of prion diseases. These aggregates don't seem to be infectious, though; transferring them from diseased to healthy animals doesn't transfer the neurological symptoms. What the causes and consequences of these protein agglomerates are remains unclear. More generally, there's still a lot left to learn about protein folding and misfolding.

PREDICTING PROTEINS

Returning to the overwhelming majority of proteins that *do* have a unique three-dimensional form, it remains surprisingly difficult to predict what that form will be given the protein's amino acid sequence. Such predictions would be immensely useful. Assessing how a potentially therapeutic drug would bind to a range of different proteins, for example, would be easier with the three-dimensional structure of each of these molecules in hand. Though structure determination has progressed considerably since our first glimpses of sperm whale myoglobin, it remains difficult, time-consuming, and capricious. The workhorse method of probing proteins with X-rays requires first coaxing them to form crystals, a craft that demands a great deal of trial-and-error tinkering, followed by interrogation with high-powered X-ray sources. Other methods exist, involving electron microscopes, for example, but none are fast or simple. It is appealing to think that, instead of actually making and measuring a protein, we might simply calculate, given the amino acid sequence, the three-dimensional structure it would adopt. Determining the amino acid sequence is easy thanks to the nature of the genetic code embedded in our DNA, which we elaborate in the next chapter. In principle, since we understand the physics of electrical forces and hydrophobic and hydrophilic interactions, we should be able to simply plug the amino acid sequence into a computer program, easy enough to write, that grinds through the requisite calculations, stopping when it has found the optimal folding of the molecular chain. In practice, however, the number of possible configurations is so enormous that even the fastest computers struggle to explore them all.

There have been lots of clever approaches devised to tackle this computational challenge—some focused on improving the algorithms for calculating forces and energies, some developing simplifications such as grouping sets of atoms together, and some exploring unconventional computer architectures. One could, for example, design a computer whose integrated circuits are tailor-made for calculating the sorts of forces experienced by amino acids, rather than the general-purpose integrated circuits of typical computers. This has been the approach taken by David Shaw, who focused the fortune he made as an investment manager to commission bespoke supercomputers devoted to the biophysical challenge of protein folding. Or one could use normal computers, but arranged in a large and haphazard array. This has been the approach of the authors of the "folding@home" program that runs in the background of volunteers' computers (anyone can sign up), using their otherwise idle moments to distribute calculations across tens of thousands of machines. Or one could try to recruit human minds. This has been the approach of researchers at the University of Washington who created a free protein folding game, called "foldit," in which players move amino acids like puzzle pieces on a screen, with the game's outcomes conveyed to the researchers. Or one could use artificial intelligence, training a computational neural network to infer patterns from known protein structures and apply them to predict new forms. This has been the approach of DeepMind, a company affiliated with Google, whose stunning performance placed it at the top of the 2020 "Critical Assessment of Protein Structure Prediction" contest. All these strategies and more have proved useful, though a quick and general method for computing the structure an amino acid chain will adopt remains elusive.

It is humbling to consider that the proteins themselves have solved the protein folding problem, shaping their structures within fractions of a second in every cell of every creature on earth. Self-assembly is awe-inspiring; it allows form to emerge from the pieces and forces intrinsic to nature's substances themselves. We'll see why this emergence can be so rapid and robust in chapter 6 when we discuss molecular randomness. But first let's explore the connection between proteins and DNA, defining the notion of the gene and setting the framework for revealing how self-assembled structures can form decision-making circuits in cells.

3
Genes and the Mechanics of DNA

We've called DNA a code, but what is it a code *for*? We've glimpsed a few of the many proteins an organism can create, but what determines this repertoire? The same concept, that of the gene, provides answers to both questions, binding together the abstract idea of biological information with the physicality of biological molecules. The power of genes as well as their limitations are inextricably linked to the physical properties of DNA, proteins, and the environment in which they exist. Discussions of genetic diseases don't often invoke the challenges of bending DNA or stuffing molecules into small spaces, but, as we'll see, such visceral concerns are important for making sense of how life works. Self-assembly is again central in this chapter as DNA and proteins, for example, join together to package genomes. Our other themes of predictable randomness, scaling, and regulatory circuits are also reflected in the handling of our genetic material, as our cells tackle challenges of size, shape, and disorder to organize their DNA.

WHAT IS A GENE?

We've seen that a protein is a sequence of amino acids stitched together with chemical bonds. The nucleotide sequence of the cell's DNA specifies each amino acid sequence. Groups of three nucleotides encode one particular amino acid. The DNA sequence TGG, for example, encodes the creation of one tryptophan, a hydrophobic amino acid. Both CGT and CGC specify the positively charged amino acid arginine. The sequence TGGCGT therefore indicates tryptophan linked to arginine. There isn't, however, a machinery that directly reads the DNA code and makes the corresponding protein. A molecule called *RNA* (ribonucleic acid) acts as an intermediary.

RNA, as its name might suggest, is similar to DNA. RNA is also a chain composed of any of four nucleotide units, three of which (A, C, and G) are the same as those in DNA and the fourth of which (U, uracil) is similar to DNA's T (thymine). A protein machine called *RNA polymerase* binds to a "promoter" sequence of DNA and then steps along the double helix like the slider on a zipper, spreading the two strands apart, reading the nucleotide sequence of one of the strands, called the *template strand*, and constructing a single-stranded chain of RNA. The process, which copies information from one form (DNA) to another (RNA), is called *transcription*, analogous to the transcription of spoken words into text or handwritten notes into type.

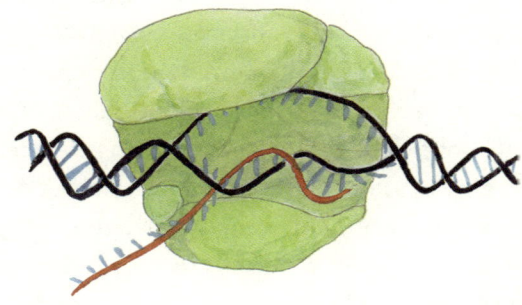

The RNA is complementary to the DNA template strand and is therefore identical in sequence to the other strand of the DNA double helix, called the *coding strand*, except for the Ts being Us. So, for example, the DNA coding sequence ATCGTT, mirrored as TAGCAA on the template strand, would be transcribed as the RNA sequence AUCGUU. Another machine, called the *ribosome*, translates the RNA strand into protein. The ribosome travels along the RNA, interacting with each three-nucleotide segment and attaching the appropriate amino acid to the protein it's constructing. The RNA sequence UGG, for example, encodes tryptophan; CGU and CGC both encode arginine. Some sequences (UAG, UGA, UAA) encode a "stop" message, telling the ribosome to halt its protein synthesis and detach from the RNA. The sequence AUG means "start here."

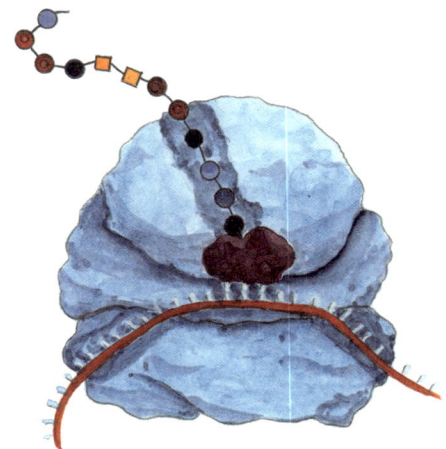

A particular segment of DNA therefore determines a protein that will be created via the processes of transcription (into RNA) and translation (into protein). Because DNA is transmitted from parents to children via sperm and egg, each such DNA segment enables the hereditary transfer of traits—the activities or characteristics of these particular proteins. Your ability to see color, for example, is made possible by three different proteins that react to different wavelengths of light, each produced in one of three types of cone cells in your retina. Differences of even a single three-nucleotide group, encoding 1 amino acid out of the approximately 350 in each of these proteins, can lead to subtle but measurable shifts in color perception. More dramatically, lacking the DNA sequence for an entire light-detecting protein results in one of several forms of color blindness.

One might think that these segments of DNA that encode proteins are what we mean by the term *gene*. This is almost, but not quite, correct.

Cells must not only specify the identity of the proteins they create but also control when this creation occurs and how much protein is made. Some stretches of DNA do not encode a protein sequence but rather affect whether other segments are read by the machineries of transcription and translation. For example, a class of proteins called *transcription factors* can bind to the promoter region at or near the starting point for RNA polymerase, diminishing or enhancing the likelihood that RNA polymerase will assemble and begin the transcription of DNA into RNA. We've seen an example of this already in chapter 2's glucocorticoid hormone receptor. Or a segment of DNA can be transcribed into RNA without subsequent translation into a protein, and this RNA can itself interact with DNA or other RNA molecules to influence protein synthesis. There are many ways in which RNA helps regulate the activity of cells, our understanding of which has in large part only arisen recently; RNA's status has been propelled from being merely a messenger between DNA and protein into a vital participant in these molecular conversations. An RNA called "Growth arrest-specific 5" made by cells sensing starvation, for example, attaches to the DNA-binding region of the glucocorticoid hormone receptor and thereby thwarts its recognition of its target DNA; the structural similarity of RNA and DNA allows the RNA to serve as a decoy.

Regulation of the processes by which genetic information is transformed into specific molecules is as important as the information itself, and it also enters the definition of what a gene is: A *gene* is a span of an organism's DNA sequence that encodes a particular, single hereditary characteristic, typically corresponding to a single protein or RNA sequence, and including noncoding regulatory sequences. It's a clunky definition, and one that's constantly changing, but life needn't comply with our desire for simple terminology. To further complicate matters, the term *gene* is still often used to mean "protein-coding DNA segment," its simpler and older meaning. Here I'll try to be transparent. The question we're now ready to ask is, thankfully, simple enough.

HOW MANY GENES DO YOU HAVE?

We can now read the genomes of all sorts of organisms, in other words, the complete sequence of As, Cs, Gs, and Ts. Because we can deduce the promoter sequences that instruct the transcription machinery to start and the terminating sequences that indicate stops, we can count the number of genes. For bacteria, we find a few thousand; each bacterium can generate roughly a few thousand distinct proteins. The bacteria that cause tuberculosis and cholera each have about 4000 genes in their genome, of which about 98% encode proteins. The genome of the *Lactobacillus delbrueckii* subspecies commonly employed to turn milk into yogurt has about 2000 protein-coding genes.

The human genome contains about 20,000 protein-coding genes. The number of noncoding genes, yielding RNA that doesn't go on to be translated into an amino acid chain, is harder to determine precisely but is estimated to be similar, around 20,000. Before you get too excited about your superiority over bacteria, because 20,000 is more than a few thousand, notice that your margin of victory isn't particularly large—there's less than a factor of 10 separating what most would consider an enormous difference in the complexity of the organisms. What's more, even among eukaryotes (organisms whose cells enclose their DNA in a membrane-bound nucleus), we're not very special. The common house mouse also has about 20,000 protein-coding genes, as do the Western clawed frog *Xenopus tropicalis* and the domestic horse. Some organisms have fewer genes. The fruit fly *Drosophila melanogaster* and the mushroom-forming fungus *Schizophyllum commune* each encompass about 13,000 protein-coding genes, and peregrine falcons about 16,000. The bread mold *Neurospora crassa* and the soil-dwelling amoeba *Dictyostelium discoideum* have about 10,000 and 13,000 protein-coding genes, respectively. Some organisms have many more genes than we do. The genome of the tiny water flea *Daphnia pulex*, about a millimeter long and nearly transparent, contains 31,000 protein-coding genes—the record, so far, among animals with sequenced genomes. Rice has about 30,000 protein-coding genes and maize (corn) about 40,000—*double* the human count—along with a few tens of thousands of noncoding genes. The number of genes tells us very little about the complexity or capabilities of organisms.

HOW BIG IS YOUR GENOME?

We've discussed your genome as a database of 20,000 protein-coding genes, but it's also a physical object, a series of A-T and C-G nucleotide base pairs that are the rungs of DNA's double-helical ladder, taking up physical space. Let's consider nucleotides first, then actual space. Your genome consists of about 3 billion base pairs. Most bacteria have much smaller genomes, typically a few million base pairs. The bacteria behind tuberculosis and cholera each have 4 million base pair genomes, and *Lactobacillus delbrueckii*'s genome size is about 2.3 million base pairs. But again, humans aren't especially remarkable or extreme in genome size. The mouse genome is similar in size to yours, while the fruit fly's is about 25 times smaller. The rice genome is also smaller, at about 430 million base pairs. (If this seems puzzling given the large gene count we stated earlier, don't worry—we'll return to this shortly.) Salamanders have especially sizable DNA, with genomes spanning 14 to 120 billion base pairs. The lungfish genome is 130 billion base pairs, and that of the flowering plant *Paris japonica* is a whopping 150 billion base pairs, 50 times larger than the human genome, making it the likely record holder for size. The single-celled amoeba *Polychaos dubiu* may surpass it with 670 billion base pairs, but this is somewhat controversial because its length was determined with outdated methods. (I'm amazed that no one has revisited this creature's DNA. If you're reading this and have a DNA sequencer and some spare time, go for it!) As with genes, there's no straightforward connection between genome size and the complexity of the organism.

Quantifying the numbers of genes and genomes brings to light a surprise and a puzzle. As we've noted, you've got 3 billion DNA nucleotide base pairs and about 20,000 genes that each encode a distinct protein. Proteins come in a wide range of sizes, but the average number of amino acids in a human protein molecule is about 400, and each of these amino acids is specified by three DNA nucleotides. Therefore, 20,000 distinct proteins require about $20,000 \times 400 \times 3 = 24$ million DNA base pairs. The human genome isn't 24 million base pairs long, however; it's 3 *billion* base pairs. The genome is over a hundred times larger than the protein-coding DNA it contains! Historically, we knew the length of the human genome before we knew its sequence of letters and the number of genes it contains; the small number of protein-coding genes compared to what we expected based on the size of the genome came as a shock. For rice the discrepancy is smaller, but it's still around a factor of 10. In general, most of a genome doesn't directly encode proteins. The puzzle, which we're still unraveling, is what the rest of the genome is doing. Some parts are transcribed into RNA but not translated into amino acid chains. These include independently functional RNA segments, as noted earlier, and RNA that is spliced out of the strand transcribed by RNA polymerase before it is translated by the ribosome. Much of the noncoding DNA, however, is never even transcribed into RNA; it can nonetheless influence the readout of genes by making up sites such as promoter regions.

Before expanding on this, let's first develop a better physical picture of DNA. We started this section with the question, "How big is your genome?" and gave a biologically accurate but physically unsatisfying answer: 3 billion base pairs. *How big* is this? Each of the two copies of your genome, housed in nearly all of your cells, would span a meter (3 feet) in length if laid out in a line. The cellular nucleus that contains the DNA is a few micrometers (a few millionths of a meter) in diameter.

Your cells stuff a meter of DNA into a space a thousandth of a thousandth of that length. Is this impressive or not? This may seem like a silly question—obviously, it's impressive. But I can roll 50 yards of yarn into a ball a few inches wide, to the amazement of no one. The central question is one of mechanics: How stiff is DNA? Is it like yarn or like steel? (Hopefully, you *would* be impressed if I stuffed 50 yards of braided steel cable into a few-inch bundle, even if it were the same thickness as yarn.)

DOES DNA BEND?

Characterizing the stiffness of materials is a topic in itself, the exploration of which could take us into detours of materials science that would distract us from biophysics. Thankfully, there's a conceptually simple model of the rigidity of polymers—long, chain-like molecules—that can give us essential insights into genome size. Imagine three different strings of different stiffnesses, each with the same end-to-end length if held taut as a straight line, but left free to form amorphous blobs.

Intuitively, we realize that the one that is most extended, as if made up mainly of long stretches of gentle curves, is the stiffest of the set (left). The most convoluted string, scrunched tight, is likely the softest (right).

Let's think about the typical distance over which the molecule looks straight—in other words, the typical distance we would travel before we start facing a different direction if we're walking along the strand. The stiffer the molecule, the longer this distance. Imagine an ant walking along an uncooked strand of spaghetti—its direction hardly changes at all as it moves along. This characteristic length for the spaghetti is very large. Now imagine a cooked strand of spaghetti, tossed onto a tabletop. Following the strand, our ant's path often bends and turns; this characteristic length is shorter, probably less than an inch or so.

Now let's replace in our minds the molecule's actual curvy path with a series of straight segments, each as long as this typical straight-path distance, connected by joints that randomly orient adjacent segments.

What we've created is something physicists and mathematicians call a *random walk*. Imagine a walker who takes a series of steps in completely random directions—one step might be to the north, the next to the southwest, the next north-by-northeast, and so on. Trying to predict where our random walker ends up after some number of steps sounds futile; who knows where she'll find herself, since the direction of each step is purely up to chance. For any individual walker, prediction is, in fact, futile. But if we imagine taking many random walks, or watching many random walkers, the *average* outcome is well defined: a random walker who takes 25 steps will find herself an average distance of 5 step-lengths from where she started; for 49 steps, 7 step-lengths; for 100 steps, 10 step-lengths; for N steps, the square root of N step-lengths away from the starting point. (This is true whether the walk is two-dimensional, as a person would take, or three-dimensional—perhaps a random swim.)

These random walks turn up in countless places. Economists describe the rapid ups and downs of stock markets as random walks. The paths of swimming bacteria and the spread of random mutations in a population are often modeled as random walks. The list of examples is ever expanding. These trajectories are paradigmatic of our theme of predictable randomness, as robust average properties coexist with the vagaries of chance. In addition, random walks, with their strange dependence of travel distance on number of steps, give us our first glimpse of the general theme of scaling. As we'll learn in part II, many physical characteristics don't simply grow proportionately to size; like our square root above, unexpected dependencies often pop up.

If we think about the DNA configuration abstractly as a random walk, our question of how stiff a DNA molecule is becomes transformed into questions of how long the "step size" is and how many steps there are. From images of DNA molecules, one can deduce that the length over which the double helix "looks" straight is about 100 nanometers, or a tenth of a millionth of a meter. In other words, replacing the actual path of the DNA with straight lines, and asking what straight-line length is appropriate, gives a value of 100 nanometers. (The technical name for a polymer's straight-line length is the Kuhn length, after Swiss chemist Werner Kuhn, and there is a precise mathematical expression for its calculation.) For scale, the width of the double helix is 2 nanometers, and the length along the ladder for a full helical turn is about 3 nanometers; the Kuhn length is large compared to the fine structure of the helix.

We can think of 1 meter of DNA as therefore being composed of 10 million straight-line steps. Our question of how big 1 meter of DNA would be if left alone, floating around in the watery environment of a

cell, becomes transformed into the question of how far a random walk of 10 million steps, each 100 nanometers long, would be. The answer: about 0.3 millimeters, or 300 micrometers. (That's the square root of 10 million, or about 3000 step-lengths, multiplied by the step-length of 100 nanometers.) That's far larger than the few-micrometer size of the cell nucleus; it's much larger even than the entire 10–100 micrometer width of a typical human cell.

You might object, perhaps, if you're aware that your DNA isn't one unbroken strand but rather is divided into 23 chromosomes. (Nearly all of your cells have 46 fragments grouped in pairs, from each of the two copies of your genome. Exceptions are egg and sperm cells, which have one copy of your genome, and red blood cells, which in humans and other mammals lack DNA.) The fragmentation simplifies the spatial challenge of packaging, but not by much: human chromosome 1, the largest chromosome, is 249 million nucleotide base pairs in length, corresponding to an overall length of about 8.5 cm and a random walk "blob size" of about 90 micrometers, which is still much larger than the size of a cell nucleus. To scale, I've illustrated below a typical human cell and its nucleus, the random blob configuration of 1 meter of DNA, and the random blob configuration of 8.5 cm of DNA (like chromosome number 1).

We should, in fact, be amazed and impressed by the packing of DNA—not because of the length of your genome, but because DNA is so stiff that it isn't amenable to being confined inside a cell. The space

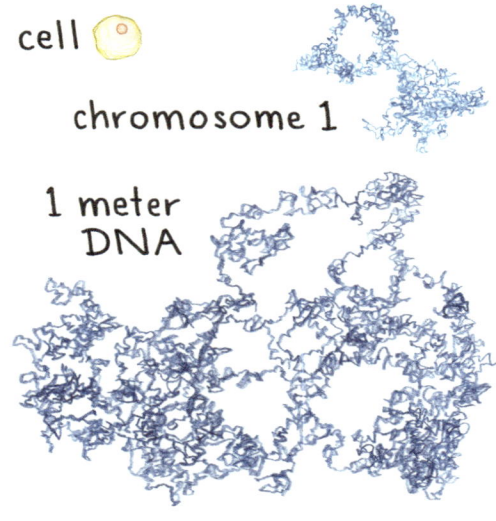

into which it's packed is far, far smaller than the space the molecule would occupy if left alone, floating in a watery world.

DNA PACKAGING

The DNA inside our cell nuclei isn't left alone like our spaghetti strands or random walks, and isn't stuffed in like clothes in a hastily packed suitcase, but rather is elegantly and compactly packaged. Much of your DNA is wrapped around little spools about 10 nanometers in diameter, made of proteins called *histones*.

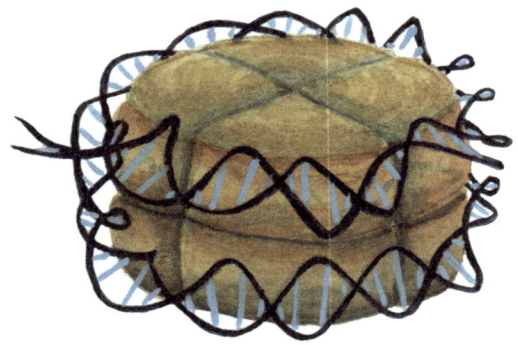

Ten nanometers is much smaller than the Kuhn length of DNA, so this wrapping requires a lot of force, provided in large part by electrical attraction between the DNA, which is negatively charged, and the positively charged outer surface of the histones. The spacing of positive amino acids matches the periodicity of the double-helical grooves, maximizing the magnitude of electrical forces. Once again, we find self-assembly at work: the physical attributes of DNA and histone proteins, especially their charge and shape, enable them to craft themselves into a well-defined, functional structure. Nearly two turns of DNA, or about 150 nucleotide base pairs, are wound around each spool. The span in between the spools varies in length, between 20 and 90 base pairs, and the whole assembly is evocatively referred to as "beads on a string."

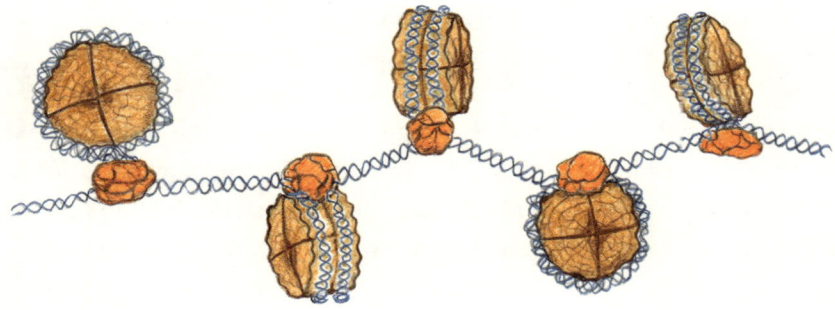

These strings of spooled DNA are further looped and packed together. What form they take has been a long-standing mystery, with a variety of structures proposed based on experiments that typically involve extracting DNA from cells or preserving cells with fixatives. The most common picture is of the beads on a string organizing into a 30-nanometer-thick fiber, with these fibers then arranged into 120-nanometer and thicker cords. Very recently, however, researchers at the Salk Institute and the University of California at San Diego, led by Clodagh O'Shea, developed a method to stain DNA in intact nuclei, decorating them with metal atoms that are readily visible in an electron microscope. Applying this approach, they did not find the expected discrete fibers, but rather uncovered chains with a broad spectrum of widths, ranging between 5 and 24 nanometers. These chains, moreover, differed in how curly or straight they were depending on whether or not cells were dividing. It may be that the packaging of DNA is less stereotypical, and more dynamic, than was previously thought.

The means by which cells stuff DNA inside themselves is more than an intellectual curiosity. The expression of genes—whether some span of DNA is actually transcribed into RNA, allowing the creation of proteins—depends a lot on the packing and organization of DNA. Regions of DNA that are wound around the histone spools or that are otherwise tightly constrained are relatively inaccessible to the machinery that reads and executes the genetic code. The exact same gene can be "on" or "off," depending on whether it is easily found or hidden away. DNA packaging, in other words, affects DNA function, and the physical arrangement of DNA is a powerful tool for regulating the activities of the cell. A variety of maladies as diverse as neurodevelopmental disorders, rare autoimmune diseases, and even cleft palate are associated with flaws not in the genes that encode the proteins that perform the associated neurodevelopmental, immunological, or skeletal tasks but rather in the DNA packaging. These flaws often involve the proteins that manipulate histone spools, increasing or decreasing their affinity for each other or for DNA, for example, by changing their charge.

In the past two decades or so, scientists have discovered that the determinants of what DNA regions are wrapped around histones are embedded in the DNA sequence itself, dictated in part by the mechanical properties of the double helix. We saw earlier that DNA is quite straight over lengths of about 100 nanometers. The exact stiffness depends subtly on the DNA sequence (the As, Cs, Gs, and Ts). Particular groups of nucleotides are less stiff than others, or by virtue of their shape prefer a slight curvature. In DNA that ends up wound around nucleosomes, these more curved or flexible regions, like little hinges, tend to be situated 10 nucleotides apart. The pitch of the double helix is also 10 nucleotides, meaning that if you stood on the twisted ladder of DNA and climbed 10 rungs, you'd be facing the same direction you were to begin with. The hinges, therefore, are all oriented the same way, allowing each span of DNA to bend toward the histone spool. Analysis of the binding between DNA and nucleosomes shows that sequences without these repeating pairs of nucleotides are less likely to be wrapped around histones. The DNA sequence itself, therefore, encodes mechanical information about how it should be packaged. DNA is a mechanical code subtly interleaved with a biochemical and genetic code—truly an extraordinary molecule!

The architecture of spools and fibers of DNA provides an example of the general theme of regulatory circuits with which cells can control their activity, turning genes on or off. As we'll see in chapter 4, many more strategies exist, enabling faster and more complex decision-making circuits.

DNA-STUFFED VIRUSES

It's not just *your* cells that are faced with the task of compressing DNA into tight spaces. Every living organism packs its DNA, and none leave it alone as a free, random-walk chain. This even extends to the not-quite-living world: viruses, little capsules of genetic material that infect living cells, hijacking their replication machinery, contain the densest-known packaging of DNA. Not all viruses contain double-stranded DNA; some contain a single DNA strand and some contain single- or double-stranded RNA. Those that do have a double-stranded DNA genome, which include the viruses that cause herpes and smallpox, must stuff this rigid molecule into a protein shell just tens of nanometers in diameter, again smaller than the Kuhn length of the DNA double helix. (Double-stranded RNA is even stiffer than double-stranded DNA. For both DNA and RNA, a single strand is much more flexible.)

In a double-stranded DNA virus, the bent, squished polymer pushes on the virus's shell, or *capsid*, trying to stretch out. When the capsid is opened, for example, when the virus infects a cell, this internal pressure helps propel the DNA into its cellular target. How can we measure the pressure of the compressed DNA? Imagine opening a closed capsid; the DNA rushes out. Now imagine squeezing the capsid from all sides, applying pressure, and *then* opening the capsid. If the external

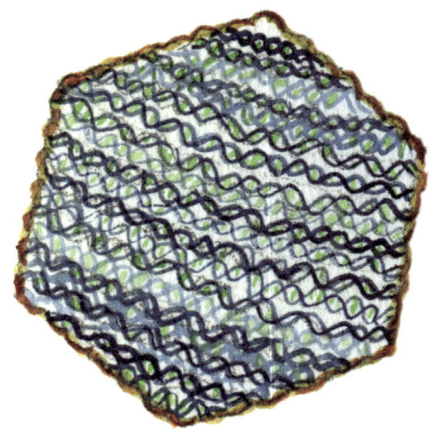

pressure is less than the internal pressure, DNA will still come out. If the external pressure is greater, the DNA will remain inside. By varying the external pressure and monitoring whether or not DNA is released, one can determine the pressure inside the virus.

That's easy to imagine, but actually doing it requires coming up with some clever experimental tricks, one of which was implemented about 15 years ago by William Gelbart and colleagues at the University of California at Los Angeles. Capsid opening is naturally triggered when a virus encounters particular proteins on the surface of its target cell. Adding these proteins artificially to a beaker full of viral capsids provides opening on demand. The viral particles are dispersed in a watery solution. Adding large molecules to the solution provides an *osmotic pressure*—a bit like bombarding the virus with all the molecules floating around it—that acts like our hypothetical squeezing of the capsids. By varying osmotic pressure and using protein-triggered capsid opening, scientists discovered internal pressures in viruses that were tens of atmospheres in magnitude. (For comparison, the air pressure in a car tire is about two atmospheres.) To get a more intuitive sense of the mechanical feats performed by these viruses, biophysicist Rob Phillips suggests envisioning 500 yards of Golden Gate Bridge suspension cable crammed into the back of a FedEx delivery truck. These huge internal pressures are valuable for the virus, helping it launch its DNA into a targeted cell where it will be replicated, initiating the generation of new viruses.

We can't understand DNA without understanding its physical properties. Shape, structure, and mechanics are inextricably tied to biological function. This statement isn't true just for DNA but for all of nature's biomolecules—a recurring theme throughout biophysics. In the next chapter, we return to the question of how a surprisingly small number of genes can guide the processes that make *you* by exploring how genes can be switched on and off—by external controls or by other genes—creating a meshwork of interactions that is again inseparable from the tangible, physical activities of life's molecules.

4

The Choreography of Genes

Chapter 3 introduced the central puzzle of genetic information: How is it that a mere 20,000 genes encodes the complexity of *you*? How do just 20,000 proteins—in other words, 20,000 tools or 20,000 components—perform the dazzling array of tasks that you're capable of, from growing to breathing to reading to reproducing? This is, of course, a human-centered way of asking these questions—we could just as well ask how 20,000 genes make a horse a horse, or how 30,000 genes make a water flea a water flea.

We're far from having a complete answer to these questions, and their exploration will keep scientists busy for decades, or centuries, to come. We have, however, uncovered intriguing general principles, motifs that illustrate how the complexity of life is encoded, and we've begun to use these principles to engineer living organisms in unprecedented ways. In the previous chapter, we viewed genes as more or less static—packaged in the space of the cell, with the potential to dictate the assembly of proteins. Now we introduce time—stimulating or suppressing the transformation of genetic information into physical activity as needed by dynamic, living organisms. Much of this choreography of genes is organized by genes themselves. We've so far considered self-assembly in a concrete, structural sense. Here we encounter a more abstract manifestation of self-assembly, as molecular activities weave themselves into regulatory circuits that make every creature a biological computer. To see what this means, we start with the notion of turning genes on and off.

GENE REGULATION

A cell, or a whole organism, can control when and whether to actually make use of any given gene—in other words, whether its string of As, Cs, Gs, and Ts is read by the machinery that transforms a DNA sequence into an RNA sequence into a protein. This control can be influenced by the external conditions that the cell or organism is experiencing, letting it activate or deactivate particular genes in response. Even before understanding regulation in detail, you could infer that something like this must exist from the fact that your body is composed of very different types of cells, though each contains a copy of the same DNA. The genomes inside a neuron, a skin cell, and a mucus-secreting cell that lines your gut are all identical. These cells, however, don't look the same, don't perform the same activities, and are not synthesizing the same set of proteins. Genes for proteins that create mucus must be dormant in neurons; genes for proteins that adhere tightly to neighboring cells must be active in your skin; your secretory cells must ignore the genes responsible for sending long-distance electrical signals. Somehow, it must be possible to turn genes "on" and "off." Let's see how this control is made possible.

Recall the transcription of a DNA sequence into RNA. The RNA polymerase machine slides along DNA like a train on a track, transcribing the sequence of a gene from its start to its stop signals, extruding a strand of RNA as it glides along. RNA polymerase isn't always bound to DNA, however. Much of the time it floats around in the watery medium of the cell; occasionally, it bumps into a specific DNA sequence that it recognizes and latches onto. As we saw in chapter 3, these sequences, called promoter sequences, are adjacent to genes or sets of genes. There's a directionality to DNA, and an RNA polymerase "reading" one strand of the DNA double helix will travel in a particular direction. Genes lie downstream of their promoters, so that an RNA polymerase that lands on a promoter sequence transcribes the adjacent genes. Controlling the tethering of RNA polymerase is the essence of *transcriptional regulation* of genes, one of the most powerful ways to control gene activity.

We first figured out mechanisms of transcriptional regulation in bacteria. Imagine you're a bacterium. You like to eat sugars, but you need to make sugar-digesting proteins to do this. You'd prefer to make more such proteins if you encounter sugar, and you'd rather not waste energy making these proteins if there's no sugar around. How can you do this? We illustrate with an actual sugar, lactose, and the regulatory machinery found in the bacterium *E. coli*, which is very similar to machineries used throughout the living world.

A gene called *lacZ* encodes part of the lactose consumption machinery. Upstream of it, as always, is its promoter region. I've drawn RNA polymerase as a green blob poised to advance and read the (blue) *lacZ* gene.

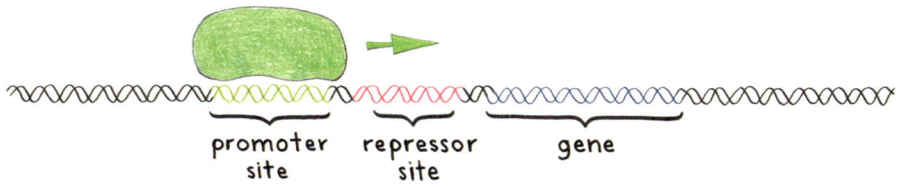

(The illustration isn't to scale; the *lacZ* gene is about 3000 base pairs long, for example, and the width of RNA polymerase spans 30 to 40 base pairs.) *E. coli* also makes a protein called the lac repressor, which binds to another stretch of DNA upstream of the *lacZ* gene. When the lac repressor (red) is DNA-bound, RNA polymerase can't attach and the *lacZ* gene is not expressed.

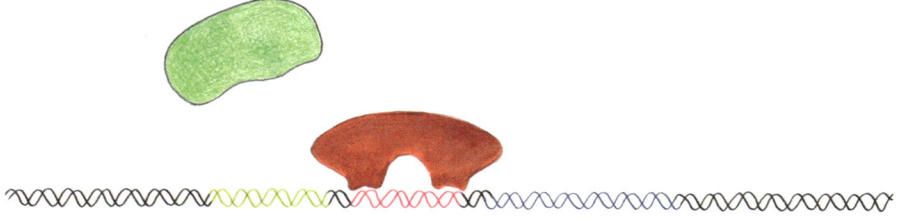

As we've seen in previous chapters, DNA and proteins are physical objects, with particular structures that guide how they work. The binding between the lac repressor and DNA has a particularly striking arrangement. The specific sets of nucleotides that the lac repressor recognizes are spaced farther apart than the repressor's width. The protein must therefore loop the DNA into a tight circle, about 10 nanometers in diameter.

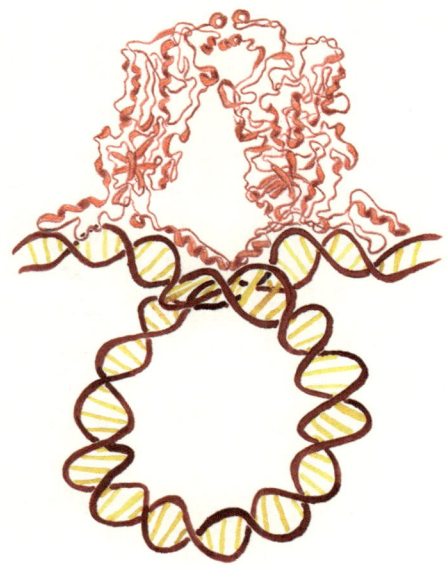

Recall, however, that DNA is very stiff. Left alone, it will be quite straight over distances of about 100 nanometers. Like a circus strongman flexing an iron bar, the lac repressor bends the DNA. The looped DNA interferes with the normal binding of RNA polymerase, preventing the readout of the genes for lactose-digesting proteins.

The lac repressor has another amazing property: it can also bind to a lactose lookalike called *allolactose* (the black circle in the illustration below); when allolactose-bound, the shape of the repressor subtly changes so that it can no longer bind to DNA. The bacteria create allolactose from lactose itself. If the bacterium encounters lactose in its environment and internalizes some of it, so that the lac repressor no longer represses, the lactose-digesting proteins are made and the bacterium can gorge on the food it has found.

Repressors like the lac repressor are common in all organisms, not just bacteria. Inhibiting RNA polymerase binding, or at least competing with it to make it less likely, is one of nature's favorite tactics

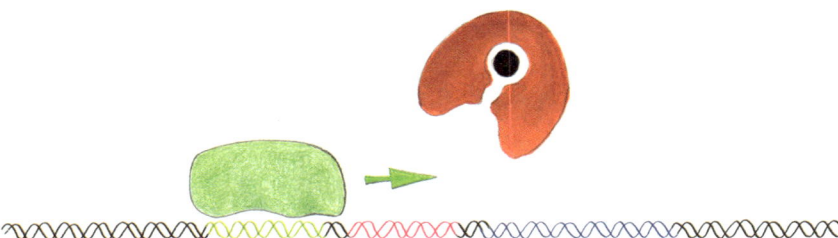

for regulating the activity of genes. As with the lac repressor and lactose, this inhibition can be coupled to an external stimulus or, as we'll see, an internal stimulus.

The opposite of repression in the tool kit of genes is activation. Especially for promoters to which the binding of RNA polymerase is weak, activator proteins with an affinity for RNA polymerase can attach to nearby DNA, again at specific sites recognized by the proteins, enhancing the likelihood that RNA polymerase will stick and initiate transcription.

Activator proteins also have a role to play in the story of lactose and bacteria. Though bacteria such as *E. coli* like lactose, they like glucose, another sugar, much more. If glucose is present, they won't waste their efforts digesting lactose even if it's available. This phenomenon was discovered in the 1940s by Jacques Monod, who balanced studies of fundamental biology with work in the French Resistance during the Second World War. At the level of its DNA, the bacterium must express the lactose-digesting genes only if lactose is present *and* glucose is absent. An activator protein makes this possible. The binding between RNA polymerase and the lac promoter region is weak, so even if the repressor is absent, the transcription of the gene is unlikely to occur. The bacterium makes an activator protein that binds to DNA only if it has also bound a molecule called *cyclic AMP*. Cyclic AMP is produced by

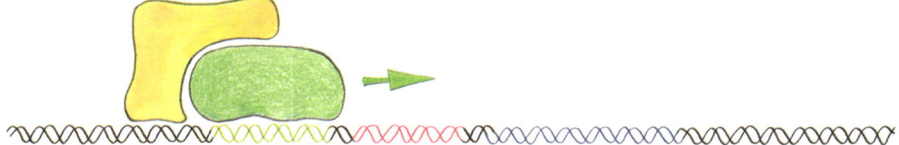

the bacterium when glucose levels are low; it's sometimes called a "hunger signal." Therefore, if glucose is present, there's little cyclic AMP, the activator won't bind, and the lactose-digesting gene won't be expressed even if lactose is present. If glucose is absent, there's a lot of cyclic AMP, the activator binds, and the lactose-digesting gene is expressed *if* the polymerase isn't being blocked by the lac repressor. It's a clever system, especially for a brainless creature a thousandth of a millimeter in size.

Repressors and activators are both referred to as *transcription factors*—things that govern the transcription of genetic information. Transcription factors are themselves proteins, and so are encoded by genes. We have a lot of them—it's not known exactly how many, but it's thought that the human genome contains over 1500 genes for transcription factors. Recall that we have only about 20,000 protein-coding genes. A sizable fraction of our genetic instruction set, in other words, is made up of the brakes and levers that regulate the readout of those instructions.

Transcription factors and the decisions they make possible are found throughout the living world and are essential for the encoding of complex behaviors by simple genes. Regulatory regions—the landing sites on the genome for transcription factors—don't even need to be adjacent to the genes they regulate. Because the genome twists and turns, a transcription factor bound to a segment of DNA can influence expression of a gene that would be distant if the DNA were laid out in a straight line but that's nearby in actuality.

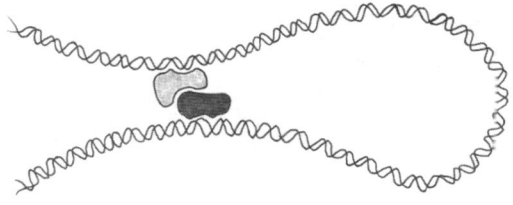

This interplay between genetic and physical distance offers even more possibilities for gene regulation and is the topic of a great deal of present-day biophysical investigation.

All the mechanisms we've explored so far involve the regulation of transcription, the first step in gene expression in which a DNA code is turned into an RNA code. Cells can also regulate *translation*, the synthesis of a protein from the RNA segment. This too can be done in many different ways, including controlling the rate at which RNA degrades, cloistering RNA in particular regions of the cell, and even creating RNA molecules that are complementary to the transcribed RNA so that the two bind together to form a double-stranded RNA that can't be translated into protein. We could spend many more pages exploring the variety of the gene regulation tool kit, but instead, let's step back and look at the universality of these tools and some motifs nature has developed to combine these tools into machines.

PORTABLE GENETIC CONTROL

We've seen how the lac system uses transcription factors to turn a gene on or off depending on stimuli from its environment, namely, the presence or absence of certain sugars. *E. coli* and other bacteria use this system to match their biochemical activity to the availability of particular foods. A researcher could easily add lactose to a flask full of glucose-starved bacteria, triggering the microbes to activate the *lacZ* gene. More sneakily, though, the researcher could add a chemical called IPTG, which is very similar to lactose except that it can't be digested. The lac repressor will bind to IPTG and therefore not repress transcription, and the cell will produce lactose-digesting enzymes even though there's no lactose to digest. The motivation for this strange subterfuge is to construct a handle for gene expression. The researcher beforehand could have inserted other genes downstream of the *lacZ* promoter, perhaps also deleting *lacZ* itself. These new genes could be genes for fluorescent proteins with which to monitor the bacterium, or genes to synthesize various useful chemicals or drugs. The researcher now has the expression of these new genes under the control of an external trigger, the IPTG.

A striking example of this genetic control is described in a paper from 2001 by Heidi Scrable and colleagues at the University of Virginia, simply titled "The Lac Operator-Repressor System Is Functional in the

Mouse." The researchers made use of albino mice, which have a mutation in a gene called tyrosinase that is necessary for pigment production. By inserting a functional tyrosinase gene (brown in the illustration) into the genome, along with its promoter sequence (green), the authors created mice with normal pigmentation: brown hair and eyes, as one would expect. The remarkable part comes from engineering control of the pigmentation genes. Though animals have a large number of gene regulatory systems, the lac system is not one of them; it's only found in bacteria. However, the researchers engineered a mouse that contains the DNA binding sequence of the lac repressor (red), situated upstream of the functional tyrosinase gene. Because the lac repressor gene doesn't exist in mammals, there's no lac repressor protein, and no repression of tyrosinase. These mice, therefore, are also pigmented (second row).

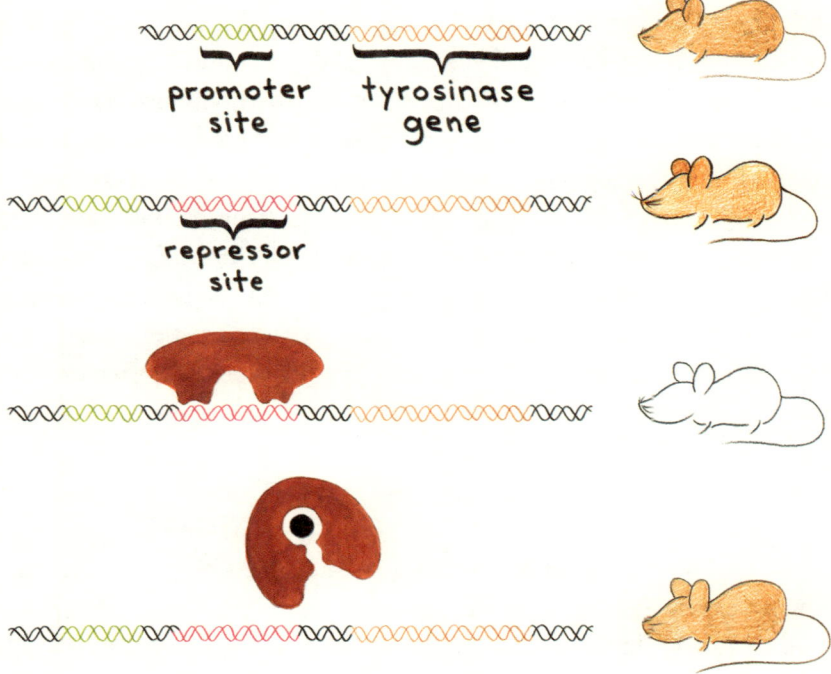

Yet another set of mice was created in which, again, tyrosinase was controlled by the lac promoter but in which the lac repressor gene was also inserted into the genome, along with its promoter, at some other location. These mice produce the lac repressor protein, which represses tyrosinase expression, so the mice lacked pigmentation (third row). If the researchers added IPTG to the drinking water of the mice, the animals produced the proper pigmentation, turning brown (fourth row). The IPTG prevented the lac repressor from repressing, just as it does for bacteria, enabling the expression of the pigmentation gene.

In addition to the almost surreal ability to change an animal's hair and eye color by adding a sugar-like molecule to its water, this outcome highlights the universality of life's machinery. The last common ancestor of mice and bacteria lived over three billion years ago. The descendants of that ancestor have evolved along separate paths ever since, giving us very different creatures—a single-celled microorganism and a palm-sized, furry mammal. Nonetheless, one can cut-and-paste a regulatory apparatus from one into the other, and it works perfectly well. As Monod himself dramatically and presciently noted decades earlier, "What is true for *E. coli* is true for the elephants."

Like the lac system, there are many others that allow organisms, or researchers, to regulate gene expression. My own lab also makes use of these sorts of engineered circuits. We don't change the colors of mouse fur, but we turn on or off the ability of certain bacteria to swim, directing them to assemble or disassemble their microscopic motors by placing a simple chemical into their water. With this tool, we can infer the role of a physical behavior like swimming on the bacteria's ability to succeed in their environment. Just a few decades ago this would have been science fiction; now it's not only possible but becoming steadily easier.

GENETIC MEMORIES

If you flip the switch to turn on a lamp, you needn't keep pressing your finger against the button for the light to remain on. The switch is toggled to a new, stable position, and it stays there until toggled to a different, also stable, position. Nature, and researchers, also often want switches like this, which set cells on a particular path once they've received a signal and keep them on that path even after the signal has gone. In plants and animals, this is especially important for the development of different types of cells. Both neurons and the glia that help neurons function, for example, descend from the same type of progenitor cell. Some signal sends a progenitor on the path to be a neuron, after which it's committed to express the proper set of neural genes, make synapses with other neurons, and perform all the other tasks a neuron is charged with. You wouldn't want it to need constant reminders not to revert to its ancestral form, or to suddenly switch to being a glia, or to turn into a confused half-neuron, half-glia. Having stable cell types requires having toggle switches for genes. Another way of saying this is that the cell needs *memory*; it needs to remember past stimuli, encoding them into the way genes are expressed in the present and future.

There are many ways to make a memory. Here's one that builds on what we've seen of transcription factors. Imagine two genes that we'll call A and B. As with lac, imagine there's a repressor for gene A. Now suppose that the gene that encodes that repressor protein is immediately downstream of gene B, so that if B is expressed, the A-repressor is also expressed. Now, symmetrically, imagine that the B-repressor gene is just downstream of A, so that if A is expressed, the B-repressor is also expressed. This mutual repression enables memory: Suppose A happens to be strongly expressed. The cell makes lots of B-repressor protein, so B will be repressed and there won't be any repression of A, consistent with A's strong expression. The cell will continue in the A state. On the other hand, if B is strongly expressed, the opposite set of events will occur, and the cell will continue in the B state. We have two stable types of behavior for the cell. We can toggle between them by, for example, flooding the cell with a lot of whatever triggers the activation or repression of A or B—IPTG, for example, if the lac repressor is used for part of this apparatus. From that point on, the cell retains a memory of this event.

There's a very general principle illustrated here, which is that genes regulate the expression of genes. In other words, feedback between genes creates particular patterns of activity. Our example uses two instances of repression (negative feedback) to create a switch. It's not just hypothetical: exactly this scheme is used throughout nature, for example, in viruses infecting bacteria that decide between an actively dividing and a dormant state. Many other schemes are also possible. We could use activators of transcription, with, for example, expression of gene A coupled to expression of an activator for gene A, amplifying whatever initially put the cell on the A path (positive feedback).

CLOCKS AND CIRCUITS

To tell time, we use clocks. Every clock is based on some periodic, rhythmic phenomenon, such as the back-and-forth swaying of a pendulum bob or the rapid vibrations of a quartz crystal. All sorts of living organisms, and even single cells, use clocks to control activities that should wax and wane with some well-defined period. The circadian rhythm is a great example. In many plants, the production of chlorophyll has an approximately 24-hour cycle, matching the periodicity of the day. The plant doesn't solely rely on external cues, subject to the whims of clouds and shadows, but has an internal timekeeper with a 24-hour period. You do as well, with body temperature, blood pressure, and of course sleepiness rising and falling roughly once per 24 hours, even if you're sequestered in a windowless, constantly lit room for weeks on end. Many animals, fungi, and even some bacteria possess a circadian clock. Sensing light helps maintain the rhythm and shifts the timings of its peaks and troughs, but the periodicity itself arises from internal oscillators.

Gene regulation provides a way for a single cell to make an oscillator, using only the ingredients of the cells themselves. We'll be a bit abstract, since the details of real cellular oscillators are rather complex, involving many interacting genes. The basic idea, though, can be illustrated with just one gene.

The simplest possible oscillator consists of a gene that represses itself—in other words, a gene that encodes a protein that represses the expression of that gene itself. At first glance, this seems ridiculous; how could such a gene ever be expressed? The answer lies in the fact that both expression and repression take time. Recall that the expression of a gene means the transcription of that stretch of DNA into a segment of RNA and then, afterward, the translation of that RNA into a chain of amino acids—a protein. After that, if the protein is to repress the gene, it must meander and find the promoter region. All this takes time. Even after the repressor binds and RNA polymerase is blocked, the activity of the gene isn't immediately quenched. The pieces of RNA that were already made can continue to be translated into protein, and the proteins that were already made can continue to do whatever they were doing. The upshot of all this is that the expression of the gene can increase for some time. The activity of the gene, therefore, persists for a while, though it represses itself. To see how this activity can oscillate, we need one other fact about proteins: they all degrade over time.

The decay of transcription factors has a large impact on their regulation of genes. A fundamental truth of the physics of molecular binding is that the greater the concentration of a molecule, like our repressor, the greater the likelihood that it will be bound to something it has an affinity for, like our promoter region. Conversely, as the free repressor proteins degrade and their concentration declines, the repressor proteins on the DNA become increasingly likely to unbind, no longer repressing transcription. The gene can then be expressed.

Putting this all together, we have the following: In the preceding circuit, the gene is initially expressed and slowly builds up the concentration of its own repressor proteins, inhibiting further creation of proteins. But the existing proteins degrade, and eventually so many have vanished that the gene can be expressed again, and the cycle begins anew. Thus, we can have an oscillator.

It's not a very good oscillator, though, and I don't know of any organism that actually uses a clock made of a single gene. The timing resulting from this arrangement is hard to tune and its periodicity isn't very precise. Both properties depend on the rate of protein degradation in a cell, which is determined by a variety of factors beyond the purview of the gene and its self-repression.

A better scheme involves three sets of genes, A, B, and C, in which A represses B, B represses C, and C represses A. I won't give a detailed analysis, but this circuit also oscillates. The amount of each of the A,

B, and C proteins goes up and down periodically. The frequency of the oscillations depends on the affinities of the repressor proteins to the DNA. We, or the cells, can tune the periodicity of the cycles by using repressors with stronger or weaker DNA binding. This A-B-C circuit, called the *repressilator*, doesn't exist in nature, at least not as a stand-alone unit. However, it was one of the first to be artificially engineered into cells, by biophysicists Michael Elowitz and Stanislas Leibler at the turn of the twenty-first century. The researchers inserted this oscillator into the genome of *Escherichia coli* bacteria and coupled it to the green fluorescent protein gene, creating cells that rhythmically switched between being fluorescent and dark. Since then, a variety of other precise and tunable oscillators have been engineered into cells.

Though a stand-alone repressilator hasn't been found in nature, similar circuits are commonplace. The 24-hour oscillator that controls the human circadian rhythm, for example, involves a handful of genes connected by intertwined feedback loops, including the repressilator motif. The resulting machinery gives rise to a clock that is robust but also trainable by external stimuli, such as sunlight, that induce various chemical changes in our bodies. This training isn't immediate, as anyone who has experienced jet lag knows. Sometimes, we'd like to reset our clock more rapidly than our bodies can accommodate. Our circadian rhythm, however, didn't evolve in a world with high-speed air travel.

Beyond memories and oscillations, there are innumerable other combinations of gene interactions made possible by the regulatory tool kit. Imagine, again abstractly, an A-B-C trio of genes in which A and B each encode an activator of C, and the expression of C is weak unless one of the activators is present. If whatever induces expression of A is present, C will be expressed, *or* if whatever induces expression of B is present, C will be expressed. We could also design some set of genes that only activate C if the A stimulus *and* the B stimulus are present, or if the A stimulus *and not* the B stimulus is present, and so on. (In fact, we've seen earlier in this chapter an example of the last configuration, in bacteria in which *lacZ* is expressed if lactose *and not* glucose is present.)

A device that can perform logical calculations on inputs—decisions based on *and, or, not,* and other such operations, and combinations of these operations—is a computer. The computers we're used to thinking about use electrical voltages—high or low, on or off—rather than the presence or absence of biochemical transcription factors, but the conceptual framework is the same. Moreover, the underlying generality is the same. With the appropriate combination of logic elements, whether they're electrical bits or genes, one can perform any computation, whether it's compressing a digital video file or deciding whether conditions warrant germination of a seed.

With logic and memory, nature can make all sorts of genetic circuits that perform all sorts of tasks. The complexity of gene-driven activities, therefore, can be vastly greater than one might think from simply tallying the number of genes.

GENES IN THE ATTIC

Another way of controlling whether genes are expressed or not involves a phenomenon we've already encountered: the packaging of DNA. As we've seen, the arrangement of histone-wrapped DNA can dictate how easily RNA polymerase can read a gene. Over the past few decades, we've learned how powerful this packaging-based gene regulation is. A tool kit of proteins adds and subtracts particular sets of atoms from histone proteins, influencing their further assembly or disassembly as dense DNA-histone fibers. This modification of histones is particularly important during the early development of an embryo, as descendants of a handful of cells permanently adopt distinct cellular identities, and in cancer, as cells change into rapidly growing and migrating forms. Genes that aren't needed by particular cell types remain packed away, as if in storage in an attic—available, but not readily accessible.

Cells can also alter DNA itself to affect its readability. Certain proteins, for example, can perform chemical reactions that replace a hydrogen atom with a carbon atom and three hydrogen atoms (a methyl group) on an A or C nucleotide. This appendage can inhibit the expression of the gene the methylated nucleotide is part of. This tiny tag on the DNA can gum up the transcription machinery and can also recruit proteins that further modify histones.

Surprisingly, it seems that these facets of genetic regulation—packaging and methylation—can be passed on from parent to child. We inherit, it seems, not just our genome from our parents—the sequence of As, Cs, Ts, and Gs—but some aspects of the genome's organization that affect how it is interpreted. This phenomenon, known as *epigenetics*, adds further complexity to the connections between our genetic code, our environment, and how we function as organisms. Studies of the Dutch who endured the 1944–1945 "hunger winter," for example, found an increased likelihood of obesity and cardiovascular disease later in life, and altered DNA methylation that persisted for decades; moreover, similar signatures of ill health were present in the survivors' children, born long after the famine, indicative of epigenetic inheritance.

We conclude this chapter by reminding ourselves that the machinery of gene regulation makes every organism a powerful, versatile computer, capable of making decisions based on the diverse stimuli provided by its environment, and orchestrating behaviors that vary in time and space. This complexity doesn't require thought, or central control, but rather comes from the nature of the genetic code itself, which contains genes as well as the means to regulate them. Again, we see self-assembly at work, with machineries that construct themselves. The elegance of genetic circuitry and predictable decision making coexists, however, with an inherent randomness that is ubiquitous in the microscopic world, the importance of which we see in chapter 6. First, however, there's one more crucial cellular component to introduce: membranes, whose architecture provides a stunning example of self-assembly separate from DNA and proteins.

5

Membranes: A Liquid Skin

You began as a single cell, a fertilized egg. This cell divided in two, these two into four, and after many more divisions, rearrangements, and changes of shape, you developed a body composed of tens of trillions of cells. Every one of these cells has a membrane at its edge. This membrane doesn't just serve as a boundary between inside and outside, but forms an arena where the cell grabs onto its surroundings, traffics chemicals, and exchanges signals with its neighbors.

Membranes exist within cells as well, as the boundaries of a variety of organelles ("little organs"). The nucleus that houses each of your cells' DNA, for example, is an organelle (purple in the illustration). Many proteins are made at an organelle (the endoplasmic reticulum, dark blue) that consists of a long, convoluted labyrinth of membrane. At another organelle (mitochondria, red), the cell synthesizes the chemicals it uses to transport energy; this organelle has a double membrane, the inner of which is folded into layered stacks. Not all cells have

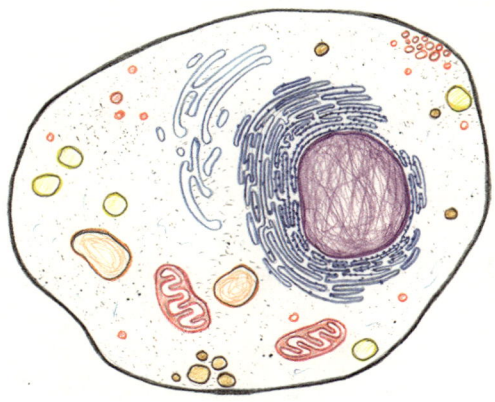

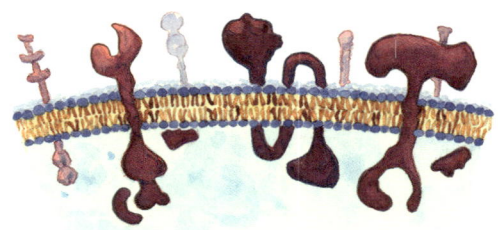

membrane-bound organelles: members of the bacteria and the archaea, two of the three domains into which all creatures are classified, lack them. Organelles, however, are prevalent in the third domain, that of the eukaryotes, which includes all animals, plants, fungi, and many single-celled organisms, such as amoebas.

At the core of every cellular membrane is a sheet just a few billionths of a meter thick made of molecules called *lipids*. Proteins are present as well, poking through the membrane, opening pores in it, or lying against it. These membrane-associated proteins account for over a third of the human genome and form the majority of pharmaceutical targets; they orchestrate a lot of biological activity. The lipids, however, are what make a membrane a membrane. Given the importance of membranes, one might expect their structure to be precisely dictated by the cell's genetic code and carefully monitored by some internal machinery. Instead, we find the opposite, laissez-faire approach: proteins make lipids and self-assembly does the rest. Physical interactions among lipids and water suffice to generate a reliable yet dynamic material. To understand the properties of membranes and how they emerge, let's first look at something more familiar.

HOW TO MAKE A MEMBRANE

Oil and water don't mix. The oil in a shaken bottle of vinaigrette salad dressing quickly coalesces into droplets. Left alone, oil molecules associate with other oil molecules and water molecules with water molecules. As we saw in chapter 2, substances such as oils and fats that separate from water are called hydrophobic ("water-fearing"); substances such as sugar and vinegar that mix with water are hydrophilic ("water-loving").

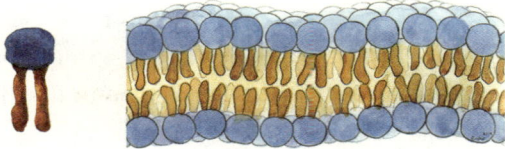

Lipids are both hydrophobic and hydrophilic. Each lipid molecule has a "head" that likes water and a "tail" that doesn't. The tail typically consists of two chains, each of which chemically resembles an oil molecule. There are other familiar substances that also have these amphiphilic ("loving both") tendencies: every soap molecule also has a hydrophilic head and a hydrophobic tail, typically just one chain, which together let the soap cling to greasy dirt as well as to the water that washes it away.

Lipids in water struggle with the contradiction imposed by their structure: their heads are happy but their tails are not. They therefore spontaneously associate with one another to shield their tails from the water, forming a two-molecule-thick sheet. This "lipid bilayer" forms the basis of all cellular membranes.

Lipid membranes have remarkable physical properties. They are essentially two-dimensional: the bilayer is about 5 nanometers thick—about 20,000 times thinner than a typical sheet of paper—while its lateral extent can be many thousands of times greater than its thickness. They are flexible, bending and curving in three dimensions, and living cells carefully control this curvature.

You might think that a membrane is rather like a plastic bag: thin and flexible. There's a key difference, though. If you take a marker and dab a spot of ink on a plastic bag, then leave it alone and check back a few minutes later, the spot will be in the same place you left it. If you were to do the same to a lipid membrane, the spot would blur and disappear, the marked molecules having meandered elsewhere throughout the membrane. Lipid bilayers and cellular membranes are fluids. Just as the water molecules in liquid water aren't fixed in relation to one another, but rather can flow throughout the fluid, lipids and proteins embedded in a membrane are mobile and can flow throughout its extent. Membranes are two-dimensional fluids. This mobility, like the other physical properties so important to membranes, follows from the nature of lipid bilayers: the lipids aren't rigidly bound to one another, but are simply associating to shield their hydrophobic tails from water. There's no impediment to molecules weaving around one another, as long as the hydrophobic core of the bilayer remains unexposed to water.

Molecular mobility is a wonderful trait: membrane molecules can rearrange themselves, interact with one another, and even form structures and patterns that help cells carry out various tasks.

A striking example occurs in your immune system, during the interactions between two cell types, known as T-cells and antigen presenting cells (APCs). APCs take up proteins from their surroundings, rip them apart, and put the resulting fragments on display by attaching them to proteins sticking out from their outer membranes. T-cells meet the APCs, make contact, and examine the fragments to determine whether they came from you or from some foreign source—perhaps a bacterium or virus. If foreign, the T-cells activate your immune system, triggering your body's defenses to fight the apparent invasion. This response to foreign protein fragments involves a dynamic molecular dance at the T-cell/APC contact. Adhesion proteins at each cell's outer membrane bind to one another and start to cluster together.

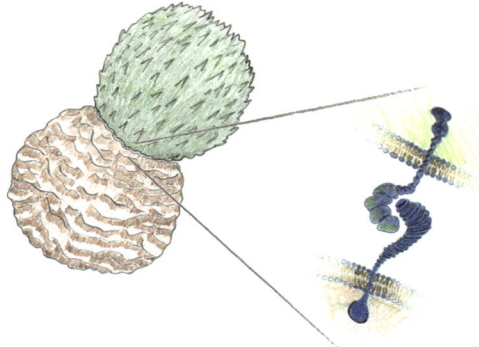

Around them, the proteins involved in the display and recognition of protein fragments—in other words, the signaling between the cells—also begin to bind to one another and group together. If we imagine the plane of this page as the interface of contact between the two cells, the initial arrangement, with adhesion and signaling proteins drawn as blue and red, respectively, looks like the illustration on the left.

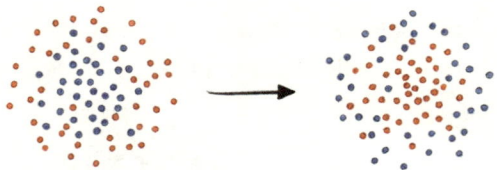

Then, over the course of a few minutes, this bulls-eye inverts, and hundreds of signaling proteins flow to the center while the adhesion proteins assemble in a ring (right).

This structure, called the *immunological synapse*, was discovered in the mid-1990s. Understanding how these spatial patterns form, and then how a T-cell translates them into the "on" signal for its activation, has been intensely studied in the years since. In addition, scientists have found that similar synapses form at the contacts between immune cells transmitting the human T-cell leukemia virus as well as the human immunodeficiency virus (HIV, which causes AIDS). These viruses, it seems, have figured out how to hijack the structure-forming machinery of cells they infect. We won't go into the intricacies of synapse formation but limit ourselves to pointing out that, if they weren't embedded in a two-dimensional fluid, T-cell signaling proteins, adhesion proteins, and many other membrane proteins in many types of cells would be incapable of performing the dynamic spatial rearrangements that nature demands of them.

Protein and lipid patterns also intersect the theme of predictable randomness. The fluidity of membranes allows their components to reorganize, but it brings with it the uncertainties of flow and disorder. Neither we nor a cell can know exactly where every lipid and every protein will be, but we can predict, on average, properties of the ensemble. Both the deeper nature of the randomness and the meaning of prediction become clearer in chapter 6.

CONES, SPHERES, AND BUBBLES

A bilayer isn't the only structure that an amphiphilic molecule can form. Consider a molecule shaped like an ice cream cone, with the hydrophilic part being the ice cream and the hydrophobic part being the cone (left). In water, it will self-assemble into a sphere (right), illustrated two-dimensionally here, to shield the hydrophobic cones.

Molecular shape is a key determinant of self-assembly. Most of the lipids in your body are "cylindrical" rather than conical; the hydrophilic heads and hydrophobic tails are similar in width, hence the ensemble of lipids forms into a fairly flat bilayer. A small fraction of your cells' lipids are not cylindrical, though—not enough to disturb bilayer formation, but enough, it's thought, to help generate curvature as the cell bends membranes into complex structures.

You might wonder whether amphiphiles could be arranged so that the hydrophilic heads are "in" and the hydrophobic tails are "out." They can, and you've created such a structure every time you've blown a soap bubble.

At the edge of a soap bubble, soap molecules sandwich a thin film of water, sticking their hydrophobic tails out toward the air. Soap films and cell membranes share some deep similarities—worth keeping in mind the next time you wash your dishes.

TUBERCULOSIS AND TOUGH MEMBRANES

In London at the start of the nineteenth century, 30% of all deaths were due to tuberculosis, an infectious disease that most commonly affects the lungs. The "white plague" infected and killed vast numbers of people worldwide, and remained the first or second leading cause of death each year in the United States through the first decade and a half of the twentieth century. Even now, about one million people die of tuberculosis annually. Tuberculosis is caused by a bacterium: *Mycobacterium tuberculosis*. A different mycobacterium, *Mycobacterium leprae*, causes the disease leprosy, another scourge of humanity that has eaten away at the skin and nervous systems of its victims for millennia, until the advent of modern antibacterial treatments. The mycobacteria are notoriously tough. We've known for about a century, for example, that *Mycobacterium leprae* and *Mycobacterium tuberculosis* can survive periods of dehydration lasting several months. This is puzzling not only because cells need water for their biochemical activities but also from a membrane-based perspective: if there isn't water near the lipids' hydrophilic heads, how can hydrophilic and hydrophobic interactions hold a membrane together, and how can the membrane hold the bacterium together?

The mycobacteria, it turns out, have very odd membranes. Their interiors are bounded by a lipid bilayer, just as in your cells and the cells of every other organism. Outside this bilayer, however, is a dense hydrophobic gel, outside of which lies a lipid monolayer—in other words, a single layer of lipids, whose greasy tails point inward and whose hydrophilic heads face the outside. Not only is this arrangement of lipids unusual, but the lipid molecules are unusual in themselves: many have a sugar called trehalose bound to their hydrophilic heads (green in the illustration). The mycobacteria and a few closely related soil bacteria are the only organisms on the planet known to have trehalose lipids. Could this be important?

I learned about these mycobacterial membranes about a decade ago, not long after starting my research lab at the University of Oregon.

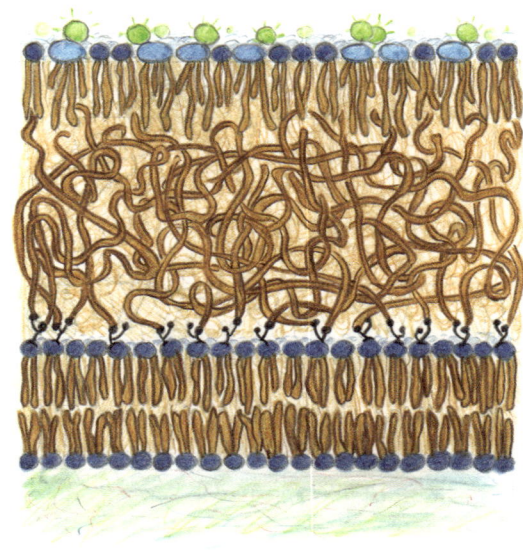

I had been working with lipid bilayers for a few years by then, mostly measuring physical properties, such as stiffness, to understand what these materials are capable of. Through some experiments involving sugars and polymers, I began to work with the group of Carolyn Bertozzi, a chemist then at the University of California at Berkeley. Coincidentally, one of her lab's major projects was to unravel the methods by which the mycobacteria create trehalose lipids and other strange molecules, both to understand the chemical tools nature has developed and to potentially disrupt them to tackle disease. It was through these interactions that I first heard about trehalose lipids, which immediately rang bells because trehalose, in other contexts, is an almost magical sugar. There are a handful of organisms, including some fungi, plants, yeast, and even certain animals, that can survive the loss of over 99% of their water. The "resurrection plant" *Selaginella lepidophylla*, for example, can withstand near-total dehydration for years, curling into a tight, brown ball that revives as an ordinary-looking green plant when hydrated. A common feature of many of these organisms is that they make trehalose, often in copious amounts. Compared to other sugars, like the familiar glucose and sucrose, trehalose is less likely to form crystals when concentrated, leaving the sugar molecules more available to interact with other substances. In addition, trehalose readily forms so-called hydrogen bonds, similar to the bonds between water molecules and between water and hydrophilic molecules, allowing the sugar to mimic water to some degree. Unlike water, however, trehalose doesn't readily evaporate. It's believed that for all these reasons, trehalose is a useful substance for dehydration resistance. In fact, there's a lot of contemporary effort to put it to use outside of organisms to preserve cells such as blood cells, and biomaterials such as proteins and vaccines, in a dry state for storage and transport. Perhaps the mycobacteria, I wondered, adapted the trehalose tool kit to link it to lipids in order to protect their outer membranes from dehydration. How could we test this idea?

We couldn't just engineer the microbes to stop making trehalose lipids and then test their resilience because we don't understand the mycobacterial machinery well enough to alter it. Even if we could, I wasn't keen on keeping tuberculosis-causing bacteria in the lab. (As we discuss later, my group happily works with the bacteria that cause cholera, but cholera is easy to prevent, hard to catch, and easy to cure—the opposite of tuberculosis.) I decided on a different approach instead, one that's been very fruitfully applied to "normal" lipid bilayers for decades: re-creating artificial, cell-free membranes on solid surfaces. Normal lipids can be coaxed to form bilayers on very clean and flat glass surfaces. Moreover, because of the hydrophilic nature of the glass and the lipids, a water layer just 1 to 2 nanometers thick separates the lipids and the glass, allowing the bilayer to retain its two-dimensional fluidity. At the cost of sacrificing some of the realism of an intact cellular membrane, this provides a controllable, convenient platform for studying bilayer biophysics.

We decided, therefore, to try to build an analogous supported membrane platform to mimic the non-bilayer structure adopted by tuberculosis lipids. We first chemically linked hydrophobic molecules to glass wafers. Then we formed monolayers of lipids at the surfaces of water-filled troughs, where the hydrophobic tails would stick out into the air, and gently transferred these monolayers onto the wafers so that the tails met the linked hydrophobic layer. We constructed the monolayers to contain specific fractions of purified trehalose lipids, among other more conventional lipids. Just like the natural mycobacteria have a dense, hydrophobic layer underlying a monolayer that contains trehalose lipids, our artificial membranes had a dense, hydrophobic layer underlying a monolayer that contains trehalose lipids.

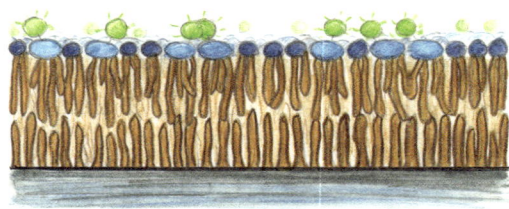

With this platform in hand, we could dehydrate and rehydrate the supported membrane. As expected, lipid monolayers composed completely of "normal" lipids didn't survive desiccation. In contrast, monolayers composed almost entirely of the mycobacterial trehalose lipids were intact after dehydration and rehydration, and even retained their two-dimensional fluidity. More remarkably, monolayers that were a mixture of normal and trehalose lipids could survive the loss of water down to a trehalose lipid concentration of about 25%. Even a minority of trehalose lipids can make a membrane resistant to dehydration! We pushed this a step further with our colleagues in the Bertozzi Lab, especially David Rabuka, who created synthetic trehalose lipids with the same hydrophobic chains as more standard lipids, differing only in the sugar at the head. (The natural mycobacterial lipids have gigantic hydrophobic chains. One could imagine these chains intertwining in some way, with this entanglement rather than trehalose driving the membrane preservation we observed.) These artificial molecules saved membranes from desiccation just as well as the mycobacterial lipid, confirming the notion that the trehalose itself is the protective agent. It was a satisfying result for our colleagues, me, and my nascent research group.

It seems that the bacteria that cause tuberculosis and leprosy have figured out a clever and robust route to stress resistance by linking sugars to lipids and of course exploiting the self-assembly of lipids into membranes to define their exterior. Could we engineer even better artificial dehydration-resistant lipids, for example, with multiple trehalose sugars per molecule, to construct easy-to-store biomaterials? Could we destroy lipid-linked trehalose to counter tuberculosis? I don't know; we'll see what the future may bring.

ORGANIZING A TWO-DIMENSIONAL LIQUID

Returning to normal cellular membranes, the two-dimensional fluidity of lipid bilayers presents a potential problem for cells: How can a cell organize its membrane, housing certain proteins together with their partners and keeping other proteins separated, if the membrane as a whole is a liquid? One tactic—taken, for example, by the T-cells described above—is to link membrane proteins to the internal scaffolding of the cell, whose struts and motors can push and pull as needed. Another possible tactic, one that's intrinsic to the physical properties of the membrane itself, involves two types of lipids. Both form fluid bilayers, driven by the goal of shielding their hydrophobic tails from water, but each prefers to be near its own type, A lipids next to A and B lipids next to B. Like oil and water, these two lipid types segregate, but the segregation is confined to the two-dimensional world of the bilayer. In the last decades of the twentieth century, scientists realized that this lipid membrane segregation could easily occur for mixtures of lipids commonly found in cellular membranes. The hydrophobic tails of different lipids can be relatively stiff or floppy, depending on the types of chemical bonds that connect their atoms. Combinations of stiff- and floppy-tailed lipids together with cholesterol (abundant in cell membranes) form bilayers that are a hodgepodge of two distinct compositions, each coexisting with the other. One composition is rich in cholesterol and stiff-tailed lipids; the other is rich in floppy-tailed lipids. Their segregation shows all the hallmarks of *phase separation*, which physicists had been studying for decades, especially its dependence on temperature. Above some critical temperature, the different lipids mix together (upper illustration, following page), while below that temperature they follow their preferences and segregate (lower illustration, following page).

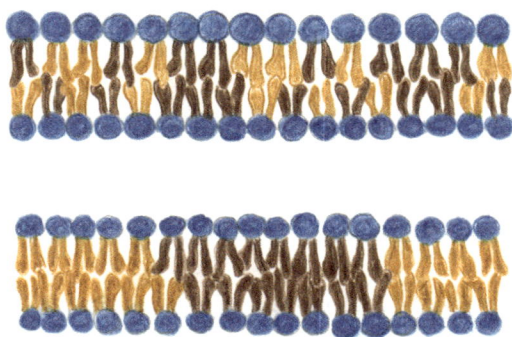

As is the case for DNA melting (chapter 1), the transition is a sharp one, and the analytic tool kit originally developed for nonbiological materials again finds applications in the living world. The picture that has emerged is that cells could make use of this cholesterol-dependent phase separation to organize their membranes. Proteins that prefer cholesterol-rich phases would be sorted into regions with other proteins of similar preference; cholesterol-poor regions would house other sets of proteins. In artificial lipid membranes, it's easy to see and study lipid phase separation. One can readily construct in the lab cell-sized spheres of lipid bilayer, which serve as tools for studying the biophysics of membranes and membrane proteins. (This is reminiscent of a soap bubble, but with water instead of air on both the inside and outside, and a single lipid bilayer as the boundary between them.) Labeling the membrane with different dyes that prefer cholesterol-rich or cholesterol-poor domains, for example, light and dark gray in the illustration, we see through our microscopes disks of one color amid a sea of the other.

It quickly became clear that this spatial organization *could* happen in actual cells, but whether it actually *does* turned out to be a difficult and contentious question that is still not fully resolved. In artificial membranes, cholesterol-rich and cholesterol-poor domains grow to sizes that are easy to see in a microscope. Moreover, one can lower and raise the temperature and watch domains appear and disappear at will. In living cells, puzzlingly, one never sees these domains, though we know that the lipids and cholesterol that make them up are the same

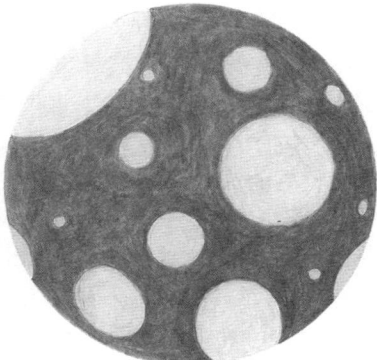

sorts used in the artificial membranes. The suggestion has been that the domains exist but are constrained by the cell's underlying scaffolding to be at most a few tens of nanometers in size. Due to the wavelike nature of light, we can't see structures smaller than a few hundred nanometers, rendering such domains unobservable. This is obviously not very satisfying; to say that something exists but is unobservable doesn't bolster one's confidence that it actually exists! Intriguingly, however, one can chemically perturb cells to create "blebs"—blisters of outer membrane detached from the underlying scaffolding. In these, one finds visibly discernible lipid phases with all the hallmarks of liquid phase separation, an observation first reported by Watt Webb and colleagues at Cornell University in 2007.

The bleb experiments lend credence to the idea that real membranes do phase-separate, but one might still argue that the cell's membranes are perhaps harshly perturbed and are not natural. Recently, scientists have observed large, visible domains in the membranes that bound an organelle called the *vacuole* in yeast cells. A group led by Sarah Keller at the University of Washington showed that these membranes in living yeast cells exhibit the hallmarks of phase separation, most importantly a critical temperature below which the domains appear. Intriguingly, the yeast cells seem to use these domains to enable the digestion of stored fats when their preferred sugars are unavailable, using the domains to concentrate the proteins involved in this process. It remains to be seen whether other cells have similar strategies, but the idea that cells harness the power of liquid phase separation to organize their membranes increasingly seems not only elegant but also true.

MEMBRANE STRUCTURE AND SELF-ASSEMBLY

This picture of cell membranes as two-dimensional fluids made possible by the self-organized lipid bilayer was cemented in the early 1970s, following decades of study of the nature of cellular membranes. The elegance of the lipid bilayer architecture is amazing: not only does it explain much of the behavior of membranes, but it shows that this behavior is a consequence of simple physical interactions. One might have expected that the tremendous biological importance of membranes would mean that cells would carefully specify the arrangements of lipids, creating precise chemical bonds to bind them together. But this isn't so: lipids are free to do as they please, just as a drop of oil suspended in water is free to form itself into a cube or a jagged star. The drop forms a sphere, not a cube or a star, simply because a sphere is the shape that minimizes the contact area between the oil and the water. Similarly, the lipids arrange themselves into a bilayer simply because that shape minimizes contact between the hydrophobic chains and water. The cell doesn't need genes to tell lipids to form bilayers; this is simply what lipids do. (The cell does need genes to encode the proteins that synthesize lipid molecules; once made, however, the lipids can organize themselves.)

As with the folding of proteins, we see the powerful general principle of self-assembly at work: simple physical concerns can orchestrate the formation of structure, allowing molecules to organize themselves. Harnessing self-assembly is not only useful for nature but also inspirational for those of us who study nature, showing that life need not be as complicated as it may initially seem—that physical simplicity may underlie biological complexity.

We've now met the essential molecules that make up every organism on earth: DNA, RNA, proteins, and lipids. This isn't a complete set of the ingredients of life—there are ions, sugars, hormones, and more that contribute in important ways—but understanding this group of universal components tells us a lot about how life works and how the living world encodes information. These molecules assemble and interact in exceptionally varied ways to generate the diversity of life around us. We'll continue to explore different types of biological structures and the physical forces that guide and constrain them, but we first dive into a major biophysical theme we've so far only hinted at: predictable randomness.

6

Predictable Randomness

Nothing is ever still. Every picture we've seen of a protein, DNA, or any other molecule is fundamentally flawed. Every lipid, for example, should be depicted as a blur of motion rather than sedately posed.

This motion is not unique to biological molecules. If I look into my microscope at a tiny glass bead about the size of a bacterium, adrift in a dish of water, within a few seconds it will meander over a distance of a few times its diameter. This travel isn't caused by flows in the water, or unevenness of my microscope stage. It's an intrinsic, unavoidable dervish that all objects perform. We see in this chapter that this motion—a consequence of fundamental physical laws—creates a backdrop to all of life's processes that is deeply alien to our macroscopic intuition, a backdrop that is fundamentally governed by randomness. Paradoxically, there is structure in this chaos, and many of the machineries of life make sense once we grasp how randomness and predictability are interrelated in the small-scale world.

THE PHYSICS OF POLLEN

This ceaseless dance of small things is called *Brownian motion*, after the botanist Robert Brown who observed and described it in 1827. Brown looked though his microscope at granules inside grains of wildflower pollen and saw their incessant motion. The motion is random: on average, there are as many steps to the right as to the left, upward as downward, with no pattern to the sequence of directions. Other microscopists had seen these dynamics before, but Brown established that they were not a consequence of life, the biological origin of their constituents, currents in the fluid surrounding the particles, or flows induced by the evaporation of liquid; rather, they were due to some sort of universal, underlying physics. To probe, for example, whether evaporation was a driver of the motion, Brown shook up a mixture of oil and pollen-containing water, with water droplets thereby shielded from evaporation by the surrounding oil; the granules nonetheless showed clear Brownian motion.

Brown established that everything incessantly shakes and wanders. But *why*? What drives this microscopic motion? Many decades passed without an answer, until Albert Einstein in Switzerland, Marian Smoluchowski in Poland, and William Sutherland in Australia each independently came up with a clear, simple, and accurate explanation in the first few years of the twentieth century. Their first insight was to take seriously the circumstantial evidence collected throughout the nineteenth century, especially by chemists, that matter is made up of discrete units (called atoms). This seems trivial from a modern point of view—we're so used to referring to atoms and molecules that it seems strange not to—but at the turn of the twentieth century, the notion of discrete building blocks for matter, rather than an endlessly divisible continuum, was contentious and not universally accepted. Einstein and others pointed out that the jostling of lots of tiny, individual water molecules colliding with Brown's granules or my glass spheres would give rise to motions with exactly the same form observed in experiments.

The second insight was in realizing the role of temperature. We're permeated by something called *thermal energy*, of which temperature, roughly speaking, is a measure. The greater the temperature, the more thermal energy an object has; the lower the temperature, the less thermal energy. As Einstein, Smoluchowski, and Sutherland realized, combining the impetus of thermal energy with the viscous drag provided by the fluid surrounding an object leads to a predictive model of random motion that matches perfectly with experimental observations. What's more, the underlying principles are universal and unavoidable; where there is temperature, there is random motion. (Only the unattainable temperature of absolute zero, −273.15°C or −459.67°F, brings stillness.) To appreciate the biophysical implications of this model, we need to be a bit more precise in describing Brownian motion. We've said that it's random. Nonetheless, it is comprehensible.

QUANTIFYING RANDOMNESS

Suppose that you walk in a straight line for 10 seconds and cover a distance of 30 feet. If instead you walk for 20 seconds, you wouldn't be surprised to have traveled 60 feet; for 100 seconds, 300 feet, and so on. We say that distance is proportional to time; to double your distance, you'd need to double the time spent walking. A graph of distance plotted against time would look like a straight line. The slope of the line would be a measure of your speed (in this case, 3 feet per second).

Sketching the paths my meandering microscopic bead might take over 10 seconds shows a variety of tangled tracks. We can't predict its

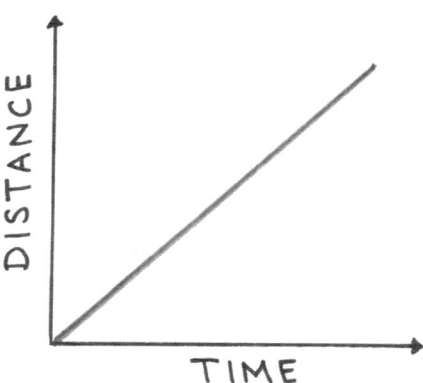

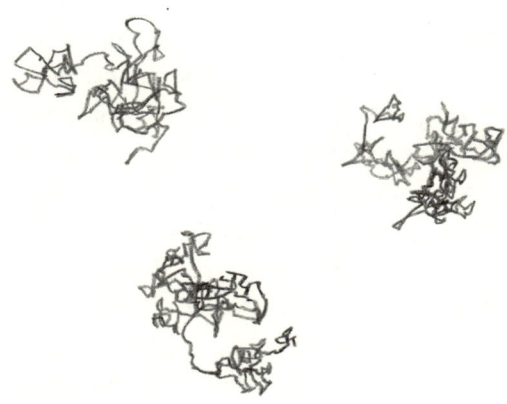

route ahead of time, nor what its final location will be. Its motion is random.

Coexisting with this randomness, however, is a sort of predictability. Just as I can't tell in advance which face a flipped coin will land on, but I know that if I flip the coin many times roughly half will show heads and half tails, I can make statements about the statistics of Brownian motion. If I repeat the 10-second meandering of my bead a few dozen times and draw a point at each of the final locations, with the starting position at the center of the page, the set of endpoints would look like the pink cloud I've illustrated.

Though the positions are random, the average distance from the starting point is well defined. How does this average depend on the travel time?

If you're experiencing déjà vu, that's great! This is, in fact, equivalent to the problem we encountered when asking how large a blob of DNA is in chapter 3. There we met the random walk, and learned that a random walk of N steps ends up, on average, the square root of N step-lengths away from its starting point. Here every instant of time during which our Brownian particle is bombarded by atoms of the fluid gives the particle a random kick, from which it takes a random step. Therefore, on average, the distance the particle covers is proportional to the square root of the time it travels. A graph of the typical distance versus time, rather than being a straight line, is curved.

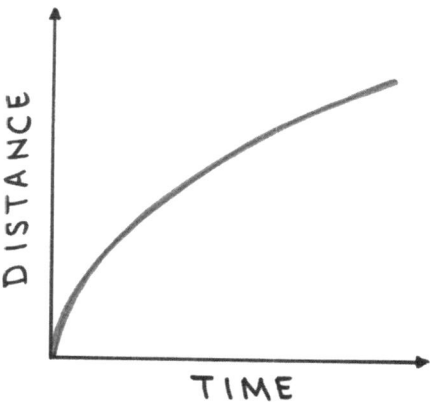

If the particle travels four times longer, it will, on average, only go twice as far. To move three times as far (on average), the travel time must be nine times longer.

In addition to time, Brownian motion also depends on the size of the particle. This makes sense—after all, we've stated that random motion is significant for microscopic particles, and we know that we don't see larger objects, like watermelons and baseballs, randomly shaking across the floor. All particles have an average displacement that grows with the square root of time, but the magnitude of this growth is greater for smaller particles than larger ones. All particles get the same kick from the ambient thermal energy, but the smaller particles respond more vigorously to it.

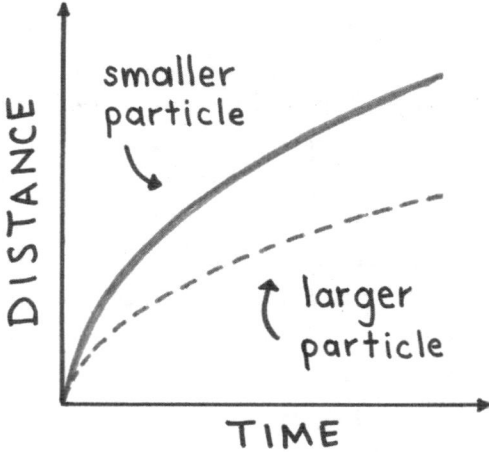

The random motion of molecules is also referred to as *diffusion*, a term commonly encountered in descriptions of dyes drifting through a liquid or gases wafting through the air. I note, however, that the common classroom demonstration of perfume spreading through the air isn't really an illustration of diffusion. The perfume molecules are certainly undergoing Brownian motion, but the main drivers of their motion across a room are air currents set up by nonuniform temperatures or by ventilation systems, people moving around, and other disturbances.

HOW TO BUILD SMALL THINGS

Beyond the obvious consequence that salts and sugars, lipids, proteins, and even whole cells are in constant agitation, Brownian motion illuminates a great many aspects of biology. First of all, it solves a nagging problem with our discussions of self-assembly in earlier chapters. As we've seen, proteins fold themselves into particular three-dimensional shapes, driven by physical interactions between the amino acids that make them up. Lego bricks, however, also have particular interactions between them, but a pile of bricks doesn't spontaneously assemble itself into some form. Brownian motion explains the difference. Being small, the amino acid chain is in constant, vigorous motion. The molecule is always jiggling about, placing some amino acids in close proximity to others, then others, then others still, until it settles into a configuration in which sufficiently strong interactions lock it into place. Similarly, thermal energy drives the random motion of lipids; they find each other and assemble into a membrane. The recipe for self-assembly, therefore, is not merely physical interactions, but physical interactions together with Brownian motion.

Similarly, gene expression and regulation also depend on Brownian motion. We've described transcription factor proteins binding to DNA, glossing over the question of how the proteins find their target DNA sequences. There's no guiding hand or train track conveying them smoothly to their destination. Rather, buffeted by thermal energy, the proteins wander through the cellular space, colliding with all sorts of DNA regions and being held for a while by those they specifically match. As with self-assembly, this strategy for running a machine wouldn't work for a macroscopic object—I can't set my office key on the floor and hope it wanders off and finds the door lock—but it's a great strategy in the microscopic world.

WHAT SETS THE SPEED OF THOUGHT?

Brownian motion also illuminates a deep connection between structure and time. As a concrete example, let's think about the connection between two neurons.

There are two types of connections. In one, called a *chemical synapse*, the two cells are separated by a space a few tens of nanometers across. (Recall that a nanometer is a billionth of a meter.) The cells communicate with each other by sending chemicals called *neurotransmitters*, the small orange circles in the illustration, across this cleft.

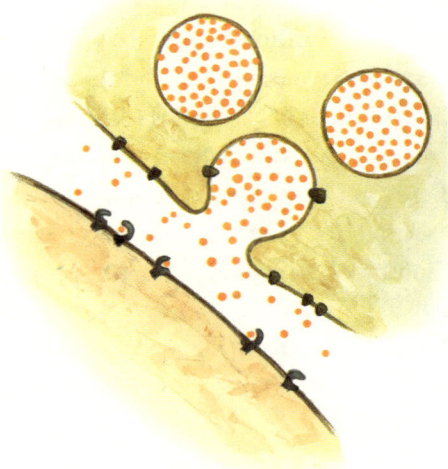

There are many different neurotransmitters and many drugs and pharmaceuticals that target their release, uptake, or degradation. Nicotine, for example, and some of the drugs used to treat Alzheimer's disease increase the levels of the neurotransmitter acetylcholine. Adenosine, another neurotransmitter, slows down brain activity, making you drowsy; caffeine blocks adenosine's receptor protein from binding it, keeping you awake. How do the neurons manage to send and receive neurotransmitters across a chemical synapse? By doing nothing more than releasing these chemicals and then letting diffusion do the work of spreading them. Left alone, the molecules meander about the cleft, occasionally running into receptor proteins on the target cell that bind them and trigger a neural response. There's no machinery needed—no nanoscale delivery van, no electromagnetic forces pushing the neurotransmitters along. The neurotransmitters are small, a nanometer or so in size, and their vigorous Brownian motion carries them a few tens of nanometers within about a microsecond.

Turning this explanation around, we can ask how long it takes information to be transmitted across a chemical synapse. If one neuron is stimulated so that an electrical signal travels along it until reaching the end of the cell, the news of its stimulation must be conveyed to the next cell in the relay—perhaps another neuron, as above, or perhaps a muscle cell. This intercellular baton passing takes about a microsecond, as we've seen—a millionth of a second. This is a rough estimate, of course. Strictly speaking, we should ask about how long it takes some threshold number of random walkers to cross the cleft rather than the average molecule. Still, whatever the criterion, the timescale is microseconds. There's no reason to expect it to take much longer—thousandths of a second, for example—and no physical way it could be much faster—billionths of a second, for example—given the physical size of the synapse.

Since I was a child, I've wondered what sets the speed of thought—why a minute feels like a minute and not a year, why we can't draw out and savor each millisecond of our experience. At a chemical synapse, Brownian motion determines the communication speed between neurons in an inescapable way. There are several other ways information can be transmitted in the brain, each with its own dynamics. Every biological information pathway, however, is governed in some way or another by molecular flows, of which Brownian motion is an integral part, helping set the rate at which our brains function.

The microsecond timescale of a chemical synapse is quite fast—obviously adequate for our needs. It's interesting to contrast this, however, with the timescale of modern computers, which is around a nanosecond, a billionth of a second, per operation. My laptop operates far faster than my brain. Rather than relying on molecules, it harnesses the motion of far smaller electrons and, moreover, directly pushes them around with electric fields. My brain is relatively slow, but its neurons are much more interconnected than the transistors in my laptop's central processing unit. The neural architecture enables a dazzling number of calculations to be performed in parallel among different sets of neurons at the same time rather than in a rigid, temporal sequence. Both the connectivity and the parallelism are helpful for conceptually difficult tasks. It's interesting to imagine what will happen when machines surpass us in both computational speed and network complexity; we'll likely reach that point soon.

CARGO TRANSPORT IN CELLS

In the example above, a neuron simply lets go of neurotransmitters, assured that they'll diffuse to their target in a reasonable time. Cells of all sorts rely on Brownian motion in similar ways. Recall from chapter 4 our bacterium that loves lactose: the lac repressor protein may or may not encounter lactose that the bacterium has picked up from its environment, and this controls whether it will bind to a particular DNA segment and inhibit creation of lactose-digesting proteins. How does the lac repressor find the DNA? Again, nothing in particular directs it. The protein meanders at random. Given its size, its random motion is vigorous enough that it will wander about 1 micrometer, the width of a typical bacterium, within about a hundredth of a second. Reaching a particular destination—for example, the repressor's specific target DNA sequence—takes longer, since most random paths don't go anywhere useful. Still, it takes only about a tenth of a second on average to reach any given point. Therefore, it wouldn't surprise us that a bacterium can reasonably make decisions informed by its environment within a few tenths of a second.

Imagine, in contrast, a typical eukaryotic cell—one of your body's white blood cells, for example, which is about 10 micrometers wide, 10 times wider than a typical bacterium. By Brownian motion, a protein would need 10 squared, or 100, times as long to diffuse a distance equal to this cell's width. Finding a specific target like a DNA binding site is even more challenging. The average time required, it turns out, is roughly proportional to the cell size cubed, and so would take $10 \times 10 \times 10 = 1000$ times as long in the white blood cell as in the bacterium. Rather than a tenth of a second, the timescale would be nearly 2 minutes—a sluggish response!

Rather than acquiesce to lethargy, eukaryotic cells take a more active approach, using motor proteins to ferry cargo. We've seen one of these motor proteins, kinesin, already; it grabs lipid-and-protein encapsulated material with one end and marches along microtubule roads with the other.

A typical kinesin walks with a speed of about 2 micrometers per second, so it could travel the length of a eukaryotic cell in a few seconds—much faster than the minutes required for diffusion. Even here, though, the cell still exploits randomness: the motor protein need not ferry its cargo all the way to its destination, but simply close enough that Brownian motion can rapidly take care of the last stage. (Upon reaching the nucleus, for example, from a starting point far away in a large cell, random diffusion can quickly bring a transcription factor to its DNA target, about a micrometer away.) The utility of molecules like kinesin is clear, but there is also a cost: the cell must expend energy to power motor proteins rather than relying on the Brownian motion that the universe provides for free.

So far, despite lots of searching, no one has found kinesin-like motor proteins in prokaryotic cells (bacteria and archaea). From a biophysical perspective, this makes sense: it's not that bacteria couldn't have evolved them, but rather that they don't need to. Brownian motion is fast for small things and slow for large things. Nearly all bacteria are small, and so they can simply and efficiently rely on this randomness for their internal transportation needs.

WHY DO BACTERIA SWIM?

Transportation outside of bacteria, including the motion of the bacteria themselves, is also deeply connected to randomness. Most bacteria can move, for example, by swimming through liquid. *Escherichia coli*, for example, has several whiplike flagella that spin one way to propel the organism forward and spin the other way to make it turn. These microbes are constantly in motion, and with a microscope one can watch them darting and dashing in a dish of water.

We might think that the bacteria swim to gobble up food, like miniature baleen whales scooping up krill in their path, but physics reveals that this is not the case. *E. coli* swims with a speed of about 10 micrometers per second, so if there were some food about a micrometer away (a distance similar to their body length), it would take a tenth of a second to swim to it. Their "foods" are sugars and other small molecules, less than a thousandth of a micrometer in size, so small that they would take only a millisecond or so to meander 1 micron. If you're a bacterium, food diffuses to you much faster than it would take you to swim to it! As physicist Edward Purcell put it, "You can thrash around a lot, but the fellow who sits there quietly waiting for stuff to diffuse" does just as well.

So why swim? Bacteria like *E. coli* measure the concentration of food in their surroundings, counting the rate at which nutrient molecules hit their receptors, and they swim toward regions of higher concentrations. Purcell again: "What it [the bacterium] can do is find places where the food is better or more abundant. That is, it does not move like a cow that is grazing a pasture—it moves to find *greener pastures*." Thanks to many years of work, we now understand *E. coli's* sensing and decision making in exquisite detail: how the detection of nutrients subtly alters the flagellar control proteins so that the little creatures swim longer on straight paths if they're in regions of increasing nutrient richness, and tumble more to change their orientation if they're headed in a less promising direction. The same sorts of mechanisms are at work in all sorts of bacteria, including many that navigate into animals, and also in many eukaryotic cells, for example, immune cells that migrate toward wound sites.

We've now met many of the pieces that make up cells, and the physical processes and motifs that orchestrate their assembly, their decision making, and their dynamics. Cells are, of course, wonderful—living, growing, reproducing entities that exist in bewildering numbers, trillions in each of us alone. Cells can be even more amazing if working together. In part II, we expand our scope to ensembles of cells, including embryos, organs, bacterial communities, and whole organisms of all shapes and sizes—again seeing general biophysical themes at play as interacting cells self-assemble, make decisions using biological circuitry, deal with randomness, and scale themselves to larger sizes.

PART II

Living Large

7

Assembling Embryos

We've now met the major building blocks of life, and we've gotten to know three of the overarching themes that govern their interactions: the concept of self-assembly, the predictable randomness of microscopic motion, and the construction of regulatory circuits. We've caught a glimpse of the fourth theme, scaling, reflected in the size dependence of Brownian motion and the long time needed to diffuse large distances. We focus much more on scaling a few chapters from now.

So far, most of our examples of these themes have been at the level of single cells or their internal machineries. These ideas also apply, however, to beating hearts, bananas, three-toed sloths, and all other manifestations of life at larger scales. Biophysical motifs shed light on collections and communities of cells, including whole organisms, and we'll uncover among them elegant examples of simplicity underlying complexity.

"THE TOTALITY OF ALL PRIMORDIA"

Rather than thinking first about groups of a few cells, or particular tissues or organs, let's fearlessly dive into what is arguably the most complex and amazing phenomenon of the living world: the development of an animal, like you, from a single fertilized egg cell. Our understanding of embryonic development has itself developed quite dramatically. Just a few centuries ago, it was widely believed that this single cell held within it a *homunculus*, a miniature but fully formed human from which the baby, child, and adult grew. In fact, some early microscopists convinced themselves that they saw these little people through their instruments' eyepieces, preformed in sperm or unfertilized egg cells. We know now that the single-celled embryo simply contains a genome—DNA from the mother and father—along with proteins, RNA, and other useful ingredients donated mostly by the mother. From this starting point, the cell divides, and divides, and divides. The descendant cells don't just split and grow, but change their positions, shapes, sizes, and gene expression patterns until reaching the set of positions, shapes, sizes, and gene expression patterns of a functioning organism.

Even with science as our starting point, the transformation from cell to animal can seem magical. Let's step back just over a century, to the late 1800s, when many of the pioneering experiments of experimental embryology took place. By watching animals develop and also by prodding, splitting, and transplanting, scientists began to map out the paths by which cells obtained particular identities and tissues acquired form. One of these pioneers was Hans Driesch, a German biologist working mostly in Naples. Driesch discovered that separating the cells of a two-cell sea urchin embryo resulted in each of the individual cells growing into a normal sea urchin. Even separating cells from a four- or eight-cell embryo would also often lead to a single cell growing into a full animal. What's more, Driesch found that gently pressing a young embryo could move cells from their standard positions (with cells that would normally be the progenitors of the top part of the animal being instead at the bottom), and they remained there when the pressure was released. Despite the shuffling, the sea urchin developed normally, as if the transported cells knew their new locations and acted accordingly. Each cell, Driesch concluded, "carries the totality of all primordia," a perspective very much at odds with a simple mechanical view. One can't scramble the gears of a watch or the pistons of a steam engine and find that they possess some deep, inherent knowledge of the new roles they need to take to make a functional machine. Driesch was so struck by the apparent contradiction between the workings of an embryo and the physics at his disposal that he abandoned the study of development altogether to become a philosophy professor, promoting the view that living things are governed by laws fundamentally different than those of nonliving substances.

Even then, Driesch's leap was extreme. In contrast, biologists such as the American Ross Granville Harrison advanced the notion that factors intrinsic to each cell and factors more broadly dispersed through the embryo together orchestrated development, a view consistent with our modern perspective and since fleshed out by a hundred years of work.

Before you get your hopes for this chapter set too high, anticipating that we will unveil the full path from single cell to complex creature, I note that embryology is not a solved problem. We can't take your genome and predict just from the sequence of As, Cs, Gs, and Ts that you're a two-armed, two-legged, hairy, air-breathing animal. We can't predict, given just the genome of a starfish, that the animal will progress from a soft, free-swimming two-fold-symmetric larva to a hard, typically five-fold-symmetric predator stalking seabeds and rocky shores. In fact, if we didn't know the organism from which the genomic DNA was taken, we could only tell that the starfish genome codes for a marine invertebrate and the human genome codes for a primate by comparison to other known genomes, not by modeling from first principles the activities of all the proteins and regulatory networks encoded by the constituent genes. Nonetheless, we can say a lot about development, thanks especially to two key features.

First, genes are remarkably similar across organisms, so learning what a particular gene does in an easily studied creature, like a mouse or a fruit fly, tells us a lot about what that gene does in another creature, such as a human.

Let's consider as an example the gene "sonic hedgehog," which encodes a protein crucial for the development of limbs and also active within expanding cancerous tumors. In a classic study published in 1980, Christiane Nüsslein-Volhard and Eric Wieschaus discovered several genes that are important determinants of the body plan of the fruit fly, naming one "hedgehog" because mutations in it give rise to spiky fly larvae. Similar genes were later found throughout the animal kingdom. Mammals, including humans, have in their genomes three genes that are very similar to the fruit fly's hedgehog. Two, desert hedgehog and

Indian hedgehog, are whimsically named after species of actual hedgehogs. The third is sonic hedgehog, even more whimsically named after a speedy blue video game character; one of the researchers involved was inspired by a *Sonic the Hedgehog* comic book belonging to his daughter.

All these proteins are remarkably similar. I've depicted the structure of part of the fruit fly's hedgehog protein (left) and part of the human sonic hedgehog protein (right). Each has a pair of tilted helices, a few short sheets, and various loops connecting them, all arranged nearly identically.

It's easy to tell a fruit fly from a person, but it's very hard to tell their hedgehog proteins apart. This similarity is also evident if we list the amino acids that make up each of these chains. I'll just write here a stretch of 46 amino acids, about a third of the chain I've illustrated, using a common alphabetical code in which each letter denotes a different amino acid, with shared units in bold:

Fruit fly:
RCKEK**LN**V**LA**Y**SVMN**EW**PG**I**RL**L**VTE**SW**DED**YH**HG**Q**ESLHYEGRAV**

a D (aspartic acid) in the human protein; both are negatively charged. The V and A in the next mismatched pair are both hydrophobic (valine and alanine, respectively). Even if the exact molecular identities of the amino acids differ, their physical attributes are in many cases similar. Nature's parsimony allows us to amplify the impacts of learning about its tools; we can be reasonably confident that the behavior of hedgehog in fruit flies is similar to the behavior of sonic hedgehog in humans, or desert hedgehog in desert hedgehogs.

The second reason we can gain general insights into development is even deeper: nature makes use of robust physical mechanisms to pattern and organize cells. These mechanisms, just like the genes and proteins involved, are also common across organisms. Let's see how they work.

KNOWING WHERE YOU ARE

Different organs grow at different locations. Wings emerge near the middle of a mosquito and antennae at the head. Your fingers sprout at the far end of your hand, not at your wrist. One might imagine that only dedicated wing-forming cells migrate toward and end up at a wing-forming zone in the middle of a developing insect—in other words, that the cells' fate is specified prior to their positioning. Or one might imagine that cells throughout the body are capable of forming wings, but only those at the appropriate location get a signal instructing them to do so. Nature, it turns out, uses both of these tactics. The second, in which spatial cues guide cell fate, is surprisingly common, and it enables an efficient encoding of instructions for the developing organism.

The existence of spatial cues has been known for over a century. Experiments like Driesch's in which the arrangements of cells were deliberately scrambled in the embryos of sea urchins and other animals, or in which some cells were transplanted from one region of one animal into a different region of another, nonetheless often lead to normal development, as if the scrambled or transported cells knew their new addresses in the embryonic neighborhood and acted appropriately. Understanding this seemingly magical sensory ability and revealing the nature and consequences of spatial cues is a more recent and still ongoing story. The basic idea, however, is straightforward and involves two of the biophysical machineries we've already met: random diffusion and regulatory networks.

The sonic hedgehog protein we discussed above isn't distributed uniformly throughout an embryo, nor is it present at some fixed concentration in some regions and absent in others. Rather, it exhibits a concentration gradient, progressing smoothly from high values around where the protein is being made to progressively lower values farther away. (Like all proteins, it decays, so the total amount of protein isn't constantly increasing.) This gradient is simply the consequence of diffusion, the random walk of molecules from their starting points, smearing out a cloud of molecules as we saw in chapter 6. Sonic hedgehog is produced at many sites in developing organisms, giving rise to many local concentration gradients. One location is the limb bud, the early precursor to each limb, which takes shape in week three of human embryogenesis. The sonic hedgehog distribution at the limb bud is most concentrated at one side, diminishing toward the other.

If you look at your left hand, palm facing you and fingers up, your thumb is on the left and your little finger on the right. Though we've never met, I can say with high confidence that this is the case, and that the ordering of your fingers isn't reversed or random. This arrangement is a consequence of the sonic hedgehog gradient: where the concentration of this protein is high, the little finger forms; where it is low, the thumb. The same holds in other animals. In the wing bud of a chick, the sonic hedgehog gradient governs the ordering of three bony digits, which progress 3-2-1 along the diffusing concentration profile. (Top: The orange blur is the sonic hedgehog gradient; the bones are as they appear in a four-day-old embryo.) Transplanting tissue from the protein-producing region of one chick wing bud to the low-concentration side of another wing bud gives two mirror-image concentration profiles from which six digits grow, 3-2-1-1-2–3 (bottom). The cells simply read the local sonic hedgehog concentration, unaware of the strange manipulations that generated it. The chick wing experiment, performed by Cheryll Tickle and colleagues at the University of Bath, used pat-

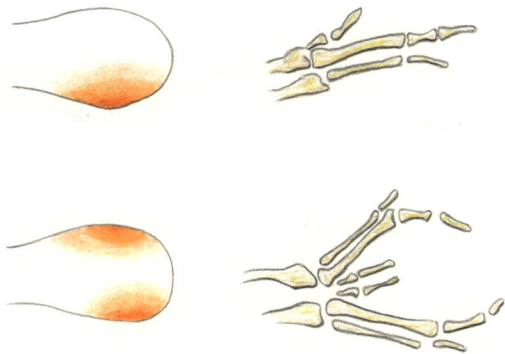

terns of digit development to probe which dinosaurs birds descended from, in addition to mapping the processes by which digits are derived. Embryonic cells measuring hedgehog protein concentrations is an ancient practice. Hedgehog gradients curate the array of fingers in your hand and the array of suckers on a cuttlefish arm, for example, though the last common ancestor of humans and cuttlefish lived over half a billion years ago. (Cuttlefish are not fish but rather cephalopods, closely related to squid and octopuses.)

Sonic hedgehog gradients drive the patterning of tissues other than limbs, playing a role in the formation of the nervous system, facial features, lungs, teeth, and more. The protein also turns up in cancers; cancer development often involves the activation of genetic processes associated with embryonic development, fostering rapid growth in unwanted ways.

Sonic hedgehog is just one of many *morphogens*, substances that govern the development of shape by variation in their concentration. Morphogens were predicted and named by mathematician and computer science pioneer Alan Turing in 1952, decades before any specific examples were discovered, in a prescient paper exploring the theoretical possibilities of such systems. Every developing embryo is crisscrossed by many coexisting and interacting morphogen gradients.

What are these morphogens doing? Either directly or via intermediaries, they act as transcription factors, turning on or off various genes, as we explored in chapter 4. The efficacy of a transcription factor depends on its concentration. This too is a consequence of physics: the binding of any molecule to any other is a constant flurry of attachment and detachment, and the probability that some transcription factor is bound to a DNA target is greater if there are more copies of the transcription factor floating around. The response function—the likelihood that the gene will be expressed, or the rate at which the protein encoded by the gene will be produced—can be a smooth function of the activator or repressor transcription factor concentration, or it can be sharp and switch-like, for example, nearly zero for low levels of an activator and at a high "on" state for activator concentrations above some threshold.

The concentration dependence of gene expression can allow surprisingly intricate patterning. Let's consider a stylized example now and add realism later. Imagine an elongated, pill-shaped embryo. This part isn't unrealistic—nearly every organism starts as a ball or an ellipsoid; we're all blobs at the beginning. Suppose the source of morphogens is at the left end, from some particular cells or materials supplied by the mother. Morphogen A diffuses, setting up a concentration gradient that decays along the length of the embryo (upper left illustration). If there's a switch-like response to morphogen A—on if A is abundant and off when it's rare—we have a steplike pattern of the output of that gene (upper right).

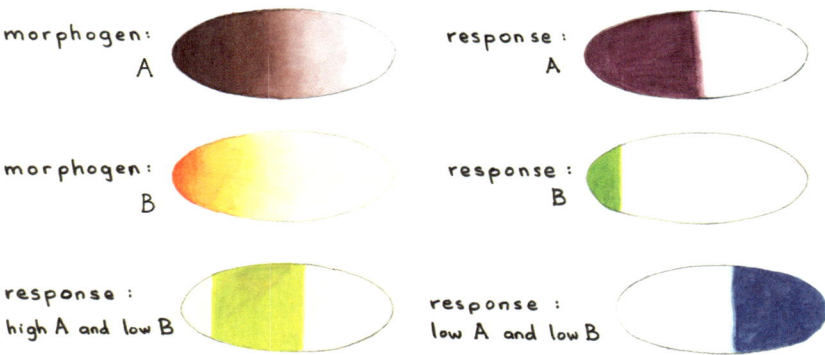

Now suppose there's also morphogen B, originating from the same location. Perhaps molecule B is larger and so its random walks aren't so vigorous; its concentration therefore has a steeper profile, and a similarly switch-like response would be localized to a smaller region.

Recall from chapter 4 that cells can create circuits of genes, integrating inputs to regulate their response. A circuit that is on if levels of A are high and B are low, and is off for any other combination, will therefore have an output profile that is on in a band just left of the middle. A "low A and low B" circuit would give a response in the right half of the embryo.

Even with only two transcription factor gradients, more than two spatial patterns for subsequent gene expression can emerge. An activated gene might encode a protein responsible for some developmental activity that should occur only in a particular region, or might encode another transcription factor that can itself interact with the first two. Suppose transcription factor C is regulated by the "high A and low B" circuit, and so is expressed in a wide band in the middle of the embryo from which it spreads by diffusion. A "high C and low A" circuit would then be activated in a narrow band, the zone just beyond where C is produced, but at which the diffusing C is still concentrated enough to be read as "high."

The narrow band demonstrates organizational precision, made possible by just a few genes. With still more transcription factors, each with an associated concentration gradient, the number of possibilities explodes. It's easy to imagine setting up specific patterns of gene expression precisely suited to the fates of the cells that will make up specific organs and tissues.

This may sound compelling in principle; remarkably, organisms actually adopt this approach in practice. We've known about such patterning in an approximate sense for decades, observing the blur of transcription factor concentrations and gene expression profiles, and in a much more precise sense in recent years for a small but growing number of genes and organisms. Our most exquisite understanding of embryonic spatial patterning comes from studies in fruit flies, again

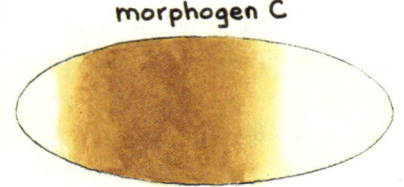
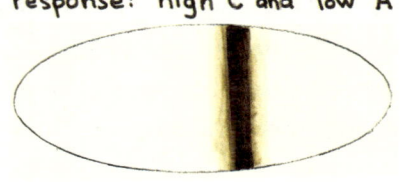

originating in the work of Christiane Nüsslein-Volhard and Eric Wieschaus. The early fly embryo, well before it develops legs, wings, or even a head, is an elongated oval, just like our hypothetical embryo above. The mother fly bequeaths RNA to her egg at one end, setting up an initial gradient of a transcription factor called bicoid that looks like our B morphogen. Bicoid binds to the promoter region of a gene called hunchback and acts as an activator. The hunchback protein also shows a high-in-front, low-in-back concentration gradient. About half a dozen other genes follow, expressed in wide swaths, combinations of which enable finer-scale expression patterns, like the seven stripes of the "even-skipped" gene, emerging just three hours after the egg is fertilized.

The patterns of even-skipped and other genes set up the organization of the fly body into fourteen segments. Different structures are fashioned from different segments. Three segments, for example, form the thorax, and each of the three grow a pair of legs, making use of hedgehog and other genes.

It's especially satisfying that this specific pattern of stripes and segments, crucial to establishing the body plan of every fly ever formed, can be both measured and predicted based on our biophysical understanding. Regarding measurement, one can survey in living embryos the transcription of a particular gene into RNA by engineering into the genome sequences that encode fusions of RNA-binding proteins and a fluorescent protein like GFP (chapter 2). Expression of the gene, therefore, generates visible, glowing proteins. One can also monitor protein levels themselves through similar insertions of fluorescent protein genes. In either case, the fluorescent glow provides a precise, quantifiable reporter of the what, where, and when of developmental activity. Regarding predictions, one can write down the equations of Brownian motion and genetic response functions and calculate patterns of protein abundance and gene activity. These computational outputs aren't yet perfect mirrors of Nature's patterns, but they do reveal stripes and gaps with sizes and timescales that are admirable representations of the reality of the fly embryo.

Looking further, the fly shows us even deeper aspects of this embryonic patterning. In our hypothetical example, we considered just two levels of gene readout, driven by low and high morphogen levels. Cells could have more finely tuned senses, giving three different responses to three morphogen levels (low, medium, high), or four, or five, or more. Having more levels seems appealing: the organism could construct fine-grained structure from even fewer ingredients. What limits an embryo's precision? Could the fruit fly detect 1000 different bicoid concentrations at 1000 different locations along its body, setting up 1000 different anatomical features from just a single spatial gradient?

This too is a question for biophysics. The limits on the precision of patterning are set by the inherent randomness of diffusive motion (chapter 6) and the randomness of molecular binding, which we haven't explicitly examined but which has underpinnings similar to Brownian motion. One can easily say on average where a cloud of diffusing molecules will be, but how well this average represents the whole cloud depends on how many molecules are present. If I flip a million coins, I'm quite confident that the fraction that land heads will be very close to one-half. If I flip six coins, I wouldn't be surprised to find four heads and two tails. The embryo, similarly, could make lots of morphogen molecules and generate a smooth, well-defined gradient from which it could reliably discern many different concentration levels. Or it could make just a few molecules, requiring less effort and energy but resulting in noisier gradients that it could only coarsely read as regions of high or low concentration. Experiments suggest that the preferred strategy is closer to the latter than the former; there aren't millions of morphogen molecules, and the fuzziness of the gradients is significant.

How exactly this statistical variation maps onto the precision of embryonic growth is not yet well understood, but we can already ask about the constraints it places on patterning. In a beautiful 2013 paper,

William Bialek and colleagues at Princeton University connected morphogen measurements and information theory to deduce how many bits of information are encoded in the early fruit fly. If a transcription factor can only be read as "high" or "low," it can be codified by one bit of information—one bit having two states, as we noted in chapter 1. If it can be read as "high," "medium-high," "medium-low," and "low," we would need two bits to encode the four possible states. We don't know how many states the fly's regulatory circuitry can discern for each gene, and so we can't directly calculate the number of bits. Bialek and colleagues realized, however, that the variability of the positions of stripes and edges as one compares individual embryos reflects the number of bits used in the patterning. Essentially, many bits imply high precision and low variability; few bits imply low precision and high variability. Analyzing images of fruit fly embryogenesis, specifically the patterns exhibited by each of the four genes just downstream of bicoid pattern formation, revealed that there are about two bits of information per gene. The four genes together are capable of defining patterns with a spatial precision of about 1%.

Once again, we're presented with an amazing feature of the living world, the generation of wonderful forms by a strikingly small set of instructions. Of course, a maggot may stretch your definition of "wonderful," but if you're unimpressed by fruit flies, keep in mind that these tractable, convenient organisms make possible the discovery and characterization of phenomena that turn out to be widespread. You, too, probably didn't require too many bits at the start.

KNOWING YOUR NEIGHBORS

In the examples of patterning we've seen so far in this chapter, we haven't really had to think about cells. We could imagine fields of morphogens or slabs of tissue without considering the discrete units that respond to or form them. For the early fly embryo, this isn't actually a simplification; the ellipsoidal embryo doesn't consist of discrete cells but rather lots of nuclei floating in a shared cytoplasmic sea. Later, membranes grow to separate the nuclei and form cells. In vertebrate embryos, including humans, distinct cells are present from the start.

In either case, once cells exist they have additional tools at their disposal for patterning.

Cells can convey and receive signals through contact with other cells. We've already seen this with membranes and immune cell signaling, in which membrane-anchored proteins reach across the gap between cells to recognize partners and trigger specific responses. Cell-cell contact plays a major role in development as well, especially in fine-scale patterning. Imagine a layer of cells that can each express genes A and B. These genes might be the defining factors for particular types of cells, A types and B types. Suppose that any cell in contact with an A cell is instructed not to itself express the A gene; it expresses B and becomes a B-type cell. Any cell that isn't touched by an A cell expresses A and becomes an A cell. Given these rules, we'd expect a mosaic that might look something like this honeycomb-like illustration, where each A cell (dark) is surrounded by a ring of B-cell neighbors (light).

This sort of patterning is commonplace. You can hear, for example, thanks to thousands of hair cells in your inner ear, so called because out of each projects a bundle of membrane-bound pillars, reminiscent of a tiny tuft of hair. Each hair cell is surrounded by a cohort of supporting cells in an arrangement that develops from exactly the patterning described above, referred to as *lateral inhibition*.

Lateral inhibition governs the layout of cells in insects' compound eyes, the specification of smooth muscle cells that make up artery walls, the generation of hormone-secreting cells in the pancreas, and more.

Though predicted since at least the 1970s, this mechanism was first clearly demonstrated in a developing animal in the mid-1980s by Chris Doe and Corey Goodman (the former now a colleague of mine at the University of Oregon) through clever experiments in which they focused a laser to destroy specific cells in fruit flies, eliminating these as neighbors of other cells that, now uninhibited, could express the genes that would turn them into neurons.

How can a cell control its neighbors' fates? For many different cell pairs in many different organisms, the key molecule is a membrane-anchored protein called Notch. Notch passes through the outer membrane of cells that express it. The external segment can attach to targets like Delta, another membrane-spanning protein, extending from a neighboring cell. The intercellular handshake of Delta on one cell grasping Notch on the other triggers a change in Notch's shape that exposes parts that would otherwise be hidden, parts recognized by other proteins that can sever amino acid chains. (It's perhaps disturbing to imagine axe-wielding proteins patrolling the cell, ready to cleave their colleagues in two as soon as they're provoked, but triggered destruction is a recurring theme in biology.) Notch is first cleaved at a site just outside the cell, releasing a large chunk of the external segment that diffuses to take part in other reactions, and then the internal part is liberated from the membrane anchor. This internal segment diffuses within the cell until it ends up in the nucleus, where it binds to other proteins and influences their binding to DNA; in other words, altering the activity of various transcription factors and thereby controlling the expression of various genes. One of the genes that the Notch fragment inhibits is Delta, decreasing its expression, so that a Delta-contacted cell won't be a Delta-expressing cell. Delta, therefore, is the A of our A-B schematic.

Notch and Delta orchestrate lateral inhibition. Notch can, however, have other partners displayed by other cells that trigger different sequences of cleavages and gene regulation. Some can cause the contacted cell to adopt the same fate as the A cell, spreading this type through a tissue. There is a rich variety of patterning-related proteins that can be present at cell-cell junctions, more than just Notch and Delta, as well as a variety of binding strengths, binding rates, and gene response functions that these proteins can show. Like a jigsaw puzzle in which a complex picture can emerge from the piece-by-piece assembly of neighboring tiles, contact-driven signals can orchestrate robust, intricate patterning from simple, local rules. Again, we see self-assembly at work, as cells contain within them the instructions that give rise to their large-scale organization.

KNOWING THE TIME

Groups of cells can organize in time as well as space, and can even use temporal cues to paint spatial patterns. We look at an example in which timekeeping enables the repetition of form. Tiger stripes, centipede legs, and the vertebrae of your spinal column all show regular features, one after the other, not exactly identical but quite similar. Consider the spine. Superficially, its bony regularity is reminiscent of the stripes of the young fruit fly, so we might expect a similar developmental origin from overlapping morphogens. In the early fly, however, the size of the animal is stable while genetic circuits build up patterns of interaction. Your spine, on the other hand, develops from segments that emerge while you are rapidly growing, as is the case for all vertebrates. Starting around the third week after conception in humans, the embryo elongates. Rather than being a smooth tube, the developing body features regularly sized chunks called *somites*, occurring in pairs along its length.

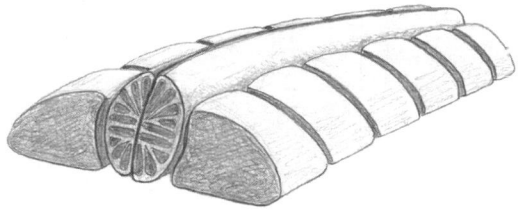

Humans form 42–44 somite pairs, though some fade away as development proceeds. Zebrafish make 30–32, mice about 65, and some snakes more than 400. Vertebrate embryos achieve robust regularity of somite spacing, sustained over dozens or hundreds of somites, by being good timekeepers.

We noted in chapter 4 that cells can use genetic circuits to build oscillators and clocks, with the ebb and flow of gene expression occurring rhythmically in time. Embryos often make use of such clocks. The first few divisions postfertilization are typically synchronized, with each cell in a two-cell embryo dividing at the same time to give four cells, each of the four cells dividing at the same time to give eight, and so on—at least for a little while, until the variety of cells becomes greater and this coordination is abandoned.

The cells of the elongating embryo that form somites also have clocks. Extracted cells in isolation have regular ups and downs of gene expression, and together in tissue these oscillations are synchronized across cells. How can the embryo turn these temporal rhythms into spatial patterns? In 1976, Jonathan Cooke and Erik Christopher Zeeman described an elegant biophysical strategy that subsequent experiments, especially by Olivier Pourquié and colleagues at the Stowers Institute for Medical Research in Kansas City, showed to be the mechanism that you and every other vertebrate animal use for somite generation: connecting a genetic clock to a morphogen gradient.

Imagine an array of cells with gene expression oscillating in sync. The rate of each cell's transcription of some gene, let's say, goes from 0 to 1 to 2 to 3, then back to 0 to begin anew. Depicting these levels as gray, green, blue, and brown, the collection of cells oscillates together, first all gray:

Then all green:

And so on:

But suppose that each cell's clock has a switch; the timekeeping circuit is active only if the local concentration of some molecule is above a threshold level. If it isn't, the clock stops. Suppose further that the controlling molecule is a morphogen, generated at the tail end of the animal, diffusing and setting up a tail-to-head concentration gradient. The clock will be running in the tail region and will be frozen everywhere forward of some point. As the embryo grows, the tail recedes further from the head, and the location of the threshold morphogen level moves steadily further back as well. The cells will have locked in the levels of gene expression that exist at the moment the threshold concentration passes by—in other words, when the clock stops. If at one location it's 2, at the next it's 3, at the next 0, then 1, 2, 3, and so on, repeating periodically. We can modify our drawing with a pink line, illustrated below, indicating the boundary to the left of which the cellular clocks have stopped; time again proceeds downward, and the tail end of the animal is toward the right. The temporal pattern is frozen in place as a spatial pattern of gene expression.

If high levels of whatever transcription is regulated by the clock generate the somite boundaries, where cells tightly bunch together, and low levels correspond to bulging somite middles, we'll have

turned our clock oscillations into repeating patterns of physical structure upon which we can construct regular structures, like vertebrae. What's more, by controlling the speed of the clock, Nature can tune the spacing of the pattern and the size of the somites. Different animals have different rates. Snakes make lots of vertebrae by having fast cellular clocks.

We now know a great deal about the specific genes involved in crafting both the clocks and the morphogen gradients. There are several genes at work, especially in the oscillator circuit, some of which encode proteins that bind to the Notch protein we encountered earlier in the context of lateral inhibition. Notch activity rises and falls, and among other things the protein helps maintain the synchrony of adjacent cells. Notch is emblematic of Nature's fondness for using a surprisingly limited set of molecules in a variety of contexts. (A joke already decades old claims there are two types of developmental biologists: those who study Notch signaling and those who don't know they study Notch signaling.)

Gradients and thresholds, contact-driven signals, and stopped clocks aren't the only physical motifs underlying embryonic development. Cells also migrate, stretch, elongate, fold over one another, vary the stickiness of their adhesion, and more, sculpting the living clay of the embryo as it grows. We're still discovering the strategies behind embryogenesis, not only enhancing our appreciation of the transformations each of us underwent from a single cell into a complex animal, but also helping us tackle transformations that don't go as we'd like, as occur in birth defects and cancerous growths.

It's too late to reassure Hans Driesch that biology is not necessarily outside the realm of scientific comprehension, but we can at least reassure each other that the field need not be abandoned. Incomplete though our current understanding is, it's clear that embryogenesis doesn't contradict the laws of physics, but rather is a beautiful manifestation of how physical properties and processes generate living forms.

8

Organs by Design

Embryos, organs, and every other assembly of cells organize themselves in response to cues patterned in space and time, as we saw in the previous chapter. Groups of cells end up as coherent entities, with distinct biological roles as well as distinct physical properties. Our fat-rich tissue is squishier than muscle, for example, as each of us can readily verify. We've recently realized that these physical properties aren't solely the consequences of tissue and organ formation but can also contribute to their causes. Development influences material characteristics that influence development, in a feedback loop of regulation that further elaborates the tool kit of self-assembly. In this chapter, we look at the roles of physical properties, like softness and stiffness, in guiding collections of cells, and then explore scaffolds for generating, someday, organs outside of bodies.

WHAT DO STEM CELLS SENSE?

Over the course of your life, you'll shed more than a ton of the cells lining your intestine. You won't mind because you're constantly generating new ones. You're also growing new skin cells, blood cells, immune cells, and more. Especially before you were born and for years afterward, you produced many trillions of cells of many different types: liver cells, muscle cells, kidney cells, and so on. All of these were formed from the division of other cells, and proceeding backward along each chain of division and specification, eventually one finds a stem cell. Stem cells are those not yet settled on a fixed identity, instead retaining the capability to generate more than one type of cell, including more stem cells, as their descendants. A single fertilized egg is a stem cell, as it has as its progeny all of an organism's varied cells. In an adult, there are stem cells with more limited potential. For example, one stem cell type generates only the cells of the blood, including red oxygen-carrying cells and immune cells of all flavors. Another variety generates the cells of the intestinal lining, including those that absorb nutrients and those that secrete mucus or digestive enzymes. Given the many possible outcomes, what determines which path a stem cell will take? Will its nonstem daughter cell be, for example, a B cell that grants immunological memory or a macrophage that gobbles debris?

Diffusing molecules form a large part of the answer. As we saw in the last chapter, cells respond to clouds of wandering molecules, tuning gene expression and other activities based on their local concentration. These molecules include hormones, growth factors, and other substances secreted by one cell and recognized by another. Diffusing molecules, however, aren't enough to specify the fates of cells. Recently, we've come to realize that an equally important signal comes from the mechanical and material environment.

Brains are soft, bones are hard, and muscle is in between. Each of these tissues is made up of cells and stuff outside the cell bodies, the latter often in the form of dense meshes of cell-secreted protein. In bone, minerals are incorporated into the protein network, but even prior to mineralization the material is hard, about 10 times as stiff as muscle, which in turn is about 10 times as stiff as brain tissue. The cells and the scaffolds they make influence the rigidity; could the rigidity in turn influence the cells?

In an elegant and influential experiment reported in 2006, Dennis Discher and colleagues at the University of Pennsylvania grew stem cells of a type that can form neurons, muscle progenitors, or bone progenitors on gels of different stiffnesses, with the liquid broth around them kept the same in every case. They found that the stem cells grown on the softest gels, similar in stiffness to brain tissue, transformed into neurons; those on intermediate-stiffness gels into muscle-progenitor cells; and those on the most rigid gels, similar in stiffness to premineralized bone, into bone-progenitor cells.

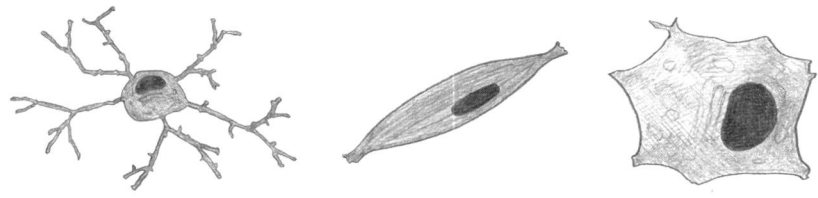

These newfound identities were evident not only in the cells' shapes—branched for the neurons, elongated for the muscle, and roughly polygonal for the bone formers—but also in the profile of genes that the cells expressed. We already know that self-assembly is amazing; collections of cells are like fabric that sews itself into clothing. But now it seems that the fabric is even more wonderful than we realized: when placed on a soft mattress, it turns itself into a nightgown; on a hard skull, it becomes a helmet.

Mechanics, not just biochemistry, provides cues for determining cell fate. Mechanics also orchestrates many other cellular activities, from the recognition of touch to the perception of sound waves to plants' sensing of gravity to distinguish up and down. The field of mechanobiology, which explores how this mechanical signaling works, has blossomed in the past two decades. Much remains unknown, but some key themes have emerged. One is the importance of channel-forming membrane proteins (chapter 2), whose configuration can be controlled by tension applied to the membrane. Links between the proteins and the internal or external environment can open and close the membrane-spanning gate. Channel proteins can also respond to stresses on the lipid bilayer; a stretched bilayer, for example, can become thinner, and the shorter hydrophobic core (recall chapter 5) can nudge the proteins to adopt an alternative conformation.

Another broad theme is that the network of physical connections between the inside and the outside of a cell can transmit information about forces. Membrane-spanning proteins can adhere, often via various intermediaries, to the external meshwork as well as the internal filamentous scaffolding. Tugging on the cell or the surroundings applies tension to the proteins, which can alter their conformations. This restructuring may, for example, expose sites that were previously hidden, triggering changes in the binding or chemical reactivity of proteins, leading ultimately to the activation or deactivation of transcription factors that orchestrate the activity of genes. Imagine that a stretched protein displays a site to which a repressor protein binds (left illustration); the sequestered repressor won't interact with DNA. In contrast, if the protein is relaxed, the binding site is hidden; the repressor is unbound, free to wander to the cell's DNA and block the expression of its target gene (right). The sketch is extremely simplified compared to the still poorly understood complexity of a real cellular response, but it gets at its essence.

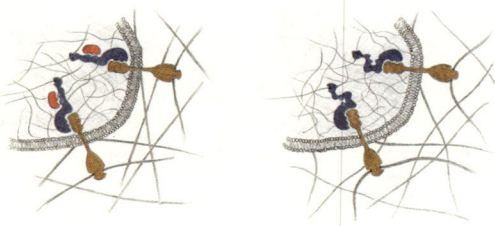

The tugging of proteins takes place even if everything appears motionless. The cell's internal machinery is never still; motor proteins (chapter 2) move, filaments grow and shrink, and the cell itself is constantly pulling. The external mesh isn't active, but its stiffness determines the equal and opposite force it exerts and thus sets the tension experienced by the structurally malleable tension-sensing intermediate.

Mechanical inputs and the material characteristics of the living environment form part of the regulatory circuitry of life and are integrated into the decisions that self-assembling cells make. The details are challenging to determine, but recent work is beginning to unravel the processes at play. Consider your skin, a layered tissue that is constantly losing cells at the outside and replenishing them due to stem cells at the innermost depths. When stretched over sustained periods of time, skin expands, growing extra cells and thereby making more skin. It's a useful response not only for the normal trials skin faces but also for accommodating reconstructive surgeries. To explore how this works, the groups of Benjamin Simons at the University of Cambridge and Cédric Blanpain at the Université Libre de Bruxelles, Belgium, examined mice in which an expanding gel had been placed under the skin. The stretching, the researchers found, led to heightened expression of genes encoding proteins involved in cellular adhesion and the cell's motor protein and filament network. Moreover, stretching induced more stem cell division as well as a greater fraction of the daughter cells being stem cells, ready to generate still more skin. The connection between stretching and cell fate decisions involved specific transcription factors that the researchers were able to identify; engineering mice to lack these transcription factors prevented the stretch-induced stem cell responses. How these particular transcription factors are coupled to the cellular scaffolding remains unknown, but at least the pieces of the puzzle are beginning to be discovered, and one can hope that further elaboration would be of use, for example, in designing therapies that call for accelerated skin regeneration.

Stiffness isn't the only material characteristic cells sense to guide their development. We're creatures of fluids as well as solids. Blood courses through our arteries and veins, and it turns out that fluid flow can induce stem cells to transform into the cell types that line blood vessels. All of our tissues, organs, and internal spaces have their own rigidity, viscosity, elasticity, and other material features whose development is intertwined with that of the cells that make them. While we try to make further sense of these connections, we push forward with ways to engineer multicellular structures, aided by our rapidly expanding insights into the role of the physical environment in guiding organ development.

ORGANS ON A CHIP

If you need a new heart, why not grow one? The dream of organs in a vat, guiding themselves into their proper forms like fruits in a garden, holds obvious appeal. Imagine replacing an injured eye with a fresh one, perhaps seeded by your own body's cells, or substituting a smashed finger with true flesh and bone rather than an inorganic prosthetic. Beyond repairing damage, assembly of isolated organs would enable the study of their development and the testing of drugs free of the complications, both practical and ethical, of studying organs inside of animals. We're still far from this vision, but our progress is accelerating dramatically, especially in the context of self-assembled cell clusters called *organoids* and partially human-assembled "organs on a chip."

We've grown animal (and plant and fungal) cells in the lab for many decades. Much of what we know about cellular biology—the filamentous network of internal scaffolding, the trafficking of cargo, and more—comes from such "cultured" cells. These tend, however, to be essentially two-dimensional: cells spread out on a petri dish, slab of gel, or other flat surface, bathed by a nutrient-rich broth.

There are obvious limitations to this approach. A layer of heart muscle cells can rhythmically pull and push, but it can't form heart-like tubes and chambers. This is more than just the trivial consequence of a flat geometry. Recall from the last chapter that cellular decisions often depend on the arrangement of contacts with neighbors and the shapes of morphogen gradients, both different in two versus three dimensions. The molecular, chemical, and mechanical cues of the three-dimensional environment are critical factors in the development of an organ.

The artificiality of two-dimensional cell clusters has been appreciated for over a century, and the earliest efforts to transcend it are nearly as old. Ross Harrison, whom we briefly encountered in the previous chapter, reported in 1906 on the growth of nerve fibers into a clotted drop of lymph fluid, seeded by a bit of embryonic frog tissue. In subsequent decades, several research groups showed that embryos of various species could be split apart into discrete cells that, if free in three dimensions, could coalesce into aggregates that recapitulate some aspects of normal embryonic form.

Over time, researchers realized that the meshwork of proteins outside of cells, called the *extracellular matrix*, is crucial to cell function, not just forming the scaffolding for tissues but also providing the mechanical and chemical cues that guide gene expression and even cell fate. In the 1980s, for example, Mina Bissell's group at Lawrence

Berkeley National Laboratory in California grew mammary gland tissues capable of secreting milk under the guidance of the appropriate matrix material, a remarkable demonstration that these collections of cells not only looked as they should but behaved as they should. Many other tissues, including cancerous growths, began to be nurtured using an expanding array of three-dimensional culture techniques. The field has taken off in the twenty-first century as our understanding of the underlying mechanisms intersects our understanding of the ideal seeds for tissues in a vat: stem cells.

Combining stem cell techniques and three-dimensional culturing methods gives us an amazing repertoire of functional, self-generating cellular assemblies, tantalizingly close to "organs in a vat," of almost any type. Such assemblies, whether stem-cell-derived or not, are referred to as *organoids*. The cells sloughed off at the surface of your intestine are replaced by the offspring of intestinal stem cells situated near the bottoms of billions of pocket-like pits. In 2009, Hans Clevers's group in Utrecht, the Netherlands, showed that a *single* intestinal stem cell, appropriately raised in a three-dimensional matrix, can divide into a community of cells shaped like a bumpy ball, with a well-defined interior space surrounded by the same sorts of cell surfaces as the intestinal interior, and with stem cells (blue in the illustration) near the bases of small pockets.

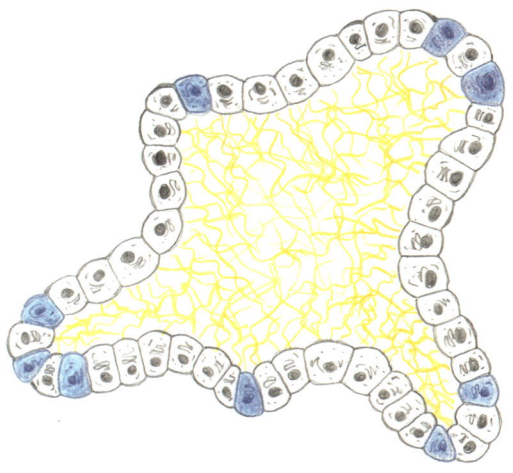

In other words, the intestinal stem cell generates an intestine-like organoid, similar enough in form and behavior to allow, for example, its use in studies of drugs targeting intestinal diseases.

Invoking the eye, Yoshiki Sasai and colleagues at the RIKEN Institute in Japan grew organoids that shaped themselves not into balls or shells but rather into the roughly hemispherical curve of the early "optic cup" (the back of the eye), initiated by stem cells of the sort that transform themselves into retina cells.

A few years earlier, in 2008, the same group demonstrated that mouse stem cells could be grown into balls of connected neurons similar in structure to the cortex region of the mouse brain. In 2013, the lab of Juergen Knoblich at the Austrian Academy of Sciences in Vienna built "cerebral organoids" that recapitulate several of the layers and structures of a normal brain, with working neurons and areas resembling the nascent prefrontal cortex, hippocampus, and more. Though they were far from functioning brains, the organoids were immediately useful. Knoblich's group investigated the puzzling origins of microcephaly, a developmental disorder resulting in a small brain, by using stem cells from a human patient as the seed for the organoid. In contrast to those derived from normal individuals, the patient-derived organoids showed fewer rounds of replication by a certain class of stem cells, driving an overall shortage of cells. Though the possibility of cerebral organoids developing sensation or consciousness is very far off, scientists and philosophers are already collaborating to map the ethical issues involved, including the question of how to assess and interpret the capabilities of a collection of neural cells.

Guts, eyes, brains—the full list of organoids developed to date is much larger than this, and it continues to expand. As noted, organoids are powerful tools with which to study development, disease, and drugs in fundamentally new ways, providing organ-like objects not ensconced in an animal body, and potentially derived from human cells. It's not much of a stretch to imagine that with further technological developments they may progress from being organoids to full-fledged organs, ready for transplantation into human recipients.

Beyond their practical utility, it's worth marveling at the biophysical lesson organoids convey. We've commented several times on the theme of self-assembly, for example, at the scale of molecules with proteins folding themselves into specific shapes and at the scale of organisms where entire bodies are generated from intrinsic rules. Here we see self-assembly being employed in a modular way at scales in between: the components of individual organs carry the instructions for their own organization. It's as if we not only have a car that forms itself from a small seed, but that a scrap of engine, bathed in the appropriate motor oil and held by the appropriate clamp, grows into a complete, growling engine or a scrap of the driver's seat, held gently, grows into a new seat; each piece on its own is, to some extent, self-sufficient. Nature harnesses self-assembly through a cascade of scales.

The self-assembly of organoids is assisted by our design of the appropriate extracellular matrix, though once set this is left alone. What if we were more hands-on, more deliberate in the operation of machinery outside the cells? Building small things, unconnected to biological applications, has been one of the great triumphs of the past half century of human civilization. The chips performing calculations inside mobile phones, for example, each contain billions of transistors squeezed together in a square inch of area, churned out by factories with astonishing speed and reliability. Our microfabrication abilities extend beyond electrical components like diodes and transistors. With materials like plastics and gels, we can fashion small channels, junctions, valves, and pumps at sub-millimeter scales—just the sorts of things one would need to bring nutrients or stimuli to groups of cells.

Merging microfabrication and cell culture gives us "organs on a chip," of which, as with organoids, there is a dazzling and expanding variety. The lab of Donald Ingber at the Wyss Institute for Biologically Inspired Engineering at Harvard University has pioneered several of these, including a "lung on a chip" in 2010. In this design, a porous, soft, very thin silicone sheet separates two chambers—one filled with air, the other filled with a blood-like, watery solution. As we discuss further in chapter 11, lungs are fundamentally an interface between air and water at which gases are exchanged. On one side of the sheet, the researchers cultured cells of the type that line the air side of the

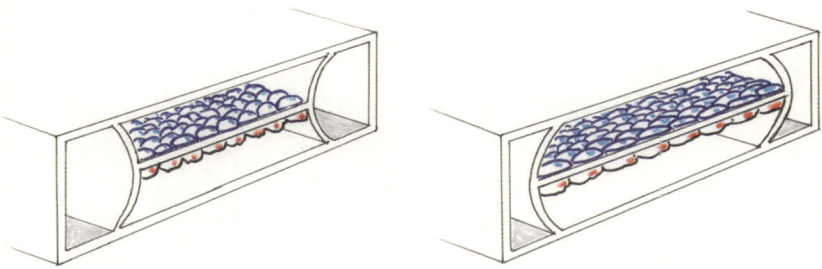

lung interface; on the other, cells that line the blood vessels. The clever part is that, at the edges of the sheet, separated by thin walls, are chambers that can be filled with air or evacuated. If under pressure, they compress the sheet and the attached layers of cells; if under vacuum, they stretch the sheet and the cells.

We therefore have a device that mimics not only the structure of the lungs but also their dynamics, stretching and contracting with, if we wish, the same rhythms as natural breathing. The chambers, sheets, and valves are all constructed by microfabrication methods that can cover a chip with a mosaic of pseudolungs, all amenable to imaging with a microscope. Ingber and colleagues showed that the uptake of particulates across the cellular boundary, a concern for both air pollution and drug delivery, is enhanced by the periodic mechanical pulsing of the cell sheets. Since then, scientists have constructed hearts on a chip, kidneys on a chip, stomachs on a chip, skin on a chip, and more, even stringing multiple organs together to build the hyperbolically named but nonetheless impressive "bodies on a chip." Much of the cell culture in these systems is still two-dimensional; integrating the three-dimensional self-assembly of stem-cell-derived organoids with the fluid handling and mechanical scaffolding of organs on a chip is an exciting area for ongoing work.

The semiartificial constructions of lab-cultured stem cells, organoids, and organs on a chip all provide us with handles on the awe-inspiring phenomena of multicellular organization. So far, however, we've treated the cells of an organism as all belonging to the same species. In the next chapter, we see that this isn't quite true—you contain multitudes of microbes—and we explore some aspects of these animal-associated microbial communities.

9

The Ecosystem inside You

You likely think of yourself as human. Your body is made up of a few trillion human cells, each enclosing a human genome, lending support to your concept of species identity. However, your body is also home to several trillion microorganisms—mostly bacteria, with some archaea and eukaryotic microbes as well—so many that if you held a vote, your human cells would probably lose. These microbes inhabit your mouth, your skin, and every warm, wet surface you can imagine, but by far the largest fraction resides in your intestines. This isn't a peculiarity of humans. All animals are hosts to a large and diverse assembly of gut microbes, and without these fellow travelers we'd have great difficulty functioning. Plants, too, have microbial partners, especially associated with their roots.

We've known of the existence of intestinal communities, often referred to as the gut microbiota or the gut microbiome, for well over a century. Our interest in them and our awareness of their importance, however, have exploded over the past two decades, driven by the technological revolution of DNA sequencing. Studying bacteria prior to this typically required growing them in a laboratory culture. Unfortunately, most bacteria stubbornly refuse to cooperate. Some, like those normally at home in the human intestine, find the amount of oxygen in the ambient atmosphere toxic. Some require especially acidic or basic conditions. Some need exotic nutrients, perhaps produced by other microbes. It's possible to meet these conditions, but it's often difficult, and moreover the solution may be different for each member of the community of interest. Therefore, for most of the time that we've been aware of gut microbes, we've known little about them.

DNA sequencing changed all this. We explore how sequencing works in part III. It suffices for now to recall from chapter 1 that we can make abundant copies of any DNA that we find. We can feed this material into a machine that reads it, returning the sequence of As, Cs, Gs, and Ts that make up the DNA fragments. This methodology can be applied to DNA from all sorts of sources, and it's radically revamped our notions of ecological diversity. In a pioneering study in 2004, a group led by Craig Venter, one of the key figures responsible for inventing modern DNA sequencing technology, scooped a few hundred liters of water from the Sargasso Sea, pureed the contents, purified and amplified the DNA (as in chapter 1), and discovered a million previously unknown genes from hundreds of novel bacteria. We've now similarly explored through sequencing environments from soil to subway stations, tips of tongues to fecal samples. (The fecal samples give us a snapshot, albeit an indirect one, of the microbes present in the intestine.)

In all these habitats we find teeming ecosystems, rich in cells and species. As we did for organs, tissues, and embryos, we can ask whether biophysical principles can help us make sense of these ensembles. In chapter 6, for example, we wondered why bacteria swim, and found an explanation in navigation to "greener pastures" of nutrients. In the tumultuous landscape of the gut, we can wonder whether similar strategies apply, or if bacteria have other motivations for moving. We can investigate whether groups of microbes self-assemble, either tangibly into physical structures or more abstractly into networks linked by biochemical exchanges. We can also apply the biophysical tools we've developed to probe the workings of the microbial ecosystem, for example, manipulating the regulatory circuits that guide bacterial decision making and observing the consequences. More than in prior chapters, we visit the frontiers of our understanding, at which answers, and even questions themselves, are still coalescing.

CATALOGING DNA

Before returning to the intestinal microbiome, I'll say a bit about two common methods to take a census based on DNA sequencing. The first makes use of a bacterial gene called the 16S ribosomal RNA (or 16S rRNA) gene. What it does isn't important here; what matters is that every species of bacterium has this gene and that some regions of the gene are identical across every species and some are different. The identical regions (gray in the illustration), when translated into RNA, correspond to sections that are crucial for the three-dimensional shape of the RNA molecule. The nonconstant regions (colored) are the record of billions of years of evolutionary variation, as different species adapted the basic rRNA architecture for slightly different purposes.

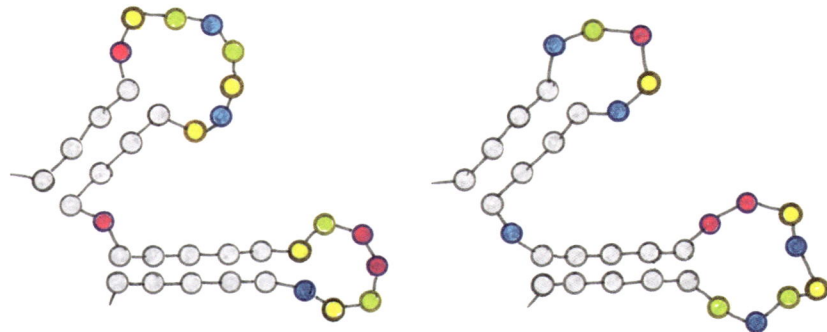

The consequence of all this is that we can conveniently use the same primer sequence (see chapter 1), corresponding to one or more of the identical regions, as the starting point for amplifying the DNA from any bacterium, getting countless copies of all the 16S rRNA genes present in our sample. Thanks to the variable regions, the complete rRNA genes are different enough that sequencing these DNA copies reveals the distinct signature of each contributing species. The 16S rRNA gene, therefore, is like a handle and a fingerprint all in one.

The drawback of 16S rRNA sequencing is that it reveals the identity of the bacteria present but nothing else. It's like having a list of names of all the people in a town, but no information about their ages, occupations, incomes, interests, or anything else that might allow you to assess what the town is like. If the 16S rRNA sequence matches that of some known bacterium, we can latch onto that information; but since most bacteria are unknown, this isn't often the case. Furthermore, closely related bacterial strains might have indistinguishable 16S sequences; it's like our list of names is a list solely of last names, not separately identifying, for example, siblings in the same family.

An alternative approach is known as shotgun sequencing. Here we amplify, break into sequence-able pieces, and sequence every scrap of DNA we collect, afterward computationally assembling them into genomes. Imagine, for example, that you had the fragments of sentences that were written on strips of paper, with multiple copies made of each sentence. Some read "a thousand times before his death"; some read "good or bad, but thinking makes it so"; some read "The fault, dear Brutus, is not in our stars"; some read "There is nothing either good or"; some read "A coward dies a thousand times"; some read "not in our stars, but in ourselves." Even if you didn't know anything about grammar, syntax, or Shakespeare's plays, you'd deduce from the overlaps ("good or," not in our stars," "a thousand times") which fragments came from the same source sentence, and you could assemble them into three distinct quotes. Similarly, we can write computer programs to determine the optimal overlap and alignment of potentially billions of DNA sequence fragments, reconstructing the full genomes from which they came. This is more complex and costlier than 16S sequencing, but it tells us what genes the members of our sample of interest contain, and therefore what proteins they can make and, in principle, what activities they can perform.

THE GUT MICROBIOME AND YOU

Thus, from the undignified starting material of a stool sample, we can glean the makeup of the ecosystem contained within you. Even from the earliest investigations, several striking features were evident. These microbial communities are very diverse; you harbor hundreds of different bacterial species. These species are special, not simply mirroring the stowaways on the food you eat or even the inhabitants of your mouth, but rather forming a group specifically acclimated to the intestine. Your gut community is unique but not wholly so; there's a lot of overlap with the communities of other people, which increases as we focus on the same geographic region and especially the same home. There's also considerable, but not perfect, overlap between the set of species in you now and the set inside you a few months ago;

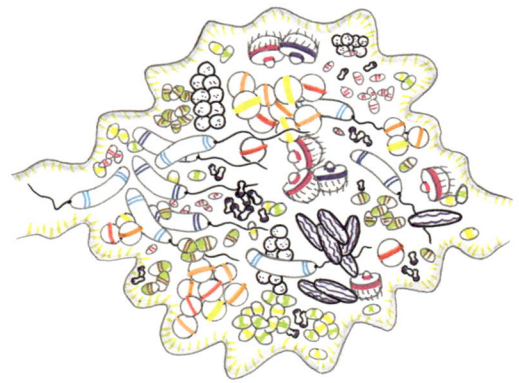

some gut bacteria are itinerant travelers, but many stick with you for the long haul.

The aspect of the intestinal microbiome that attracts the most attention, however, and that has made "gut bacteria" a household phrase, the subject of newspaper articles and advertisements, is the correlation between the composition of the gut microbiota and a wide range of complex diseases. Disorders as varied as diabetes, inflammatory bowel disease, gastrointestinal cancers, and even neurological ailments, such as multiple sclerosis and Parkinson's disease, all seem to be associated with altered makeup of the microbial communities of their sufferer's guts, compared to healthy people. These differences aren't characterized by the presence or absence of just one or two species. Unlike "classic" diseases like tuberculosis (caused by the bacterium *Mycobacterium tuberculosis*) or bubonic plague (caused by the bacterium *Yersinia pestis*), it seems that a broad shift in the abundance of tens or hundreds of species somehow takes place in the course of these more mysterious diseases.

Of course, correlation isn't causation. It's difficult to determine whether aberrant microbial community structure is a symptom of disease or whether it contributes to its origin, and the two options are not mutually exclusive. Still, the potential for a causal arrow from microbial ensembles to devastating diseases spurs an intense amount of effort aimed at elaborating these links and, one hopes, deliberately curating the intestinal ecosystem to foster health and fight disease.

At least for some disorders, there does seem to be an arrow one can draw from gut bacteria to health. Pernicious, recurrent infections by the often toxic bacterium *Clostridium difficile*, for example, have proved amenable to treatment with fecal microbiota transplantation. As the name implies, the procedure involves transferring fecal matter from a healthy donor to the patient, overwhelming the patient's aberrant native community with the new microbial immigrants. Fecal microbiota transplantation also shows tantalizing efficacy for inflammatory bowel disease and other disorders, though the results to date are highly variable. It's deliberate that the word *transplant* in the procedure's name echoes organ transplantation. The gut microbiome is, in a sense, an organ, serving a physiological role in its host despite not being composed of the host's own cells.

These issues of health and disease raise many questions. Could we replace the "fecal" part of fecal transplantation with a pill containing a carefully chosen set of bacteria, both to reduce the odds of unintentionally infecting a patient with harmful microbes via the transplant and to make the transplantation less aesthetically unpleasant? Would the pill even require bacteria at all, or simply nutrients preferred by the microbes we'd like to boost? (These questions spur a whole industry, that of probiotics.) What are the bacteria actually doing that affects, or is affected by, the health of their host? How can we perturb or even perhaps "reboot" the gut microbiome?

CAN WE UNDERSTAND THE GUT MICROBIOME?

We don't know the answers to these questions. The field of microbiome research is thrilling and vibrant. It is also a mess. Contradictory, inconclusive, and overhyped studies abound, along with many gems. Rather like the Wild West, adventurers and fortune seekers both good and bad rush in, and laws and sense come later. Many factors contribute to this chaos. First, as mentioned, the gut microbiome is highly variable. Tolstoy's assertion that "all happy families are alike; each unhappy family is unhappy in its own way" applies fairly well to conventional organs. Healthy hearts, for example, all resemble one another in their anatomical structure and in their rhythmic contractions. In fact, we can use structural aberrations or arrhythmic electrical signals as reliable indicators of disease. The gut microbiome, in contrast, doesn't have a stereotypical membership list. Different healthy people have quite different gut microbiomes, leaving us with the challenge of discerning the subtle statistical features that distinguish health and disease.

Second, fecal samples, as useful as they are, are an indirect reporter of what's present in the gut. Strictly speaking, they're an indicator of what is *not* in the gut, and it's a rather strong assumption to consider fecal samples as representative of what's left behind. In a tour de force experiment, a team of researchers led by Eran Segal and Eran Elinav at the Weizmann Institute of Science in Israel compared the microbiome of fecal samples with that obtained by more invasive, direct sampling of the intestine, finding that the former is clearly not the same as the latter, in both humans and mice. In the same study, the scientists fed mice and people supplements containing commonly used, commercially sold, probiotic bacteria, and found quite limited and highly variable colonization of the intestine by these "desirable" microbes. Examining what goes into or out of the gut is relatively easy, but it may not tell us enough about what's actually within the gut to make sense of the intestinal microbiome.

Third, it's difficult to do controlled experiments that untangle cause and effect, or that clearly distinguish meaningful changes from random variation. Every food we consume is a source of nutrients that affect some microbes differently than others; every room we step into is a potential source of new microbes. These incessant perturbations are hard to get rid of. We can't raise groups of people that are identical in every respect except for their intestinal microbes and see how they progress toward different diseases and disorders. We can't easily compare existing groups, for example, people from different regions, and be sure that differences in health are attributable to their gut microbes and not to many other confounding variables.

With animals, we can achieve a much greater degree of control. Mice, zebrafish, and other animals can be raised "germ-free," devoid of any microbes, and then either maintained in this pristine state or introduced to particular microbial species. Such studies have revealed that germ-free animals exhibit a wide range of abnormalities; resident microbes seem crucial to the training and stimulation of immune cells, the proliferation of cells that line the gut, and more. The control afforded by these experiments has in several cases enabled discovery of specific chemical factors by which microbes influence their hosts. The lab of my colleague at the University of Oregon, Karen Guillemin, who has pioneered germ-free zebrafish studies, found that germ-free fish larvae have a paucity of the insulin-producing beta cells of the pancreas, a defect that can be reversed by bacterial colonization or by the application of a specific protein that certain gut-native bacteria secrete. In humans, type 1 diabetes is characterized by destruction of beta cells; the zebrafish finding may point to previously unimagined paths toward beta cell regeneration. Microbiome-assisted development has been observed for several other organs and tissues, for example, bone growth in infant mice. Gut bacteria, it seems, have figured out a variety of communication pathways that link them to their host. Conversely, the animal host listens to its microbes, especially for inputs into how it should be developing early in life. In a sense this is surprising; it seems risky for animal development to depend on nonanimal partners, especially flighty and variable creatures like bacteria. On the other hand, animals evolved in a world already occupied by microbes. Their presence is unavoidable, so relying on microbes may have been nearly as risk-free as relying on the laws of physics. This last statement is a controversial one—I'm not even sure I believe it—but it's fair to say that our concepts of animal development are being altered by our expanding insights into microbial communities.

Returning to the pessimism of a few paragraphs ago, however, I will point out that raising germ-free animals is quite difficult. Mice can be kept germ-free to adulthood, but it requires considerable effort and expense. Zebrafish are easier, but still not easy; my lab and the labs of my colleagues struggle with this routinely, as bacteria and fungi exploit every possible chance to leap into a germ-free fish. Moreover, zebrafish can't be kept germ-free to adulthood (at least, not yet), because they need live food for adequate nutrition, food that brings with it its own microbes. As a result, a remarkable number of much-touted studies, especially in mice, are based on analyses of single-digit numbers of animals. It's a bit like trying to predict the outcome of a national election by polling a single-digit number of voters—the variability and complexity of the system demand much wider sampling.

For all the reasons we've seen and more, many microbiome-related findings are not as robust as one might hope. Links between human obesity and the composition of the gut microbiome, for example, have faded in strength in the decade since their first announcements. This isn't to say that such links don't exist—gut microbes certainly participate in many digestive processes and influence functions such as fat absorption that are tightly connected to obesity—but the roles of varied microbes aren't as simple as one might hope. Similar pitfalls occur for other features of health and disease. It may seem strange that science is prone to such missteps. Among many scientists, however, these issues of reproducibility are under increasing scrutiny in many fields.

THE GUT MICROBIOME AND ME

My goal here, however, is not an exhaustive survey of what we do and don't know about the gut microbiome, nor is it to examine the structural problems of contemporary science, fascinating though both topics are. Rather, it is to ask whether a biophysical perspective can help us understand the gut microbiome. We wonder, for example, whether the architecture of bacterial colonies can be illuminated by concepts of self-assembly, and whether general strategies for bacterial navigation emerge amid the confined, churning landscape of the intestine.

Even more than for the other topics in this book, the answers to the questions we wish to ask are unknown; making sense of our intestinal ecosystem is very much a work in progress. This quest is, in fact, the major focus of my research lab at the University of Oregon; so in addition to commenting on what some general principles may be, I'll describe how I dropped nearly all my other research to pursue the idea that there may be physics in the strange substance of the gut microbiome.

My decision was admittedly odd. I'm a physicist by training and a professor in a physics department. As we've seen, physics is more than magnets and quarks and lasers. There's a lot of physics in the living world, but even among biophysicists it's not obvious why one might consider physics relevant to the messiness of guts, in contrast to the precise choreography of protein folding or the mechanical rules of DNA packaging. Why would I, and a small but growing number of other biophysicists, gamble on the biophysics of the microbiome being something worth exploring?

Imagine a tropical forest, dense with plants and animals. If you knew that there exist organisms called monkeys, leopards, elephants, and trees, but were somehow unaware that trees are stationary, that monkeys climb trees but elephants don't, that leopards hunt monkeys but leave elephants alone—if you were unaware of any aspects of the behavior, locations, size, and mobility of the forest's organisms, it would be hopeless to try to construct an accurate picture of how the forest ecosystem works. If you were unaware that on a rocky shore the tide comes in twice a day, that starfish are mobile predators, and that sea lions swim in to snack, you'd similarly struggle to understand the tidal zone ecosystem, no matter how much of the creatures' DNA you swabbed off the rocks or scooped up from the seawater. This seems obvious for macroscopic ecosystems, and the lesson that structure and dynamics matter is a very general one.

As mentioned, however, most of our information about the gut microbiome comes from DNA sequencing methods that are blind to its layout and activity. I became increasingly fixated on this lack of biophysical understanding about a decade ago, coincident with a global explosion of interest in the intestinal microbiota and many conversations with my aforementioned colleague Karen Guillemin. Around the same time, I became entranced by developments in microscopy, especially a method called *light sheet fluorescence microscopy*, which makes possible fast, three-dimensional imaging of large fields of view ("large," by microscopy standards, being many tenths of a millimeter). My research group, therefore, built its own light sheet fluorescence microscope and pointed it at the intestines of live zebrafish larvae, realizing that we could capture images and movies that would span the entire gut, yet be precise enough to see single bacteria. No one had done this sort of imaging before in any vertebrate animal. We took advantage of the optical clarity of the larvae and their amenability to germ-free derivation, exposing fish devoid of any intestinal microbes to just one or two species to see what, in the absence of confounding complexity, the organization and behavior of these microbes would be.

If we were unlucky, I reasoned, we'd find featureless swarms of bacteria, the same regardless of species, and the same as what we'd see in a beaker or a test tube. We'd perhaps occupy ourselves with measurements of gut bacterial growth rates or some such boring but possibly useful tasks.

Thankfully, Nature was kind to us. Even from our first observations, it was evident that a wonderful variety of forms are present. Some bacteria swim freely; some clump together. Some prefer the forward part of the gut; some are mostly behind. There are clear candidates for physical features that could influence how an intestinal ecosystem works and that might help us figure out how to tinker with competition and cooperation among gut microbial species.

Nearly all my lab's explorations with zebrafish larvae involve native bacteria, found in the zebrafish gut. However, for the first vignette that illustrates the physical backdrop to microbiome dynamics, I'll describe an experiment using a nonnative species, *Vibrio cholerae*, the bacterium that causes cholera. *Vibrio cholerae* has been studied intensely for over a century, and while cholera is not the globally devastating scourge it was a hundred years ago, it still kills around 100,000 people each year. I had barely thought about cholera until an unusual meeting at Biosphere 2, the site of an ill-fated attempt in the early 1990s at operating a sealed, self-contained experimental ecosystem. There I met Brian Hammer, a microbiologist at the Georgia Institute of Technology, and Joao Xavier, a microbiome expert at Memorial Sloan Kettering Cancer Center in New York City. We weren't locked into a concrete building and forced to grow our own food, but rather were attending a workshop organized by the Research Corporation for Science Advancement, a small private funding agency. (Biosphere 2 is now a tourist attraction and meeting site administered by the University of Arizona, which also runs experiments in the project's glass-domed and no-longer-sealed buildings.) Through conversations, we realized that the means by which *Vibrio cholerae* invades a human intestine are not at all well understood. It's not an empty space the bacterium wanders into, but rather a gut densely packed with trillions of resident microbes amid which *Vibrio cholerae* must somehow gain a foothold.

For years, Brian has been studying an amazing tool that *Vibrio cholerae* and many other microbes have, a syringe-like device called the type VI secretion system with which bacteria stab adjacent cells and inject toxins. We wondered whether *Vibrio cholerae* might be using this system to help invade the gut, and whether our combination of light sheet microscopy and zebrafish larvae would enable us to assess this. In addition, no one had ever watched *Vibrio cholerae* colonize and compete within a live animal gut; who knew what thrilling things we might find? Brian's group engineered several strains of *Vibrio cholerae*, including one variant in which the type VI secretion system genes were always on, so the bacteria were always ready to stab (left illustration), and one variant in which the syringe apparatus was defective, so the bacteria were incapable of stabbing (right). Joao's group conducted petri dish experiments to examine and visualize bacterial-bacterial killing, itself quite beautiful to watch. We looked inside the fish.

As the simplest possible intestinal invasion experiment, we precolonized initially germ-free fish with a single native bacterial species, and after 24 hours added to the surrounding water one of the *Vibrio cholerae* strains. When potentially invaded by the stabbing-defective *Vibrio*

cholerae, the native bacteria looked just like they do when on their own, abundant and forming dense colonies. In contrast, the natives were annihilated after introduction of the always-stabbing invader; within a day, the populations would drop by more than a factor of 100 on average, often disappearing completely. By imaging, we could watch bacterial colonies losing their grip, steadily receding down the intestine until eventually expelled from the fish. At this point, we were excited to see the type VI machinery making a clear difference, and we assumed that the armed *Vibrio cholerae* were killing the native bacteria, dislodging them. We kept looking, however, and turned our attention to what the zebrafish itself was doing. Just like your intestines, the larval zebrafish intestine pulses periodically, squeezing to churn and move its contents. Savannah Logan, the graduate student leading the experiments, observed that the intestines of fish colonized by the always-stabbing *Vibrio cholerae* showed much stronger contractions than germ-free fish, or fish colonized by the other strains. Analyzing the images, the magnitude of the contractions was about 100% larger in fish harboring bacteria with the active type VI machinery. It looked, therefore, like bacteria weren't stabbing their competitors, but stabbing their host.

Proving this required more genetic engineering and more microscopy. A part of one of the proteins that constitute the bacterial syringe was already known to be toxic to eukaryotic microorganisms, like amoebas, via disruption of the filamentous scaffolding present in all eukaryotic cells but absent in bacteria. Brian's group engineered yet another *Vibrio cholerae* strain that lacked this piece but that formed an otherwise functional syringe. Colonizing fish and performing the same invasion assay, we found normal levels of gut contractions and normal, robust levels of the native gut bacterial species. *Vibrio cholerae*, therefore, was not defeating its bacterial competitors by killing them but rather by using its syringe to poke its host, antagonizing the fish to strengthen its intestinal contractions in response and thereby expel aggregated resident microbes. Conveniently, *Vibrio cholerae* itself was unaggregated, motile, and individualistic. This marked the first finding that any bacteria could use their syringe machinery to manipulate animal physiology. More broadly, it highlighted that the physical landscape of the gut is crucial for governing the gut microbiome, through mechanisms that are fundamentally unknowable if one looks solely at DNA sequences or test tube experiments.

Will our findings help cure cholera? I doubt it. Of course, the traditional answer, and the requisite answer in any news report or press release, is to say yes, or at least imply that cures for every even tangentially related ailment are just around the corner. For cholera, however, there's a more important issue than the long and unpredictable chain that connects basic laboratory science to practical treatments: cholera is already easy to cure. The treatment, except in the most severe cases, is water with salts and sugars. The shockingly large number of people who die each year of the disease is a sad testament to the inadequacy of sanitation and public health systems throughout the world. Why, then, should we care about *Vibrio cholerae's* type VI secretion system? Aside from its intrinsic interest, what excites me is that *Vibrio cholerae* is just one of many bacterial species with this machinery. Tens or even hundreds of species in your intestinal microbiome have a type VI secretion system, and so understanding its role in the gut may help us understand what determines your microbiome composition. Manipulating the type VI secretion system in a range of bacteria might give us a long-sought path to altering the gut microbiome, reshaping it to foster health.

Nearly everything we've looked at in the zebrafish gut has revealed a strong biophysical signature—some way in which the physical aspects of behavior or response, whether swimming and navigating, the formation of three-dimensional colonies, or the manipulation of intestinal forces, are a major determinant of outcomes. As another example, we found that weak doses of a common antibiotic can induce normally motile bacteria to elongate and entangle and normally aggregated bacteria to form fewer but larger clusters, in both cases leading to severe drops in the intestinal population as the overly cohesive microbes are pushed around by intestinal forces. We suspect that this could be a mechanism behind large and mysterious antibiotic-induced changes in the human gut microbiome, uncovered through DNA-sequencing-based methods, which is especially a concern because low levels of antibiotics are commonly found as environmental contaminants. This project, like many of ours, was a collaborative effort between my lab and that of our close colleague Karen Guillemin, and was executed primarily by Brandon Schlomann, a PhD student in physics, and Travis Wiles, a postdoctoral researcher in biology, both happily blurring boundaries between subjects.

Watching bacterial behaviors is great; controlling them could be even better. Our general theme of regulatory circuits resurfaces. In chapter 4, we encountered tools that could activate or deactivate specific gene circuits; recall the color-changing mice. The aforementioned Travis Wiles engineered such handles into the genome of a zebrafish-native gut bacterial species, allowing control of swimming and chemical sensing. These swimming bacteria move through fluid by rotating a corkscrew-shaped, taillike flagellum that extends from one end of the cell. The flagellum and its motor are formed by the self-assembly of many different proteins, including a pair called PomA and PomB (green in the illustration of the base of the flagellum) that form part of the motor. Without PomA and PomB, the flagellum forms normally, but the motor can't generate any torque with which to rotate the flagellum and propel itself. A switch, therefore, that in response to an external chemical cue turns off the *pomA* and *pomB* genes in a bacterium that normally expresses them, or that turns on the genes in a bacterium in which they're normally silent, allows us to control whether these microbes swim or don't swim in the gut. (In our implementation, the external cue must always be present, like a button that

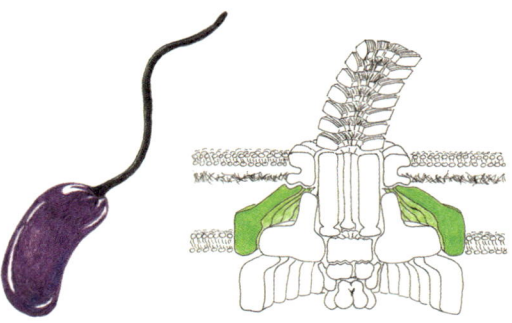

must always be held down for a light to be on. It's not a memory-enabling switch, therefore, but what an engineer would call a momentary switch.)

A switch offers more insights than a simple deletion or a constant activation of a gene. If we delete genes involved in motility, for example, and do not find these bacteria in the gut, it could be because swimming is necessary to persist in the intestine, or it could be because swimming is necessary to reach the fish and colonize it in the first place. Turning off swimming or other behaviors after bacteria have colonized allows us to delineate roles in the specific context of living in the gut. Turning off motility, we found, led to large drops in the populations of these microbes, as they were helpless against intestinal flows that transported them out of the fish and couldn't grow rapidly enough to counteract their losses. More surprisingly, the animal itself could sense these behavioral changes. Using zebrafish engineered to produce green fluorescent protein whenever genes in an immune circuit turned on, we saw large immune responses when the fish was colonized by the normal, motile bacteria, as we expected from earlier observations, but very low responses when the bacteria couldn't swim. The textbook picture of immune cells simply binding to proteins on the bacterial surface can't explain this; the external appearance of the bacteria is unchanged. Motion matters, we suspect, by allowing bacteria to push closer to the boundaries of the gut and make contact with sensory cells. This remains to be proved, but regardless we suspect that the behaviors of bacteria in a gut are as important as the behaviors of animals in a forest for governing the activity of the ecosystem.

We're not the only people excited by the potential of genetic circuit engineering. Many researchers have realized that microbes with memories could record intestinal conditions, with their state after they pass through the gut indicating whether they were exposed to particular toxins, nutrients, or other chemicals along the way. Coupling genetic switches to circuits that make specific biochemical agents could enable the delivery of therapeutic drugs only if and when certain stimuli are present, using the decision-making capabilities of cells to supplant traditional pills with more sophisticated agents.

Returning to physical perspectives on the microbiome: other labs have also been uncovering biophysical drivers of gut microbial population dynamics. The lab of Terence Hwa in the physics department at the University of California at San Diego, for example, has built artificial devices that mimic aspects of the pulsatile flow of the human gut, recapitulating the natural ratios of abundances of characteristic bacterial species. Hyun Jung Kim at the University of Texas at Austin, makes stretchable gut-on-a-chip devices, similar to the lung-on-a-chip platforms of chapter 8, with which to study mechanical couplings to cultured intestinal cells. Other groups have looked at the chaining together of bacteria by antibodies, the mechanical roles of fiber and osmotic stresses, and more. I should stress that there are far more studies exploring the biochemical and genetic properties of gut microbes rather than their physical attributes, and those sorts of investigations are particularly well suited to discovering the means by which microbes communicate with each other and their host. Bacteria can synthesize unusual proteins, fats, hormones, vitamins, and even neurotransmitters. Undoubtedly, biological, chemical, and physical principles are all simultaneously at work in the gut microbiome.

THE SELF-ASSEMBLED ECOSYSTEM

We can think of the dynamics that emerge from the interplay of bacterial architecture and gut mechanics as another manifestation of our theme of self-assembly. There are even more unusual and abstract ways in which self-assembly emerges in microbial communities related to general properties of ecosystems. I mentioned earlier that your gut is home to many different species of microbes—hundreds, in fact. You'll also find an abundance of species in a bucket of seawater or a spoonful of soil. This diversity is puzzling. In fact, it's a classic conundrum in ecology, named "the paradox of the plankton" by G. Evelyn Hutchinson in 1961. The problem is this: Imagine some bunch of species and just one type of food available. There will always be one species that's better than the others at consuming the food and reproducing, becoming more abundant than the others by a margin that can only grow with time.

Eventually, we'll find an environment that's completely dominated by that species; rather than diversity, a monoculture. With a few different types of food present we can have a few coexisting species, but unless there's an amazing variety of food with an exquisite matching to particular species' preferences, highly diverse ecosystems should be impossible.

Nature, however, scoffs at this argument, and despite what theory dictates, routinely generates and maintains cacophonous diversity. Even in theory, however, there are many resolutions to the paradox of the plankton. One is spatial structure; if different creatures dwell in distinct zones, they can coexist even if their nutrient preferences are the same. Another is temporal structure; populations can oscillate out of sync, for example, so that one is dominant at one time and another at another time. Other resolutions relate to metabolism. In our simple picture above, an organism simply takes in food and reproduces. In reality, there are many intermediate steps. Molecules are converted into other molecules, broken apart, and joined together as they are consumed. Especially for bacteria, some of the intermediate molecules are secreted into or taken up from the surroundings. There are many more types of molecular nutrients available, therefore, than what we'd guess just by looking at the starting or ending points. Microbes feed each other through this chemical cross talk. Because of this, broth containing just a single type of nutrient can reliably sustain dozens of bacterial species. A few years ago, Alvaro Sanchez and colleagues at Yale University put together hundreds of microbial communities derived from samples of soil and plant leaves, feeding each on just one nutrient, and demonstrated not only that several species can routinely coexist but that the compositions of the resulting ensembles are predictable and reproducible.

Our theoretical understanding of ecological diversity is also advancing, offering deep insights into why coexistence can be more common than previously thought. One can write down mathematical equations describing the growth, death, and interactions of species in an ecosystem. Rather than considering particular values of the parameters in these models, one can assess the range of outcomes for many instances of randomly chosen parameters to get a sense of what sorts of properties are likely or unlikely to emerge. Again, our theme of predictable randomness shows up; rather than assessing the average properties of random walkers, we assess the average properties of random ecosystem models, evaluating, for example, how often they give outcomes in which all species are present or some number are extinct. This approach was pioneered by ecologist Robert May, who in classic and highly influential work in the 1970s concluded that the diversity and stability of ecosystems do not go hand in hand. Rather, addition of species to an ecosystem makes coexistence of the species less likely. Many theorists have continued along these paths, for example, Pankaj Mehta and colleagues at Boston University, who showed that coexistence can emerge beyond the destabilizing point theorized by May, though among particular subsets of the interacting species rather than all of them together.

Other theoretical approaches more explicitly relate nutrient use to the rise and fall of populations. Often called *consumer resource models*, these date to classic studies from the 1960s and 1970s by Robert MacArthur and other ecologists and gave rise, for example, to the paradox of the plankton noted above. Resolutions to the paradox, we now realize, can come from many sources. Simply adding constraints to nutrient use, even without cross-feeding, can be sufficient for species with similar nutrient preferences to coexist. The constraints aren't complex; imagine you can eat potatoes, carrots, and peas, but the total amounts are limited by the space on your plate, so any increase in the potato portion must be offset by a corresponding decrease in the area occupied by carrots and peas. Mapping onto metabolism, the vegetables represent different digestive enzymes, tailored to different nutrients, and the plate is the overall rate at which an organism can produce enzymes. Ned Wingreen and colleagues at Princeton University found that a mathematical description of this sort of constrained resource usage finds a surprisingly large set of parameters for which coexistence occurs, essentially because there are many ways to gently consume multiple foods that are equivalent in overall intake to ravenously consuming one food.

The field of theoretical ecology is vast, and the few paragraphs above certainly don't span its breadth. I picked this handful of examples in part because they illustrate important recent insights, but also because they illustrate yet again cross talk between physics and biology. The mathematics of Robert May's methods was that of *random matrix theory*, developed by physicist Eugene Wigner in the 1950s to make sense of the energy levels of heavy atoms by considering random parameter sets for the governing quantum mechanical equations rather than intractable exact solutions; May realized that the formalism could be translated to ecological systems. Pankaj Mehta and Ned Wingreen are both physicists, and their work draws upon theories of phase transitions and other physical systems to illuminate the mysteries of ecology.

Returning to my own work, my lab's observations in zebrafish have convinced me that physical structure and mechanical forces play major roles in the dynamics of the gut microbiome. There's no way we could have predicted the outcomes of *Vibrio cholerae* invasion, or antibiotic responses, by making measurements in a petri dish. The physical environment of the gut is as central to this ecosystem as the rocks and tides are to the coast. Of course, it's possible that everything we see is an idiosyncrasy of zebrafish and that our experiments don't teach us anything about humans or other animals. I think this is unlikely, though, not only because of the parsimony of nature that we've commented on before but also because the core underlying mechanisms are widespread throughout the animal kingdom. Every gut pushes material by mechanical contractions. Aggregation, in a gut or elsewhere, is one of the most common bacterial behaviors. It is difficult to imagine, therefore, that these dynamics would somehow disappear when applied to your own intestinal community.

Of course, the size of the gut, the numbers of species present, and the magnitudes of flows differ greatly between a larval fish and you. How do these attributes depend on the size of an animal? Can we think of your gastrointestinal tract as a scaled-up version of a fish's or a scaled-down version of an elephant's? We don't yet know. Despite enormous interest in the human gut microbiome and a sizable body of work on a few model animal species, there has been much less investigation of the gut microbiota of different animals. These questions, however, lead us to the even more general question of what "scaled" actually means. It turns out that there are general rules governing the variation of physical forces with size and shape, with dramatic consequences for differently sized organisms, and it is to these rules that we next direct our focus.

10

A Sense of Scale

Living creatures span a staggering range of sizes. A blue whale, a few tens of meters from head to tail, is about 10,000 times longer than an ant. Since van Leeuwenhoek in the seventeenth century, we've known that the ant and animals like it are not the extreme but rather the midpoint of nature's scale. Another factor of 10,000 separates the ant from the smallest bacterium. No less impressive is the variety of shapes that accompany this variety of sizes. Large organisms don't simply look like blown-up versions of small ones.

We wouldn't mistake the spindly legs of a rhinoceros beetle with the stumpy limbs of a rhinoceros, even if the former were stretched to the size of the latter. Photosynthetic algae are bulbous and compact; none bother with the riotous branching that trees are so fond of, despite their shared aim of capturing sunlight and carbon dioxide. These differences extend to behaviors as well as shapes: the back-and-forth stroke of a tail fin is typical of sharks, but we never see it while watching bacteria swim. We'll see why this is shortly.

As in earlier chapters, we can ask whether the amazing diversity of life coexists with underlying rules, and again the answer is yes. The size, shape, and behavior of animals are all intertwined, with relationships governed by the physical forces they encounter and the environments they inhabit. A powerful concept that guides our understanding is *scaling*. Forces, flows of energy and matter, material characteristics, and geometric properties all have particular dependencies on size—they *scale* with size in particular ways—and these relationships dictate the forms and actions that life can harness.

HOW BIG IS A HORSE?

To tie size, geometry, and other characteristics to the ways animals and plants work, we're going to make use of a few mathematical tricks. One is a very liberating sort of inexactness.

How big is a horse? You already know the answer to this. We don't have to look up measurements of horse heights or books on equine anatomy. We needn't worry about whether I mean the distance from hoof to shoulder, or head to tail, or something else. We all know that a horse is about 1 meter in size. It's bigger than 0.1 meters (about 3 inches) and smaller than 10 meters (about 30 feet), whatever horse we picture in our minds. Whether our hypothetical horse is 1.0 or 1.5 or 2.53 meters tall matters if we're making a sweater for it, but not if we want to understand why a horse's bones are thicker and its metabolism slower than a 0.1 meter mouse's. Life spans vast extremes of sizes, and the relationships between size and function aren't determined by minor details.

Let's think of several different living things and note their sizes. To start, an ant is about a millimeter long, or 0.001 m; a typical virus is about 0.0000001 m in diameter.

It's tedious to write all the zeros in these numbers, and it's hard to keep track of how many zeros there are supposed to be. We therefore make use of scientific notation, denoting the powers of 10 in the number we're considering. Consider the number 100, which equals 10×10, in other words, 10^2. The number 10,000 is $10 \times 10 \times 10 \times 10$, or 10^4. Similarly, $1,000,000 = 10^6$ and $10 = 10^1$. What is 10 to the zero power, or 10^0? It's one fewer factor of 10 than 10^1, and so is 10^1 divided by 10, which is 1. Therefore, $10^0 = 1$. (By similar logic, *any* nonzero number raised to the zero power is 1.) What is 10^{-1}? We divide by another factor of 10, so $10^{-1} = 10^{0}/10 = 1/10$, which is 0.1. Similarly, $10^{-2} = 0.01$, $10^{-6} = 0.000001$, and so on.

You've almost certainly encountered scientific notation before. I describe it here for completeness but also to highlight the patterns by which we can construct relationships among the numbers. In the classes I teach for non-science-major college students, I often ask, "What is 10^{0}?" and nearly everyone answers "1." Few, however, can explain why. I ask them to imagine telling a friend that $10^{0} = 1$, having the friend reply, "I don't believe you!," and then trying to persuade them of this mathematical relationship. Simply stating "That's the rule" isn't (or shouldn't be) a convincing argument, but describing patterns among numbers is. Better still, by understanding the patterns, you can construct the rules for yourself, whenever needed, without relying on rote memorization. It's liberating.

Returning to our list of living things, here's mine, with "order of magnitude" (i.e., powers of ten) values for characteristic sizes:

Object	Size
a molecule of DNA (width)	10^{-9} m
an antibody	10^{-8} m
an influenza (flu) virus	10^{-7} m
an *E. coli* bacterium	10^{-6} m
a red blood cell	10^{-5} m
a human egg cell	10^{-4} m
an ant	10^{-3} m
a blueberry	10^{-2} m
a rat	0.1 m
a horse, or a human	1 m
a blue whale (length)	10 m
a redwood tree (height)	10^{2} m

You can make your own list, spanning its own wide range of powers of 10. How do the physical forces that animals and plants deal with vary, as we look up and down the ladder of sizes? Let's start by considering swimming.

WHY CAN'T A BACTERIUM SWIM LIKE A WHALE?

With elegant up-and-down strokes of its tail, a whale glides through the ocean. Sharks and many other fish similarly make use of a back-and-forth motion, in those cases left to right rather than up and down since their tail fins are vertical, but nonetheless *reciprocal*, meaning that the stroke in one direction retraces the path of the opposite stroke. If you peer through a microscope at a swimming bacterium, paramecium, or other microorganism, you never see that sort of swimming. You'll find a dazzling variety of motions—corkscrewing rotations of flagella, bulges that propagate along the cellular body, and more—but never a back-and-forth, reciprocal motion. Let's see why this is.

Every creature that swims through water pushes fluid as it moves. There are two reasons that pushing fluid is difficult. One is inertia: just as kicking a stationary soccer ball requires a force to accelerate it, it takes force to take an otherwise motionless bit of water and increase its speed. The second is viscosity: the honey pushed by a spoon drags against the honey alongside it, and it takes force to overcome this resistance. The ratio of these two required forces, the inertial force and the viscous force, is called the Reynolds number, after the fluid dynamics pioneer Osborne Reynolds, who became in 1868 the second ever "professor of engineering" in England. Every situation involving fluids has a Reynolds number, and the Reynolds number gives a concise way of characterizing the flow. High Reynolds number flows are turbulent—the inertia-dominated flows eddy and swirl, resulting in all the bits of water colliding chaotically with one another like little soccer balls. In contrast, low Reynolds number flows are smooth—the viscosity-dominated flows gently decay around the moving object. ("High" and "low" numbers are in comparison to 1, the Reynolds number at which the inertial and viscous forces are equal in magnitude.) We can calculate the Reynolds number given the properties of the moving object and the fluid it's in. High speed, large size, and low viscosity give a large Reynolds number; low speed, small size, and high viscosity give a low Reynolds number.

For a bacterium in water, with a size of about 10^{-6} m and a speed of about 10^{-5} meters per second, the associated Reynolds number is about 10^{-5}, or 0.00001—very, very low. For a cruising whale, the Reynolds number is about 10^8, very large, and 10,000,000,000,000 times greater than that of the bacterium. (You can see why we care only about orders of magnitude; worrying about whether the bacterium is 1×10^{-6} m long or 2.61×10^{-6} m long would be utterly irrelevant given the 13 powers-of-10 difference in the Reynolds number that we care about.) The bacterium and the whale, therefore, live in very different fluid worlds: the bacterium's is smooth and the whale's is turbulent.

In a classic 1977 paper, "Life at Low Reynolds Number," physicist Edward Purcell explained that this fact has surprisingly deep consequences for how aquatic creatures can and cannot move. At high Reynolds numbers, flows are irreversible, meaning that if we move an object one way through a fluid and then retrace its path to bring it back to its starting point, the fluid doesn't return to its original configuration. In other words, if you pour cream in your coffee, move your spoon through the liquid mixing together the coffee and cream, and then move your spoon back to its starting point tracing back the same route, the cream doesn't unmix. Being large, fast animals, we're very familiar with the high Reynolds number world; this irreversible behavior is so commonplace, we hardly think about it. (The spoon and coffee, by the way, have a Reynolds number around 1000; their motions drive the molecules of water and the oils and particulates of cream and coffee in chaotic, turbulent swirls.)

At low Reynolds numbers, flows are reversible. If I took the same cup of coffee and magically increased its viscosity by a factor of a million, its behavior would be dominated by viscous forces and hence would be reversible. Sweeping the spoon one way might seem to mix the cream and coffee, but by retracing the opposite path, each bit of liquid would also retrace its path, unmixing as the spoon returns to its starting point. When finished, we'd see a compact blob of cream in the coffee, just like it looked immediately after we poured it. Demonstrating this is one of my favorite in-class activities, which I perform with a rotating cylinder of very viscous corn syrup and dye, mixing and demixing as if by magic. (A classic video of this effect by fluid dynamicist G. I. Taylor is available online; see the references for a link.) What does this have to do with bacteria? Increasing viscosity gives us a low Reynolds number, but so does reducing size or reducing speed; a bacterium in water, as we noted, lives in a world of very low Reynolds numbers.

Purcell realized that because of the reversibility of low Reynolds number flow, microorganisms simply cannot swim using back-and-forth motions. It's not that they haven't found the genes to make it possible or haven't developed the right biochemical reactions, but rather that if you're small, the laws of physics make it impossible to get anywhere with reciprocal motion. If a bacterium waves some sort of stiff appendages one way and thereby moves forward . . .

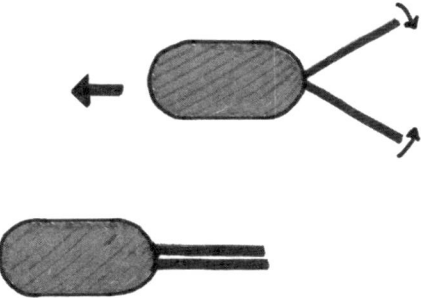

. . . it will move backward the exact same distance when it moves the appendages back:

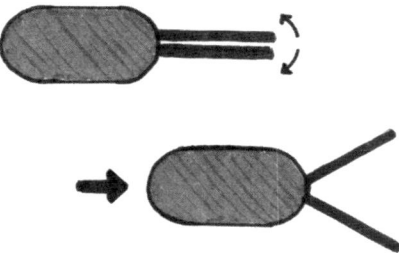

This is true even if the speeds of the back-and-forth motions are different, as long as the paths taken by the appendages are the same. Purcell named this the *scallop theorem* after the mollusks that move by clapping the two halves of their shell open and closed, a motion that would be futile if they were microorganisms.

What, then, do microorganisms do to swim? Anything they want, as long as it doesn't involve reciprocal motion. A common tactic is rotation of one or more helical flagella:

As long as the motor doesn't reverse, the movement of the flagellum never retraces its path backward and the creature steadily swims. Some microorganisms propagate bulges or kinks along their body, again making sure that the distortions don't retrace themselves. Another strategy is to wave hairlike cilia attached to a surface, making sure that the backward stroke isn't the opposite of the forward stroke. The cilia sway as they move one way . . .

. . . then bend as they move back:

Though you're not microscopic, you make use of this motion a lot: cilia line the airways of your respiratory system, their motions propelling microbe-trapping mucus and transporting it away.

The world seen by the whale, therefore, is fundamentally different from that seen by the bacterium. Crossing the orders of magnitude in size that separate the two requires much more than shrinking or growing, but also changing the very form of the creatures' behavior.

SHAPES AND SCALING

As we've seen, differences in size can influence how animals must function. So too can differences in shape, and the two concerns are linked. We can develop insights into the complex shapes of animals by first considering some seemingly simple aspects of geometry. Suppose we have a square and we double the length of its sides. The area of the square has increased by a factor of 4, which we can see by drawing . . .

. . . or from the math that the original square's area is $L \times L = L^2$, where L is the length of a side, and the new square's area is $(2L)^2$, or $4L^2$. If we have a triangle, each of whose sides has the same length, and we double this length, the area again increases by a factor of 4; tripling the length of a side, the area increases by a factor of 3^2, in other words, 3×3, or 9.

This also holds for triangles with unequal sides, provided we stretch each side by the same factor.

In all the above cases, the area is proportional to the square of the length. Another way to state this is that the area scales as L^2, which we can write symbolically as $A \propto L^2$. Increasing the lengths by a factor of

2 means that we increase the area by a factor of 2^2, or 4. Replacing L with $3L$ means that we increase the area by a factor of $3^2 = 9$. Turning L into $4L$ increases the area by a factor of $4 \times 4 = 16$.

This probably seems quite basic. After all, you might say, we learned about the areas of simple shapes in elementary school. There is a subtle lesson here, however, that is not often stated. We don't need to know mathematical formulas for the areas of shapes. *Any* shape that doubles its lengths while keeping its form unchanged will quadruple its area. That's what area *is*; it's the geometric property that scales as L^2. A circle whose radius increases by a factor of 5 has an area that is $5^2 = 25$ times larger than it originally was, and there's no need to invoke the equation for the area of a circle. A sphere whose radius increases by a factor of 10 has a surface area that has grown by a factor of 100. The blob on the left has 4 times less area than the blob on the right, whose lateral extent is 2 times larger:

The *volume* of an object scales as length cubed, in other words, $L \times L \times L$, or L^3. You can convince yourself of this by drawing boxes (or other shapes, if you're ambitious) and showing that doubling all the lengths increases the volume by a factor of $2^3 = 8$, tripling the lengths increases the volume by $3^3 = 27$, and so on. Again, this is independent of what the shape is. If we multiply the radius of a sphere by a factor of 4, its volume increases by a factor of $4^3 = 4 \times 4 \times 4 = 64$. Halving the lengths of a three-dimensional blob, but keeping the blob shape the same, would give a new blob with one-eighth the volume.

Finally, we note that if we stretch or shrink an object, not altering its shape, the proportionality between any of the lengths doesn't change. Scaling up a triangle in a way that doubles its height also doubles its width. All lengths scale as L, which seems odd to write but which is good to keep in mind. Similarly, all areas are proportional to other areas; if we expand a shape so that its cross-sectional area increases by a factor of 4.7, its surface area also increases by a factor of 4.7.

The generality of all these aspects of scaling, whether of length, area, or volume, mean that we can apply them to questions of size and form of even the most complex organic shapes, as we'll see shortly. First, let's briefly revisit bacteria. We learned in chapter 9 that there are at least as many bacterial cells in your body as human cells—perhaps disturbing from the perspective of a census, but less so in terms of space. A typical bacterium is about 10 times smaller in width than a typical human cell. Its volume, therefore, is about $10^3 = 1000$ times smaller. Despite their large numbers, the microbes are dwarfed in volume by the human cells in your body.

Are the shapes of big animals similar to small ones? We can assess this in a more rigorous and quantifiable way than just visual observation. As we've seen, if shapes are similar, then their volumes scale as length cubed and their areas scale as length squared. We can flip this statement around: if a collection of animals shows volumes that are proportional to length cubed, or areas proportional to length squared, it tells us that in a general sense their shapes are similar. The technical term is that they show *isometric scaling*. If the animals' volumes, for example, aren't proportional to the cubes of their heights, perhaps becoming disproportionately chunky when large or not showing any consistent relationship at all, we know that nature has discarded isometry, hinting that other concerns are at work. The challenge, then, is to assess the scaling behavior of real-life animal forms. One can do this with equations, but the easiest and most insightful approach is visual, with a tool that forms our second mathematical trick: logarithmic graphs.

Suppose we make a graph of volume versus length of a side for cubes. The usual way to plot this would look like the graph on the left, swooping upward with a cubic form.

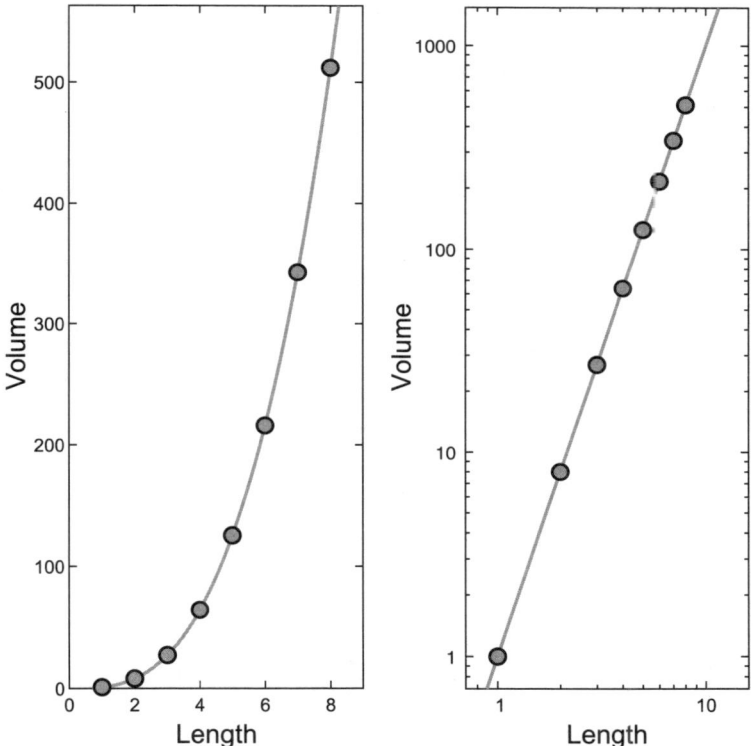

If we instead plot the same numbers on a different sort of graph paper, in which the equally spaced divisions are powers of 10 (right graph), we find something remarkable: the points lie on a straight line. What's more, the slope of the line—the number of "vertical" powers of 10 divided by the number of "horizontal" powers of 10—is 3. If we had plotted surface area rather than volume, we'd again find a straight line, but with a slope of 2. Quite generally, if y is proportional to x^p, where p is some exponent, plotting y versus x on logarithmic axes gives us a straight line of slope p. We can simply read off the scaling exponent from the graph. This ability to discern scaling relationships visually, from the tilt of the trend on an appropriately constructed plot, delivers all sorts of insights into animal shape and form.

A COCKROACH IS A COCKROACH IS A COCKROACH

Returning to animals and armed with our graphical tools, we can ask the burning question, Are cockroaches isometric? It may seem a silly thing to ask, but the answer can reveal what mechanisms of development shape the growing animal, how mechanical stresses affect different members of a species, and even how new species evolve. Rather than plotting volume, researchers often consider the mass of an animal, which is easy to measure by weighing. Most animals have similar densities of their constituent cells and tissues, so mass is roughly proportional to volume. I've reproduced here a plot of mass versus leg length for several different cockroaches, big and small.

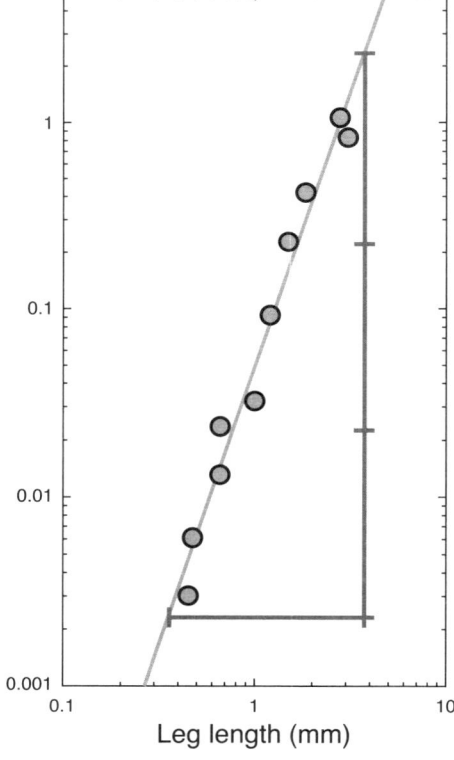

The data aren't mine—as fond as I am of nature, I can't stand cockroaches—but come from a 1977 paper by biologist Henry Prange, examining the mechanics of exoskeletons. The points are exceptionally well fit by a straight line. Of course, animals are not as well behaved as cubes or spheres, and the data are scattered about a line rather than lying perfectly on it. Still, the best-fit slope is 2.95, almost exactly 3, indicating that mass scales as length cubed just as we'd expect for isometric objects. (I've included on the graph horizontal and vertical lines with equally spaced tick marks, making it more evident that the slope is 3.) The large cockroaches are just like scaled-up versions of the small cockroaches; the size changes, but the shape is essentially the same. The cockroaches, in other words, are isometric.

Here's another example of isometry, this time across many different animals. All mammals have lungs with which to bring into their bodies fresh, oxygen-rich air. We can ask how the size of the lungs depends on the overall size of the animal. By "size," we could mean surface area or we could mean volume. In the next chapter, we consider surface area, by far the more interesting of the two measures, and how it's deeply connected to the ways lungs work or sometimes fail to work. Here we consider the more boring measure, volume, as a prelude.

We might expect isometry from mammalian lung volumes if, for example, every cell needs a similar volume of oxygen at each breath, so that the total volume of air is proportional to the total volume of cells in the body; in other words, lung volume scales with animal volume. Or we might guess that large animals need disproportionately more or less oxygen volume per cell, in which case the volume of the lungs wouldn't be proportional to the volume of the body. In the first case, plotting one volume versus the other on a logarithmic graph gives a slope of 1; in the second, it doesn't. Let's make the plot, taking values from a 1963 paper on lung physiology. As before, we plot the mass of each animal rather than its volume. Across mice, monkeys, manatees, and more, we find a slope of 1 (1.02 to be exact), indicating isometric scaling of the lungs. As mammals get bigger, their lung volume gets bigger by the same proportion. It's tempting to conclude that the typical cell in each animal is satisfied by roughly the same volume of air, as suggested above. It could be, however, that lung volume isn't the only factor that sets the availability of oxygen, and in fact we'll see in

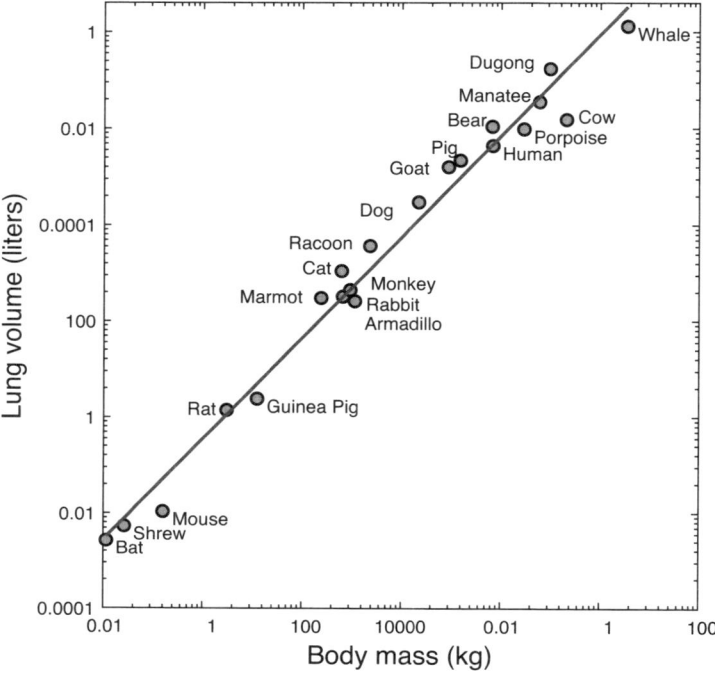

each of the next two chapters nonisometric scaling governing oxygen transport and metabolism. It's worth keeping in mind that logarithmic graphs and scaling exponents don't in themselves provide explanations, let alone correct ones. They can, however, point us toward insights that might not otherwise be apparent, as the next example shows.

WHY ELEPHANTS CAN'T JUMP

As we've seen, nature sometimes obeys isometry. It doesn't always do so, however! The elephant and the elephant shrew look very dissimilar, despite sharing a long nose. Many measures fail the test of isometry, sometimes revealing other rules. To illustrate, let's consider the large family of cloven-hoofed grazing mammals known as bovids, which includes massive water buffalo, midsized goats and sheep, and antelopes ranging down to the foot-tall (30 cm) dik-dik. In their excellent book *On Size and Life*, Thomas McMahon and John Tyler Bonner plot on logarithmic axes the diameter versus the end-to-end length of the humerus, a leg bone, of lots of bovids. I've reproduced the graph here. If these animal skeletons were isometric, these data points would have a best-fit line with a slope of 1. (Diameter and end-to-end length are both measures of length and should therefore be proportional to each other for shapes that stay the same.) The graph instead has two striking features.

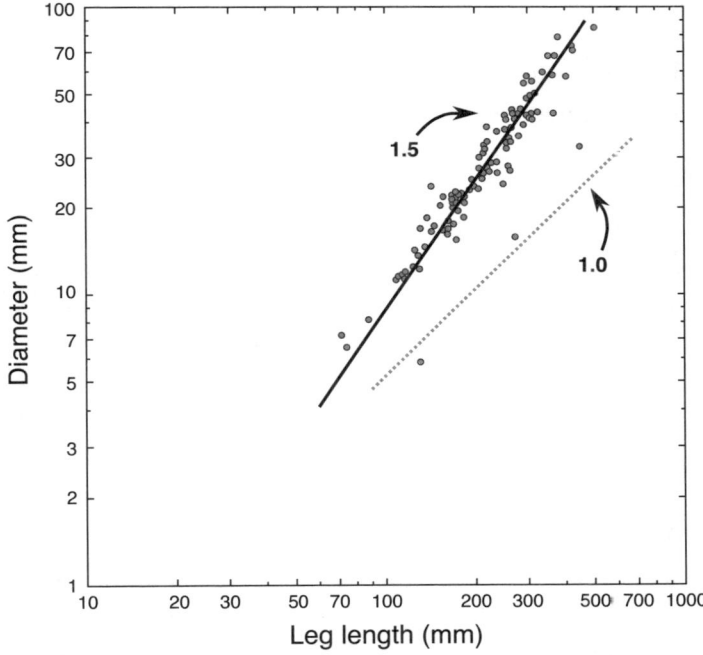

First, there *is* a relationship between diameter and length—the points aren't just randomly arrayed across the page. Second, the relationship is *not* isometric. The best-fit line has a slope of 1.5, unmistakably different from 1.0. (I've drawn a dotted line of slope 1 for comparison.) The bovids, therefore, have abandoned isometry, with the larger bovids sporting disproportionately thick bones compared to their smaller cousins. If we double the length of a bovid bone, we must more than double its diameter; in fact, the diameter must increase by a factor of $2^{1.5}$, which is about 2.8. The bovid skeletons are not similarly shaped versions of each other but have some other rule governing their form.

What is this rule, and why an exponent of 1.5? The physics behind it has to do with gravity, tugging animals and everything else downward, and the strength of bones, holding the animals up. The force of gravity is proportional to the mass of an object. If the mass increases by a factor of 10, the force of gravity increases by a factor of 10. The strength of a bone, specifically the maximum force it can withstand, depends on its cross-sectional area. This isn't a peculiarity of bones but rather a general statement about the mechanics of any sort of beam. Given these laws, our bovids would face a terrible problem if they were isometric. If we doubled, for example, all the lengths of an antelope's anatomy while keeping its shape the same, its volume would increase by a factor of 2^3, or 8. The animal's mass, and therefore the force of gravity acting on it, would also increase by a factor of 8. However, the cross-sectional area of the isometric animal's bones, like any other area, would only increase by a factor of 2^2, or 4. The strength of the bones increases less than the weight it must support. There are two possible responses to this. One is for the big animals to give up on having bone strength (relative to their weight) similar to the small animals, and to adopt correspondingly different lifestyles. The other is to have disproportionately thick bones, abandoning the similar shapes of isometry in favor of bones that are wide enough for their strength to balance their gravitational demands. The bovids take the latter path. Rather than similarity of shape, McMahon and Bonner and others posit that *elastic similarity* governs the bovid bones—the amount the bones may bend, relative to gravitational stresses, is roughly constant across different species. It turns out that a diameter-to-length scaling exponent of 1.5 corresponds to elastic similarity; if the bones become extrathick in just this way, their resistance to bending is like that of the smaller bones, relative to the force that's bending them. In other words, if the small bovids can walk, run, and graze, the large ones can too, in similar ways, thanks to their skeletal architectures scaling according to elastic similarity rather than isometry. (There's a lot of complexity that I'm glossing over, as well as a justification that bending rather than other failure modes is the critical factor for skeletons.) Of course, the bovids don't make logarithmic graphs or perform mechanical calculations. Rather, the evolutionary demands of being a bovid selected for animals whose bones obeyed the geometric relationship illustrated above. Those with smaller bone diameters were too weak to withstand gravity; those with diameters that scaled even more sharply than length to the 1.5 power perhaps expended much more energy on growing chunky bones than was necessary for a bovid lifestyle. The bovids, without knowing it, took a path constrained by biology and physics to reach an elegant endpoint of nonisometric scaling.

I mentioned, however, that there's another path that large animals can take as they consider the balance between bone strength and gravity: giving up on having the same behaviors as small animals. This is, in fact, what large animals in general do. Plotting on logarithmic axes the diameter versus the length of leg bones for a wide array of animals—cats, dogs, horses, elephants, and more—gives a scatter of points with a crude slope that's steeper than the isometric value of 1.0, but not as steep as 1.5. The bones of an elephant are truly impressive: my students and I have gathered in awe around the femur of an elephant named Tusko that I've carted into class. It's massive: a meter long with the girth of a small palm tree. (Tusko, by the way, lived a mostly miserable life in early twentieth-century circuses until ending up under far more pleasant care at the Woodland Park Zoo in Seattle. After his death, his skeleton was donated to the University of Oregon, where the elephant lives on, teaching students about bones.) Alongside Tusko's leg bone I show a coyote femur, far smaller at just a centimeter wide and 12 centimeters long. The elephant's femur is 9 times longer and 16 times wider than the coyote's. Tusko's bone is thicker than isometry would dictate (which would be just 9 times wider), but not as thick as elastic similarity would demand (a factor of $9^{1.5}$, which is 27, times wider). We can't fault the elephants for not maintaining elastic similarity with the smaller animals. If they did, bones would make up about three-quarters of their body mass, which would likely cause all sorts of anatomical complications. The consequence, however, is that the bones of the large animals are not as strong, relative to the animals' weight, as those of the small animals. The coyote can jump; the elephant cannot. The extra stress would shatter its bones. It's for this reason that elephant enclosures in zoos are often surrounded by simple, narrow moats that many animals—but not the elephant—could easily leap across.

The observation that bigger animals have disproportionately thick bones, and the assessment that they're still not thick enough to match the relative strength of small animals, are both quite old. Galileo, in fact, wrote about bone scaling in his 1638 book, *Two New Sciences*. One of the characters who animates the book speculates, "I believe that a little dog might carry on his back two or three dogs of the same size, whereas I doubt if a horse could carry even one horse of his own size." Though Galileo is renowned for performing insightful experiments, I don't think he actually tried this one.

The realization that animals' sizes, forms, and activities are all interrelated, and that these relationships are often set by the physical forces manifest in the environment, gives us deep insights into the natural world. There's a unity to all the bovids in their elastic similarity, for example, that's even more profound than their superficial similarities would imply. The back-and-forth swaying of a shark's tail is a testament to its liberation from the fluid constraints that apply to smaller creatures. The repeated emergence of scaling relationships as we study the living world across its wide range of sizes elevates the notion of scaling to a broad guiding theme. Scaling concepts provide a coherence that unifies the diversity of animal shapes and behaviors. Scaling also helps explain why such diversity exists, as organisms evolve amid physical forces whose manifestations and magnitudes are different at different sizes. In the next chapter, we focus on scaling issues related to surfaces, especially the challenges of breathing.

11

Life at the Surface

You have lungs. Ants don't. Why?

The ants' cells, like yours, need oxygen. Rather than lungs or any analogous organ, they make do with a few tiny holes along the sides of their bodies, one pair per body segment, connected to tubes on the inside. It's not because of some marvels of cellular engineering that small insects can bypass the need for lungs. Instead, it's because of scaling, and the physics and geometry of surfaces. We explore in this chapter the interfaces between inside and outside, or between one space and another, universal features of which govern the shape of an elephant's ears, the challenges of a baby's first breath, and more.

ABANDONING ISOMETRY

Every creature exchanges oxygen and carbon dioxide at some surface. The rate at which it can do so is limited by the surface area available. There's nothing complicated behind this—gas molecules going from one environment, such as the air, to another, such as the interior of a blood vessel, have to cross the boundary between them, so the flow of molecules is proportional to the surface area of that boundary. The overall demand for oxygen is set by the creature's volume (deferring some complications we consider in the next chapter), as each of its cells consumes oxygen to perform the chemistry of respiration. For a small creature, such as an ant, the surface area of its internal tubing suffices for gas exchange with its tissues, and the animal's outer surface area suffices for capturing the oxygen that its volume needs.

Imagine, however, magically expanding an ant, isometrically. If we double all the lengths of an ant, making it 6 millimeters long rather than 3, its volume increases by a factor of eight, as we saw in the last chapter. The number of cells also increases by a factor of eight, and the total amount of oxygen it needs increases by a factor of eight. The surface area, however, increases by a smaller factor, namely, four. As we increase size further or imagine even larger animals, the discrepancy gets even greater. A human-sized ant, 1000 times longer, would demand 1000^3, or 1 billion (10^9), times as much oxygen, but would have only 1000^2, or 1 million (10^6), times as much surface area with which to supply it. At human size, one can't exchange enough oxygen and carbon dioxide at the surfaces of simple scaled-up tubes, nor can one absorb the requisite amount of air through a passive outer surface. What was easy to do when small becomes impossible when large, simply because of geometry.

All large animals and plants, you included, deal with this challenge by abandoning isometry. Instead of keeping shapes similar, larger organisms adopt forms that have much more surface area for gas transport, enough to keep up with the increase in volume and cellular mass.

Small photosynthetic bacteria are smooth and round. Trees, in contrast, are highly branched, with an abundance of leafy surfaces at which to exchange carbon dioxide and oxygen. You, too, have branches, but on the inside: you have lungs, the airways of which divide into smaller and smaller airways in a cascade of ever-finer features, resulting in an enormous internal surface area. It's often noted that your lungs have the volume of a few tennis balls and the area of a tennis court.

The actions of your organs are also rooted in scaling. Your lungs inflate and deflate, driving gas in and out of your body. The ant, in contrast, can simply rely on the holes at its surface. Though ants and other larger insects can expand and contract parts of their bodies to actively pump air, the dominant mode of gas transport—and the only mode present in smaller insects—is passive diffusion. This won't suffice for larger creatures; as we saw in chapter 6, diffusing over long distances is very slow, and so you'd suffocate if you stopped pumping. Similarly, your heart vigorously drives oxygenated blood through another branching network spanning your body. The ant doesn't bother; internal gas exchange is also made simple by its small size. A little creature can harness the predictable randomness of Brownian motion, as air that enters its superficial holes meanders through passageways that penetrate the body. Invoking the insights of chapter 6: the random walk of a particle, such as a molecule of oxygen, travels on average a distance that grows as the square root of its travel time. Inverting this relationship and using the language of scaling, the travel time scales as length squared. Being 1000 times larger than the ant demands 1000^2 more time for diffusion, in other words, an extra factor of one million. Rather than patiently enduring this millionfold slowness to oxygenate our tissues, large creatures like us carry oxygen in blood and force the blood through a circulatory system, bringing it close enough to each cell for diffusion to quickly carry it the rest of the way.

There is, by the way, a way to be big and to have sufficient oxygen without building respiratory organs with large surface areas, and that is to live in an environment with a highly oxygen-rich atmosphere. Such a place doesn't presently exist, but there were times when it was the norm. In the Carboniferous period about 300 million years ago, for example, the ambient oxygen concentration in the air was 50% higher than it is currently, and we find in the fossil record abundant evidence of giant insects. A prehistoric dragonfly with a 2-foot (60 cm) wingspan would have trouble with today's relatively impoverished air.

Surfaces influence many aspects of animal form. Moose are bigger in colder regions. Bears are larger as well; polar bears, closely related to brown bears, are more massive than their southern cousins. This general relationship that animals of a species tend to be larger at colder latitudes has been noticed for centuries. The likely explanation is surface area. If you're warm blooded and live in a cold place, your body's surface area is a liability—that's where you lose heat. Because surface area scales as length squared, and volume as length cubed, the ratio of surface area to volume decreases with size. If an animal's internal heat production is matched by the heat loss through its skin and we isometrically double the animal's size, it now produces eight times as much heat from its eight times greater mass, but its rate of heat loss has only gone up by a factor of four. It could therefore overheat or, more realistically, it would require and consume fewer calories to sustain its body temperature. The larger animal can therefore survive more easily, giving rise to an evolutionary advantage for increased size. All other things being equal, being big helps in cold places.

An animal in a hot climate, on the other hand, must worry about overheating, and it benefits from having more surface area at which to dissipate heat. The ratio of surface area to volume is greater for smaller size, so all other things being equal, being small helps in warm places. Of course, one could instead abandon isometry, as with the elephant's giant-surface-area ears; but within species, changes of form tend to be less dramatic—hence the general observation about sizes and latitudes, named *Bergmann's rule* after a nineteenth-century biologist.

So far, our examples of surface-related principles relate to the form of animals. Surfaces also affect behavior, and what creatures can and cannot do.

WALKING ON WATER

At the top of a tranquil pond, water striders and many other insects effortlessly dance on the liquid, as easily as you might stroll on a lawn. Why aren't you similarly able to walk on water? The magic of a water strider lies not in the makeup of its legs but rather in its size. The insect's abilities are a consequence of scaling, specifically the scaling associated with a force called *surface tension*.

Surface tension arises at any liquid surface. No matter the liquid, the molecules making it up attract one another. That's an inherent attribute of liquids—if molecules don't attract one another, they'd very likely form a gas. Every water molecule wants to be next to other water molecules. Every oil molecule wants to be next to other oil molecules. Any liquid molecules at a surface, such as the surface of a pond, have about half as many neighbors as do molecules in the bulk. To anthropomorphize for a moment, the surface molecules are unhappy, and the liquid as a whole minimizes its surface area to ensure as few unhappy molecules as possible. Furthermore, the liquid resists any process that increases its surface area; the resulting force is known as surface tension. For soap bubbles, or liquids floating in space, or droplets in an oil and water vinaigrette, surface tension pulls the fluid into spheres because a sphere is the three-dimensional shape with the smallest surface area for a given volume. Water in a bucket, or water in a pond, has gravity and the container walls as additional constraints, and a flat interface with the air minimizes the available surface area. Whatever the context, we can think of every liquid surface as constantly pulling, contracting itself to have the smallest surface area possible given the constraints of its volume and whatever else is acting on it.

Now we can understand why the water strider can stride on water. Its legs push the water's surface, driven downward by gravity tugging on the insect. The legs are hydrophobic; the water molecules don't have any particular affinity for them, preferring instead to be near each other, again minimizing the overall surface area as much as possible. The insect's spindly legs deform the water surface; the water responds with the force of surface tension, trying to push the interface back toward a flat shape. If we imagine the leg moving downward as it alights on the pond, pulled by the force of gravity, the deformation steadily increases, and the upward force of surface tension also increases.

One of two things then happens: If we reach a deformation at which the force of surface tension balances the force of gravity, the insect stays above the water, not breaking the liquid surface, held up by the affinity of the water for itself. Or, if the gravitational force outweighs the maximum force that surface tension can provide, the surface breaks and the insect is submerged. Thankfully for the water striders, their evolutionary history has led them to the first outcome. This fluid-based support is easy to demonstrate, by the way, with a metal paper clip gently laid on a water surface. As long as both are very clean, the clip will be supported by the fluid despite being much denser. If you push the paper clip under the water's surface, however, it will sink—surface tension only applies at the surface.

So far, this argument explains why the water strider can walk on water, but it isn't clear why it fails to apply to you. After all, even though the force of gravity acting on your body is far greater than that on the water strider, your contact with the liquid surface is also larger than that of the insect. Shouldn't the upward force be greater, too? It is, but it's not enough to keep you standing atop a swimming pool. The reason once again lies in scaling. The force of gravity, as we noted in the last chapter, is proportional to the mass of an object and therefore scales with volume, or length cubed. The force of surface tension scales not as length cubed, or even length squared, but just *length*. A cube 1 inch on a side, perched on water, has a 4-inch length of perimeter defining the contact zone.

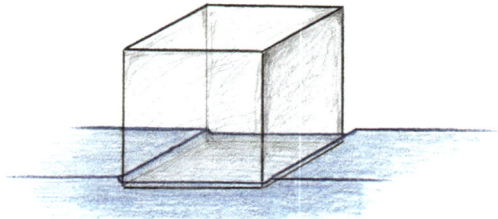

A 2-inch cube has 8 inches of perimeter, a factor of two larger than the original cube. It is along this perimeter that the water's surface is curved and extended compared to a flat interface, and so it is the scaling of the edge length that governs the scaling of surface tension forces. A creature 10 times as long as another, all other things being equal, has 1000 times the gravitational force pulling it downward, but can only muster 10 times the upward force from the fluid. A small creature may be fine at a liquid surface, but if we imagine it growing, it very quickly reaches the point at which gravity is far too strong to resist. This crossover occurs at a size of a few millimeters. Below this, it's not hard for a creature to support itself by surface tension on water; above this size, it's hopeless.

Certain ants provide another illustration of the consequences of surface tension. Fire ants are an aggressive set of species with a painful, burning sting, hence the name. They are native to tropical areas prone to heavy rains that can flood their habitats, at which point they need to stay above the water and also stay together as a colony. Though ants are denser than water, an individual ant is small; like the water strider, it can stay atop water using surface tension. A group of ants clinging to one another to stay together, however, faces a problem. As the group grows, its mass increases more strongly than the surface tension force pushing it up via the same scaling that we've just seen. For more than a few dozen ants, gravity overwhelms surface tension and the group will sink. It helps only slightly to form a two-dimensional raft rather than a three-dimensional blob; the mass of the flat raft scales as length squared, which still quickly surpasses the length-to-the-first-power scaling of surface tension. The ants, it seems, are doomed by the physics of scaling to either disperse from one another or drown. They've devised, however, an ingenious solution to this dilemma: air bubbles. An ant's surface is hydrophobic, and an individual ant can clasp a bubble to its body like a mother might hold an infant. A raft of ants together hangs on to a large bubble of air; the buoyant bubble counteracts the gravitational force pulling the insects downward. Since surface tension is fundamentally incapable of doing the job, the ants make use of the low density of air to manipulate the gravitational side of the equation.

Other creatures manipulate surface tension more directly. You, in fact, are one of them. Our next example takes place within you, every time you breathe.

BREATHING IS HARD WORK

On August 7, 1963, Patrick Bouvier Kennedy was born, five and a half weeks early, to President John F. Kennedy and First Lady Jacqueline Kennedy. Within two short days spent struggling for breath, he died.

The tragedy was mourned by millions, and though the context of being a child of the president of the United States was special, the cause of the baby's death was frighteningly commonplace. Young Patrick died from infant respiratory distress syndrome, or IRDS, the leading cause of death among premature infants. IRDS is a problem of surfaces.

For every breath you inhale, you expend a lot of work to inflate your lungs. Lungs are often depicted as elastic sacs, like rubber balloons, stretched by muscles. Your lungs aren't just balloons, however; they're *wet* balloons. The cells that line each of the hundreds of millions of tiny air sacs that together make up your lungs coat themselves with a thin layer of liquid mucus (blue in the illustration). Inhalation stretches the rubbery tissue and also increases the area of the liquid surface, that is, the area of the interface between the inflating air and the fluid film. As always, the liquid "wants" to keep this area minimal and thus will oppose its growth.

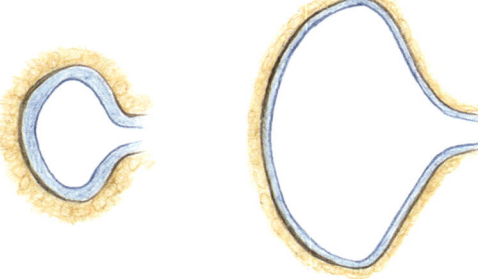

Both the stretching of the lung tissue and the expansion of its liquid surface take effort, and in fact the energy required is comparable for the two. One can measure this by pumping freshly dissected lungs with air or water. Filling with water, there's no air-water interface and hence no surface tension to fight, so the work required is solely due to stretching the lung tissue. Filling with air, we stretch the tissue and expand the air-water interface, which takes about twice as much work as filling with water. In other words, about half the effort of breathing is due to surface tension. This isn't too surprising—as we've noted, your lungs have an enormous surface area.

The high price of breathing would be even higher if the lung surfaces were lined with pure water. All liquids pull, but some pull more strongly than others. Among common liquids, water's surface tension is one of the largest—about twice as strong as oils or alcohols, a consequence of strongly attractive forces between its molecules. We can lower the surface tension of a water interface by adding a tiny bit of soap. You can demonstrate this with the paper clip that we placed on a water surface a few pages ago—with just a touch of dish soap, it sinks, its weight overwhelming the force that the soapy interface can muster. Soap accomplishes this feat by virtue of its molecular structure. As noted in chapter 5, each soap molecule has one end that's hydrophobic and one that's hydrophilic, just like lipids. The soap, therefore, gladly goes to the water's surface, where the hydrophobic tails jut out into the air. The surface is then no longer the domain of water molecules that would rather not be there. There's no longer such a large energetic cost for surface area, and surface tension is greatly reduced.

Returning to lungs, nature cleverly mitigates the cost of their expansion by simply adding soap. The secretions of the cells lining the lungs technically have the more impressive name of *pulmonary surfactants*, but "soap" is a better description of this mostly lipid substance. This is another manifestation of self-assembly: without any external guidance, the secreted soap organizes itself at the liquid-air interface, forming a layer of molecules that assists the entire organ.

The crucial ability to dramatically lower the surface tension of your lungs isn't one you've always possessed: pulmonary surfactant is made rather late in embryonic development. Premature infants, depending on how early they are, enter the world with a shortage of surface-modifying secretions or even with none at all; breathing may be a struggle or even impossible, as the muscles fight to defeat surface tension.

Patrick Kennedy wasn't the only child of his time to succumb to infant respiratory distress syndrome. In the United States alone, IRDS claimed 25,000 infants per year in the 1960s. By 2005, however, the annual mortality rate had plunged to less than 900. IRDS still occurs, just as before, but it is straightforward to treat: we squirt soap into the infant's lungs. Technically, of course, it's pulmonary surfactant, either extracted from animals or chemically synthesized, but this detail shouldn't distract us from the treatment's wonderful and effective simplicity, based not on complex biochemistry or genetics but rather on the physics of breathing. Self-assembly underlies this simplicity: the surfactant molecules put themselves in place, each positioned at the two-dimensional interface between liquid and air, on their own. Understanding surface tension saves lives.

Surfaces and the scaling associated with them govern many aspects of the living world and also illustrate how scaling connects the microscopic (for example, the structure of lipid molecules) and the macroscopic (for example, the mechanics of expanding your lungs). As in the previous chapter, we've seen how scaling helps us make sense of phenomena we observe in ourselves and other organisms. Scaling isn't a panacea, however, and we turn next to an example we're still struggling to make sense of.

12

Mysteries of Size and Shape

In the preceding chapters, we've seen examples of scaling relationships—connections between properties like bone strength and measures of overall size. These relationships often transcend simple proportionality; in surprisingly many cases, they are well described by dependencies of one measure on another raised to a power or exponent, a form generally known as a *power law*. Scaling illuminates many features of the living world. There are many examples beyond what we've seen so far, spanning sea, land, and air. Measuring the swimming speed and stroke frequency of a vast range of aquatic organisms, we find that the two are linked by an exponent whose value can be explained by hydrodynamics. Among animals that run, from cockroaches to horses, there is a power law scaling relationship between the energetic cost of locomotion and body mass, set by physical mechanisms deeper than the peculiarities of particular gaits. Flight speed, power, and mass are all governed by aerodynamic scaling laws that unify not only flying creatures but jet planes as well. We could explore all of these and more, but instead we'll take our successes for granted and look at something more puzzling. (If you're curious about these examples, I've listed some readings in the references.) We'll consider a scaling-related mystery that, if we could solve it, might help explain what sets the tempo of heartbeats, the biodiversity of forests, and even the limits of life spans.

ENERGY AND SIZE

Our mystery is old, controversial, and surprisingly easy to describe. Every organism extracts energy from chemical bonds. These bonds are present in the food consumed by animals, the nutrient molecules scooped up by bacteria, and the sugars photosynthetic creatures craft using sunlight. The released energy is directed toward all the tasks associated with life: growth, development, motion, reproduction, and more. The rate at which energy is used is known as the metabolic rate, the term *metabolism* referring to all the varied chemical reactions conducted by cells in the course of their activities. This rate isn't constant—when we sprint, we use energy much more rapidly than when we sleep—and it's never zero. Even at rest, every creature is using chemical energy at a rate called the *basal metabolic rate*. Different organisms have different basal metabolic rates. A resting elephant, for example, consumes more calories per minute than a resting mouse. How much more? More generally, how does the basal metabolic rate of an animal scale with its mass?

We might expect that no sensible relationship would hold across the diversity of life, with every creature's metabolic rate set by the uniqueness of its anatomy. Or we might expect that the basal metabolic rate would be proportional to body mass—an animal with twice the stuff inside it uses twice as much energy every minute.

In fact, neither of these is true. There *is* a relationship between basal metabolic rate and body mass, but it's not one of simple proportionality. For bone shape (chapter 10), we see a trend of disproportionately thick bones for larger animals. For basal metabolic rate, we find a trend of disproportionately lower energy use for larger animals. Basal metabolic rate is often expressed as the rate of oxygen consumption, because oxygen is required by the chemical reactions that drive metabolism. A resting mouse uses oxygen at a rate of about 40 milliliters per hour. An elephant, whose mass is 100,000 times greater, doesn't use 100,000 times as much oxygen as the mouse, but less than 10,000 times as much. A gram of resting elephant consumes energy at a rate about 20 times lower than a gram of resting mouse, on average. Lifting an evocative description from biochemist and writer Nick Lane, "an elephant-sized pile of mice would consume twenty times more food and oxygen every minute than the elephant does itself." This isn't a peculiarity of mice or elephants. In general, the larger the creature, the lower the metabolic rate per gram of body mass.

How much lower? In a 1932 paper that launched a thousand studies, the Swiss American physiologist Max Kleiber considered a range of animals from doves to cattle, plotting their metabolic rates and body masses on logarithmic axes. As we saw in chapter 10, this graphical form allows us to clearly see relationships described by exponents. A hypothetical basal metabolic rate that is proportional to mass (i.e., mass to the first power) would show up as a line with a slope of exactly 1. Kleiber didn't find a random scatter of points, nor did he find a slope of 1. Rather, the metabolic rate rose with body mass with a well-defined slope, and therefore a scaling exponent, of 0.75. Increasing an animal's mass by a factor of 100,000, for example, increases energy consumption by a factor of $100,000^{0.75}$, which is about 5600.

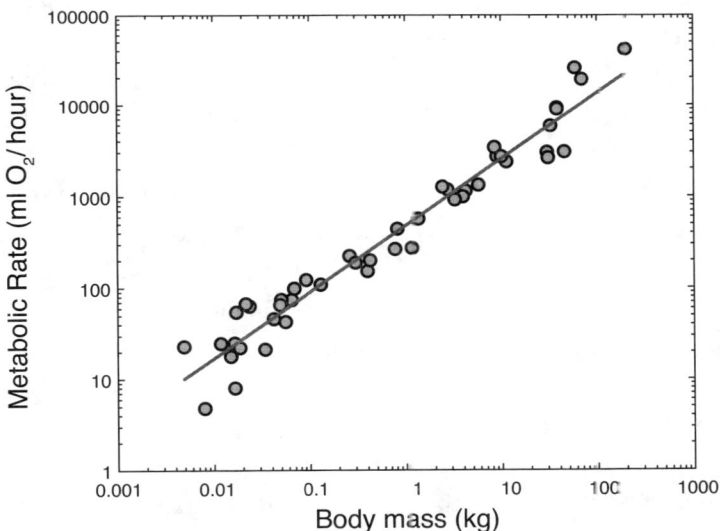

Many subsequent researchers added data points from more animals, expanding on Kleiber's set of 13. To illustrate, I've plotted about 50 points from a 2003 dataset of over 600 mammals, randomly chosen but roughly evenly spread across the wide range of masses, from the sooty mustached bat (5 grams, about the weight of a piece of paper) to the common wildebeest (200 kilograms). The linearity of the points is suggestive of some simple underlying law.

No one knows, however, what this law is. In the 90 years since Kleiber's observations, elevated to the status of "Kleiber's law" in countless textbooks and research papers, there has been no shortage of explanations, discussions, and arguments. These have not, however, converged to any agreement about what drives the relationship.

The simplest model that one might construct, beyond direct proportionality to mass, is that basal metabolic rate is proportional to the surface area of an animal. Echoing chapter 11, we might expect that the outflow of energy as heat depends on the external surface and that this must be balanced by the body's overall energy consumption rate. After all, nearly every activity of cells, tissues, and whole organisms ends up dissipating energy as heat, a consequence of universal laws of thermodynamics. This possible equivalence of metabolism and surface area predicts a basal metabolic rate that scales as mass to the ⅔, or 0.67, power. (If you're comfortable with algebra, this is because surface area scales as length^2, and mass as length^3. Inverting the latter relationship, length scales as mass$^{1/3}$, so surface area is proportional to (mass$^{1/3}$)^2, or mass$^{2/3}$. If you're not comfortable with algebra, pretend you didn't read this.) This surface area model doesn't match the data, as we've seen; 0.67 is notably different from 0.75. A mass$^{2/3}$ scaling does, however, show up in several, though not all, studies of animals *within* the same species. In fact, 50 years before Kleiber, the German scientist Max Rubner examined the energy consumption of seven dog breeds and found a mass$^{2/3}$ scaling. Oxygen consumption in guinea pigs shows the same behavior. As in chapter 10, the data suggest that isometry is a good rough guide if considering individuals of the same species, but that the differentiation of species requires more drastic changes.

If surface area doesn't explain Kleiber's law, what does? One possibility is that Kleiber's law isn't really much of a law. Basal metabolic rate is a difficult quantity to measure, so we perhaps shouldn't place too much confidence in the data points. Furthermore, the points aren't arranged in a tidy row but are scattered quite widely about the best-fit line; we perhaps shouldn't trust its slope too much. In fact, the best-fit exponent for the 600 mammals noted earlier is 0.73, not exactly equal to the appealingly elegant ¾ (i.e., 0.75). Several expansions and reassessments of the data cast additional doubt on simple conclusions. In 2001, Peter Dodds, Dan Rothman, and Joshua Weitz at the Massachusetts Institute of Technology evaluated several datasets and concluded that values for small animals, less than about 10 kilograms (22 pounds, roughly the mass of a bobcat), were well fit by a scaling exponent of 0.67. Larger animals deviate upward from this trend. If one insists on fitting a single line to all the data points, an exponent around 0.71 emerges, which can be pushed closer to ¾ by evaluating a higher fraction of large mammals. Considering birds alone, Peter Bennett and Paul H. Harvey in the 1980s found a scaling exponent of 0.67, consistent with basal metabolic rate being determined by surface area. Studies of reptiles, however, give an exponent of 0.80.

Different sets of creatures yield different metabolic scaling relationships, calling into question the very notion that there is any sort of universal principle governing all of them. On the other hand, one might argue that these variations are to be expected, due to the natural peculiarities of organisms with different behaviors and life histories, and it is precisely the overall slope—the slope that emerges if we toss all of their data points onto one graph and squint to blur the discrepancies—that matters for insights into general principles. The latter point of view would be more compelling if we could state what those general principles might be.

For decades, debates about Kleiber's law simmered; but they were brought to a boil in 1997 by physicist Geoffrey West at Los Alamos National Laboratory and ecologists James Brown and Brian Enquist at the University of New Mexico, all three of whom shared an affiliation with the interdisciplinary Santa Fe Institute. West, Brown, and Enquist proposed a creative explanation for metabolic scaling, and moreover one that they claimed leads inexorably to a ¾ scaling exponent. In their picture, the key determinant of metabolism isn't the energetic demands of an organism's cells but rather the physical constraints of the circulatory and respiratory systems that supply them. As with the surface area argument and its associated ⅔ scaling, geometry is central to the theory. Unlike the familiar geometry of surface area, however, West, Brown, and Enquist's model made use of a different sort of mathematics known as fractal geometry.

188 • Chapter 12

FRACTAL BIOLOGY

There is an almost rhythmic complexity to the shape of cracks in a sidewalk, veins on a leaf, or frost on a window that we can sense is absent in the standard shapes we learn about as children. Simple geometric shapes look different if we change the scale at which we examine them. Consider a circle. As we zoom into its edge, it looks increasingly straight.

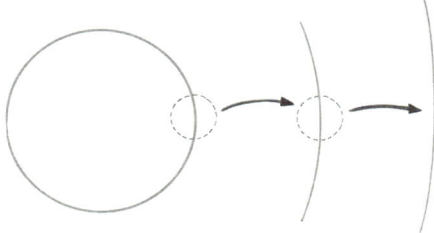

Now consider instead a shape formed by iteration after iteration of branching, taking a single line and splitting it into two, then each of these two into two more, and so on.

The final shape, after infinitely many levels of branching, looks the same no matter how much we zoom in. It is "self-similar."

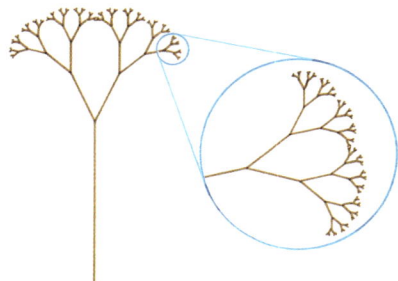

Branching is just one of many ways to create self-similarity. A rich area of mathematics called *fractal geometry* examines the properties of self-similar shapes—for example, how they can be generated by processes like diffusion, how they can fill a given space despite a sparsity of substance, and how they behave not as one-, two-, or three-dimensional objects but rather as objects with fractional dimensions (hence the name *fractal*). Many features of the natural world are well described by fractals. Fine twigs growing from a thin branch look roughly like thin branches growing from a thick branch, which look roughly like thick branches growing off the trunk of a tree; an approximate self-similarity connects features at different scales. As mathematician Benoit B. Mandelbrot noted, "Clouds are not spheres, mountains are not cones, coastlines are not circles." In other words, many of the shapes that exemplify nature are fundamentally different from simple, standard geometric forms. Their self-similar character isn't just a minor modification to spherical, conical, or circular shapes, but is their inescapable essence. Mandelbrot pioneered the analysis of such shapes and coined the term *fractal*. (His middle initial, B., the joke goes, stands for "Benoit B. Mandelbrot.") Though self-similarity of real systems doesn't proceed to infinitely fine scales, fractals nonetheless provide a powerful framework for understanding nature. The exact pathways of blood vessels may differ in their details from animal to animal, but all can be thought of abstractly as fractal branching networks.

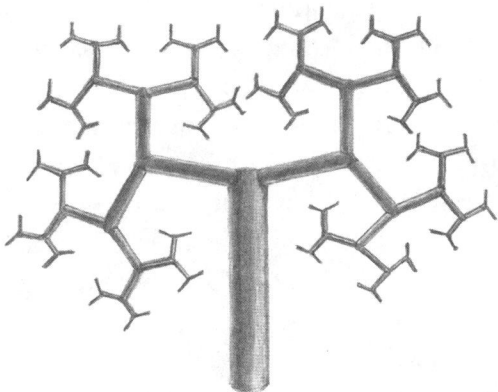

West, Brown, and Enquist recast the mystery of metabolic scaling into a question about fractal networks of pipes. These networks might supply oxygen, blood, nutrients—whatever. Rather than asking what metabolic rate an animal or plant needs, West and colleagues boldly asked what flow its supply network could sustain. This, in turn, is a question of physics and geometry. It's harder to push fluid through a narrow pipe than a wide one. More precisely, a pipe with half the cross-sectional area of a larger pipe needs four times as much pressure to drive fluid at the same flow rate. Branching into ever-finer pipes, as in a network of blood vessels, leads to an ever-greater energetic tax on the organism, the magnitude of which depends on the geometry of the array of tubes. A larger animal needs more levels of branching to connect the cellular scale with the whole-body scale. West, Brown, and Enquist expressed this mathematically and argued that the most efficient self-similar network would scale with body size in such a way that an exponent of ¾ naturally emerges for the scaling of metabolic rate with mass, that is, Kleiber's law.

West, Brown, and Enquist's paper, published in 1997 in the prestigious journal *Science*, made an enormous splash. It has been cited by over 4000 research papers since it was written. Finally, it seemed, we had uncovered the biophysics underlying Kleiber's law.

Sadly, however, the story doesn't end so simply. Determining the energetically optimal network geometry is mathematically challenging, and several researchers found subtle but important flaws in West and colleagues' proof. Just as problematically, the fractal supply model rests on several underlying assumptions—related, for example, to details of the branching architecture—that are not necessarily true. (We don't know enough to know whether they are true or not.) We're still left with a fascinating model, but rather than one that can proclaim Kleiber's law emerges from well-understood basic principles, it's one that tells us Kleiber's law is probably equivalent to a collection of other possible laws of nature, encouraging us to investigate their validity.

Perhaps the greatest contribution of West, Brown, and Enquist's work was that it reinvigorated our thinking about metabolism and spurred many biologists, physicists, and mathematicians to turn their attention to the topic. Some of the thousands of subsequent papers were celebratory. Some were critical. Some applied fractal models to other systems—more on that in a moment. Some pushed the theory in surprising directions. In 1999, for example, Jayanth Banavar (currently a colleague of mine at the University of Oregon), Amos Maritan, and Andrea Rinaldo developed a mathematical model that doesn't require self-similarity at all. They examined the relationship between the rate at which a branched delivery network can deliver material and the volume occupied by the network itself, and argued that the most compact network is the one for which the delivery rate scales as the ¾ power of the animal's total volume (or mass). The idea, then, is that animals evolve to minimize the volume required by their circulatory systems, thereby giving Kleiber's law. Applying this model to animal metabolism again requires assumptions whose validity several researchers have questioned, but the ideas are nonetheless intriguing.

Where do we go from here? How can we understand Kleiber's law and, more broadly, understand metabolism? As mentioned, one possibility is that Kleiber's law isn't really a law and that the ¾ exponent requires wishful thinking to believe in. Another, also mentioned, is that a rough ¾ law is a signature of aspects of shape and form that we still don't understand. Yet another is that Kleiber's law isn't a fundamental aspect of how metabolism must work, but rather emerges as a consequence of the many other biophysical laws that organisms are subject to. For example, imagine that animals were made only of muscle and bone, with a kilogram of muscle and a kilogram of bone, each having its own characteristic and constant fuel requirements, but with the fuel consumption rate being greater for muscle than bone. If the proportion of muscle and bone remains the same regardless of overall size, the animal's overall metabolic rate will simply be proportional to its total mass. However, we saw in chapter 10 that bigger animals have bulkier bones. The overall proportion of bone mass is greater for larger animals to help compensate for the weaker scaling of bone strength compared to weight. The animal's overall metabolic rate would therefore scale more weakly than mass to the first power, giving an exponent below 1.0 for reasons of bone mechanics rather than anything intrinsically related to energy consumption or nutrient use.

Of course, animals are more than just muscle and bone. Every organ and tissue has its own signatures of size and shape, currently understood to varying degrees. Together, these might all intersect to give the metabolic characteristics of organisms, described by simple exponents or perhaps more complex relationships. This perspective has been nicely described by biochemist and author Nick Lane and was precisely formulated by Peter Hochachka's group at the University of British Columbia, Canada. Unfortunately, Hochachka and colleagues' initial model had significant mathematical flaws, and it remains to be seen if this approach can be molded into a rigorous and convincing theory.

WHO CARES ABOUT METABOLIC SCALING?

All we've concluded in this chapter is that we can't conclude anything. You, dear reader, may be wondering why I've dragged you through all this if not to reach a satisfying end. Or you might wonder why *anyone* cares about a puzzle that has so steadfastly resisted solutions. In part, the simplicity of the question—How does metabolic rate depend on mass?—and the hints from data of universal laws are the main drivers of interest. Beyond this, though, the solution matters to more than questions of animal energetics.

How much forest does an elephant need? The answer is set in part by the foliage it must graze on to sustain its metabolism. Shifts in species abundances and changes in available space—driven, for example, by disease, poaching, or deforestation—can ripple through an ecosystem in ways that depend on the metabolic needs of larger or smaller creatures. More grandly, one can ask what the metabolism of the whole ecosystem is and whether there are rules that transcend the details of particular constituent creatures. Brown, West, and others have argued for a "metabolic theory of ecology" along these lines. One might hope that a better understanding of metabolic scaling will lead to better strategies for conservation and land use.

Metabolism influences fundamental features of our existence, such as life span. Our elephant from early in this chapter lives much longer than the mouse. The life expectancy of a typical African elephant is over 60 years. The average mouse, even if it escapes being eaten, lives about 2½ years. The contrast represents a general trend: life span on average increases with body mass throughout the animal kingdom. A plot of life span versus mass for mammals shows a rough power law scaling with an exponent of approximately ¼, again for reasons that remain mysterious. The fact that the Kleiber's law exponent of ¾ is a whole number multiple of ¼ may reflect a shared origin, as the network models imply, or it may be coincidental. The elephant also differs from the mouse in its heart rate; the elephant's heart beats about 30 times per minute, the mouse's about 600. Again, this illustrates a general trend of heart rate slowing with size, and yet again we find a scaling exponent, $-¼$, that's a multiple of ¼. (The negative exponent means that the relationship slopes downward on logarithmic axes.) The opposite exponents of heart rate and life span together have an intriguing consequence. Multiplying heart rate and life span—beats per minute times the total number of minutes lived—gives the total number of beats the heart performs over the animal's existence. The $+¼$ and $-¼$ cancel to give a number of heartbeats that is constant regardless of body mass. (More precisely, $mass^{¼} \times mass^{-¼} = mass^{0}$, a flat line with no slope at all.) The mouse, the elephant, and all mammals great and small each get about a billion heartbeats allotted to them. Humans are an anomaly, by the way. Our life span is about twice what's normal for an animal of our mass—a typical 100 kg mammal lives about 30 years—so we've got about two billion beats at our disposal. These relationships regarding heart rate, life span, metabolism, and more might all be coincidental, or they might point to connections that, if we understood them, could give us insights into some of the fundamental aspects of being alive.

Perhaps most surprisingly, understanding nature's metabolic scaling laws might help us understand unnatural sorts of scaling, such as the activity of cities. Geoffrey West and others, building on the fractal model sketched above, have argued that cities, like living organisms, are governed by the properties of networks. These may be meshes of highways, roads, and lanes; arrays of pipes and wires moving liquid and electrical currents; webs of financial transactions; or social networks connecting people. The data suggest that unexpected scaling laws might govern the dependences of income, road coverage, patent filing, and more on city size. The data are, however, noisy, and their trends are even more contentious than those of animal metabolism. Nonetheless, the possibility of deep principles underlying urban life is tantalizing, especially because the continuing urbanization of the earth's human population calls for better design and management of cities.

As fascinating as the implications of understanding metabolic scaling may be, you might still be frustrated by the present state of confusion. That's fine—I am, too. A fundamental problem is the vast imbalance between abundant theory and sparse data. Papers written about Kleiber's law outnumber the measured data points! Science is much more than theorizing. In fact, what makes science science is that ideas are tested by their ability to make predictions about the real world. There is no shortage of theories that are mathematically sound, elegant, and wrong. I run a research lab, performing experiments rather than working solely with equations and computers, in large part because I want to see, and be surprised by, the way nature actually behaves.

It's hard to experimentally test metabolic scaling ideas. Though we've imagined several times so far animals getting smaller or larger, real elephants and mice don't come with knobs to dial their masses up and down. There are a few clever approaches that get at this obliquely. Frank Jülicher, Jochen Rink, and colleagues in Dresden, Germany, for example, recently examined flatworms that grow or shrink dramatically depending on the availability of food, with a mass range spanning more than a factor of 1000. Strikingly, their measured metabolic rates scaled as body mass to the ¾ power, just as Kleiber's law would predict, and in contradiction to the ⅔ exponent noted earlier that typically characterizes animals of the same species. The researchers were able to attribute this scaling form to mass-dependent changes in how the animals' cells store fats and sugars. Whether this is a general path to Kleiber's law or a peculiarity of flatworms is still to be determined. Our mystery remains mysterious. Its resolution awaits new data, or perhaps new animals.

The scaling relationships we've encountered throughout the past few chapters reflect connections between the cellular machineries encoded in genomes and the large-scale constraints of physical laws. In part III, we see how and why to read and write genomes, potentially redesigning organisms in ways that influence both small- and large-scale function. We might imagine that understanding biophysical scaling relationships enhances our ability to make genetic predictions, or that the outcomes of genetic engineering inform our understanding of scaling laws. Both are likely to be true.

PART III
Organisms by Design

13

How We Read DNA

In the preceding chapters, we've seen that life involves the interplay of physical objects and physical laws—molecules, cells, and organs governed by principles of self-assembly, regulation, randomness, and scaling. The two categories of objects and laws aren't really separate. A protein is a protein, for example, because of the self-assembly of its constituent amino acids, driven by the incessant dance of Brownian motion to explore the possibilities of three-dimensional shapes. An organism's genome ties together tangible stuff and the forces that shape the stuff, encoding the amino acid sequences and regulatory motifs that physical interactions will mold. Changing the genome, therefore, changes the resulting organism.

So far in this book, my aim has been primarily to illuminate the intersection of physical principles and the natural workings of life. Now in part III, we dive into the application of our newfound knowledge to alter how living things function, in ways both subtle and radical. The physical nature of biological matter is crucial not only to the implementation of new technologies but also to their implications. What we infer from differences in DNA sequences, for example, or what outcomes are possible or impossible if we alter such sequences, depends on the processes that orchestrate the readout of genes, the architecture of the proteins formed within our cells, the forces that guide and constrain self-assembly, the randomness inherent in the microscopic environment, and other such biophysical concerns. Keeping in mind the physical context of life will help us form a realistic vision of the impacts of present and future biotechnologies, distinguishing between possibilities that are likely or unlikely, feasible or far-fetched.

In chapter 15, our explorations lead us to methods for rewriting genomes. Before writing, however, we must be able to read, discerning the sequence of As, Cs, Gs, and Ts that defines a given set of DNA. We have achieved this ability, creating dazzling technologies that harness the physical characteristics of this vital molecule and the constructive power of self-assembly and microscopic randomness.

WHY IS DNA HARD TO READ?

Reading DNA in itself provides deep insights into life with thought-provoking, practical consequences. For example, detecting unusual sequences (i.e., mutations) that correspond to increased likelihoods of disorders such as cancer may spur preventive measures. Relatedly, sequencing the cells within cancerous tumors can reveal particular genetic signatures that suggest particular treatments. These applications and many others require maps of a 2-nanometer-wide, 1-meter-long molecule.

We know from chapter 1 that DNA is a chain-like molecule, composed of just four possible units. Why, then, isn't it simple to read its sequence? If I write the word *molecule* on a page, you can immediately see that its sequence is M-O-L-E-C-U-L-E. The problem with DNA is that it's small. Each nucleotide is about one-third of a nanometer long, a nanometer being one-billionth, or 10^{-9}, of a meter. Not only is that small on a human scale, it's too small for any microscope that uses light. Light, like radio waves or X-rays, is an electromagnetic wave, a traveling undulation of electric and magnetic fields. For visible light, the wavelength—the distance between the peaks—is a few hundred nanometers, the exact value depending on the color of the light. The laws of optics dictate that the smallest features one can discern are about the size of the wavelength of light one uses to see them, no matter what lenses, mirrors, or microscopes one builds. Any features finer than this are blurred together. Even if we were to label each A, C, G, and T with a different color or its own distinct flag, our image would blur together each nucleotide's marker with those of a thousand of its neighbors. Trying to simply read out the sequence would be hopeless.

You might guess that the way out of this is to use something other than visible light to form an image, but this approach fails. Smaller waves certainly exist. X-rays, for example, have wavelengths between about 0.01 and 10 nanometers. Electrons behave as waves, and when guided in electron microscopes have wavelengths that are tenths of nanometers or smaller. DNA nucleotides would in principle be resolvable with either of these probes, but multiple problems pop up in practice: X-rays are challenging to focus; the high energies of X-rays and electron beams can be destructive; and the different DNA nucleotides are almost identical from the perspective of X-rays and electrons, so we couldn't determine the sequence even if we were able to form an image. You might find it odd that X-rays aren't helpful, since we learned in chapter 1 that these waves were used to determine the double-helical structure of DNA. This, however, involved shining X-rays through a whole crystal of DNA, a grid of many trillions of identical molecules. The interaction of the waves with all these DNA strands reveals the twisted ladder structure they all share; it doesn't discern the sequence of any individual strand.

PIECING WORDS TOGETHER

Reading DNA, therefore, is difficult. Fifteen years after the 1953 revelation of the double helix structure of DNA, Ray Wu and Dale Kaiser managed to decipher 12 of the nucleotides of a viral genome. Its full genome contains about 48,000 nucleotides. Five years later, Allan Maxam and Walter Gilbert determined the sequence of 24 nucleotides that make up the DNA binding site of the lac repressor (chapter 4). This took two years of demanding work, a rate that would translate to 250 million years for sequencing the full human genome. Clearly, one would hope for better methods.

Better methods emerged, and over the next several pages we look at them in some detail, not only because of their importance to the modern world but also because their very existence is a testament to the power of studying the physicality of DNA and other biological molecules. Properties like size, stiffness, and electrical charge, and themes like self-assembly and randomness, serve as the backdrop of this very practically focused chapter.

Two clever approaches to determining DNA sequences arose around 1977, one by the same Maxam and Gilbert, and the other by Frederick Sanger and colleagues. Sanger's method was the easier of the two and soon came to dominate DNA sequencing, becoming the method of choice for over two decades. I begin our description of how to read DNA with Sanger sequencing.

Imagine that, rather than reading one by one the letters M-O-L-E-C-U-L-E, you instead had many truncated copies of the word in which you could only make out the last letter: "??L," "?O," "?????U," and so on. Noticing that all the three-letter fragments end in L, all four-letter fragments in E, and so on, you deduce that the complete word is MOLECULE. Conceptually, this is the essence of Sanger's method as well as some of the others—reimagining the task of reading as an identification of distinct pieces rather than progression along a strand.

We've already encountered, in chapter 1, several of the steps in the recipe for this approach. Imagine a fragment of DNA—not a whole genome, but something more tractable, perhaps a few hundred nucleotides long. The *polymerase chain reaction* (PCR) generates countless copies of the fragment, and heat separates the two halves of each double helix. Ignore one of these halves for now; imagine millions of identical, single-stranded DNAs. Keep in mind that DNA polymerase generates a perfect complement to any single-stranded DNA by stitching free nucleotides together to synthesize a new partner strand. Now for the new idea: Imagine that the researcher again uses DNA polymerase to replicate DNA as in normal PCR, but spikes the stock of free nucleotides with a small fraction of defective units—slightly different As, Cs, Gs, and Ts that can still be attached to the strand, but to which new nucleotides can't be linked. Each nucleotide addition is a roll of the dice: If DNA polymerase latches onto a normal free nucleotide, the elongation of the strand progresses. If it grabs an altered nucleotide, the elongation terminates with this final unit. Because the addition of terminal nucleotides is random and rare, the researcher ends up with many DNA strands all beginning with the same starting point but extending different lengths.

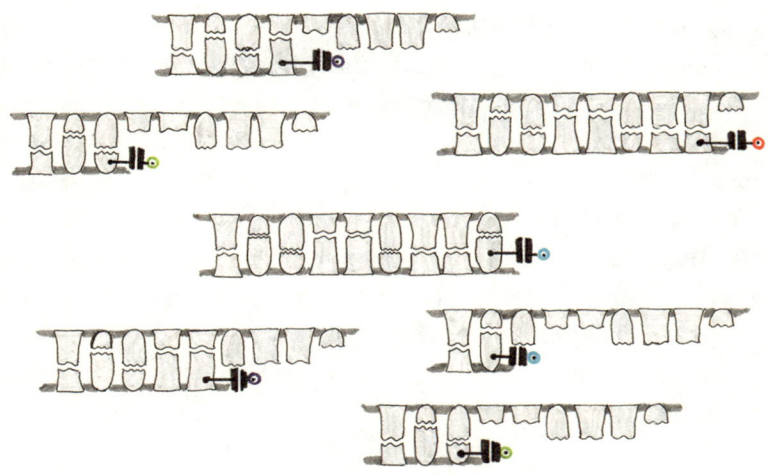

So far, it seems like the consequence of the altered nucleotides is just to botch PCR, but these terminal units are designed not just to thwart DNA elongation but also to emit light—a different color for the not-quite-A, not-quite-C, not-quite-G, and not-quite-T. There are other possible markers than color. Sanger sequencing at first used radioactive labels and didn't use PCR (which hadn't been invented yet), but rather employed bacteria to clone DNA. Our description here is of its later, more efficient variants, but the principles are the same.

Continuing, a final melting step separates each DNA molecule into single strands, with the variable-length fragments labeled at their ends, like our pieces of words with the last letter visible. The researcher still doesn't know how long a given fragment is, however, and the fragments are too small to observe with visible light. As we saw in chapter 1, PCR makes use of one of DNA's important physical properties: its melting, that is, its separation into single strands above some critical temperature. Sanger sequencing builds on this by exploiting another physical property of DNA: its electrical charge.

DNA has a negative electrical charge, as we noted in chapter 3 when discussing its wrapping around histones. As a result, one can move it with electric fields, dragging the molecule toward positive electrodes and away from negative ones. In plain water, pieces of DNA move with similar speeds regardless of their size. Larger fragments have more charge and hence a greater electrical force pushing them along, but also experience more drag from the fluid. The physics governing the scaling of force and drag with fragment size turn out to be complex and subtle, but the ultimate result is a near cancellation of effects that make mobility roughly independent of length in a simple liquid. The situation changes, however, in a slab of gel. In a gel, like the edible gelatin of Jell-O, long, chain-like molecules are pinned to one another, forming a porous, three-dimensional meshwork permeated by water. The single-stranded DNA must snake through the pores to travel through the gel—the technical term is *reptation*—which requires a series of contortions that take considerably more time for longer DNA strands than for shorter ones.

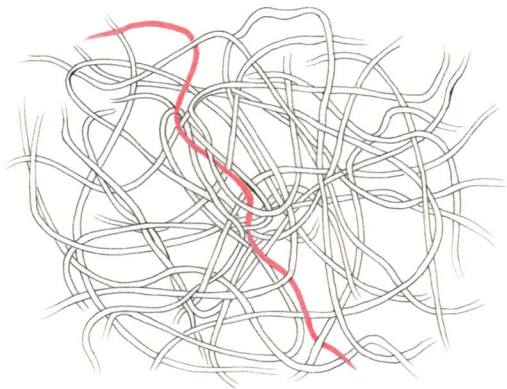

The predictable randomness of Brownian motion is crucial to this transport. Without it, DNA would get stuck in the gel no matter how strong the electric field; if the two ends of a strand fell into different pores, the molecule would be draped across the barrier like a towel on a clothesline, never to be free. Thanks to Brownian motion, however, the DNA is constantly jiggling and reorienting, coming loose from one hole to pass through another. The statistical predictability of microscopic randomness gives a well-defined, and mathematically tractable, speed to the molecule's travel.

The final result of the duplication, end-labeling, electric-field-driving, and dragging-through-gel of the DNA is a straightforward readout of nucleotide sequence. All the fragments of a given length have the same color labeling their end. Strands that are 27 nucleotides long end with C and are red, let's say, with a terminal-modified C. Likewise, 28-nucleotide-long strands ending with T are blue, let's say, with a terminal-modified T. And so on. None of the 27-nucleotide-long strands are blue, since all fragments of that length, being clones of one another, must end in C and all terminating Cs are red. (If you've conscientiously worried for the past few pages about the ignored half of the original double-stranded DNA, which seems like it would contribute another set of molecular fragments, fear not: a judicious choice of primer guides DNA polymerase to work on just one of the single strands in Sanger sequencing, so the other is not replicated at all.)

All the DNA fragments start out moving through a thin tube of gel together, but they separate as they travel at different speeds. Watching a particular point along the tube, the researcher sees a red pulse go by, then a blue one, then another blue one, then a green one, and so on, from which it follows that our original piece of DNA had the sequence . . . CTTA. . . . We have read our DNA.

Enhanced by a proliferation of technological improvements, Sanger sequencing and its variants, often referred to in retrospect as first-generation sequencing methods, were able to read around 1000 nucleotides per day by the mid-1980s. It's more common to see this written as 1000 bases per day and to refer to the length of genomes in terms of the number of bases. For our purposes, the distinction between nucleotide and bases is irrelevant. (If we want to be pedantic, an A, C, G, or T nucleotide consists of an adenine, cytosine, guanine, or thymine base joined to a sugar called *ribose* and a collection of phosphorus and oxygen atoms called a *phosphate group. Nucleotide,* in other words, means base + ribose + phosphate, and it's the ribose and phosphate groups that are linked together to make a strand of DNA.)

The sequence of an entire genome is revealed by stitching together all the sequence fragments. In 1982, we had the complete 48,000-base genome of the bacteria-infecting virus that Wu and Gilbert caught a glimpse of in 1968. The full genome of the yeast *S. cerevisiae* (12 million bases) was mapped in 1996 and the nematode worm *C. elegans* (100 million bases) in 1998. The most tantalizing target, of course, was *Homo sapiens.* Though it was clear that Sanger sequencing could in principle tackle the human genome, applying it to a genome billions of bases in size was in practice a massive technological challenge. The task called not only for improvements in biochemistry, related to terminal nucleotides, for example, but for advances in our tools for physical manipulation of DNA—melting, pushing, detecting emitted light, and more, all to be performed robustly and rapidly.

In 1988, the US Congress authorized funds for what became known as the Human Genome Project, set to launch in 1990 with an estimated time to completion of 15 years and an estimated cost of $3 billion. (For scale, total annual federal non-defense-related research spending in the United States in 1990 was about $23 billion.) Much like the Apollo moon program of the 1960s, the Human Genome Project evoked notions of exploring new frontiers, this time of the inner world within cells rather than outer space. The Human Genome Project was publicly funded, organized mostly through the US National Institutes of Health and the Department of Energy, though with considerable international collaboration. In 1998, however, a privately funded group led by biotechnologist Craig Venter announced a plan to independently sequence the human genome, and moreover to do so faster and at a lower cost than the Human Genome Project. A race was on! Both groups succeeded and in 2001 jointly announced a 90% complete version of the human genetic code. In 2003, they reached 99% coverage, declaring the achievement of an essentially complete genome two years ahead of the original schedule. Work continued to fill in the few remaining segments that defied initial efforts due to difficulties such as repetitive sequences, and by 2004 they had a 99.7% complete genome.

You might wonder, *Whose* genetic code was sequenced? Both projects used composites: DNA from several individuals so that some fragments came from one person, some from another, in total giving a generic picture. In practice, however, the majority of the genetic material in each project came from a single individual. For the Human Genome Project, this was an anonymous man probably from Buffalo, New York, and for Venter's effort, this was an anonymous donor later revealed to be . . . Craig Venter himself! These people are certainly not representative of the whole of humanity; to understand our species, we'd like a statistical portrait, which would require sequencing many more humans' genomes. Similarly, if I were to develop cancer, my doctor would like the genome of *my* malignant cells, not a generic average. Addressing these limitations calls for much faster and cheaper technologies. Luckily, these were just around the corner.

READING MANY WORDS AT ONCE

The $3 billion price tag of the Human Genome Project works out to about $1 per base pair. This is a stunning achievement given that less than a human lifetime earlier we didn't even know the structure of DNA; nonetheless, it's far too expensive for routine applications. In the early twenty-first century, several clever new techniques emerged, spurred in part by public funding for sequencing innovations. These approaches are known collectively as next-generation, second-generation, or high-throughput methods. In Sanger (i.e., first-generation) sequencing, one replicated fragment is sequenced at a time. Mixing fragments together would be disastrous, as we'd lose the unique correspondence between the length of a truncated subfragment and its terminal nucleotide. Second-generation methods are inherently parallel, capable of assessing many fragments at once and in many cases reading DNA strands as they grow. We look at a few of these second-generation sequencing approaches now. Though diverse in their details, all these methods share a theme: leveraging physical attributes of DNA, DNA-associated materials, or both.

Pyrosequencing is a next-generation method made possible in part by the awesome abilities of fireflies. As we've noted, DNA polymerase stitches new nucleotides onto DNA strands. A careful accounting of the atoms that make up the nucleotide and the atoms that make up the strand shows that they don't quite match up. The reaction that links a nucleotide to a DNA chain liberates a tiny molecule called *pyrophosphate*, made of two phosphorus atoms and seven oxygen atoms. A protein included in the pyrosequencing soup transforms pyrophosphate into ATP, a molecule used by cells for many different activities. One such activity is a light-emitting chemical reaction performed by proteins called *luciferases* that use ATP as their fuel. (The Latin *lucifer* means "bringer of light.") Creatures such as fireflies, click beetles, and luminous fungi naturally make luciferases. As we saw with jellyfish-crafted green fluorescent protein in chapter 2, the variety of life provides a variety of tools that we can repurpose in creative ways.

Pyrosequencing therefore works as follows: As with Sanger sequencing, one starts with DNA fragments, replicated to give many copies and melted into single strands. Again, DNA polymerase will grow the second, complementary strand from a given single strand. Imagine for the moment that there is only one molecule of single-stranded DNA, anchored to a beaker. The researcher pours in a soup that contains luciferase and the other ingredients but that includes only one type of nucleotide—A, let's say. Detecting a pulse of light implies that A must have been added onto the strand by DNA polymerase, and so A must be the next nucleotide in the DNA sequence. Darkness, in contrast, implies that A wasn't added and isn't the next nucleotide. The researcher removes the A-containing soup and repeats the process three more times, with C, G, and T. One and only one of the four will yield a burst of light. The letter is now known. Repeating the process gives the next letter, the next, and so on. The DNA is read as it's being formed.

I haven't explained how this procedure could be parallelized. Moreover, it seems to require perfect sensitivity—production of a single pyrophosphate must lead irrevocably to a single luciferase emitting a single, exceptionally faint pulse of light that we must detect perfectly. If any of these steps fail, we miss a letter of our genetic code. These two issues, parallelism and robustness, are addressed by the same physical tactic: bunching many identical DNA fragments together.

As in Sanger sequencing, one first fragments DNA into small pieces, less than 1000 bases long. The DNA is melted to separate the two strands of each double-stranded fragment (chapter 1). A short sequence of chemically modified bases links each single-stranded DNA to the surface of a microscopic bead, with DNA and beads mixed together with such an excess of beads that it is very unlikely that any bead will have more than one DNA molecule bound to its surface.

The beads and DNA are in a watery solution. Mixing with oil under agitation or flow gives water droplets not much larger than the beads, encapsulating no more than one bead within.

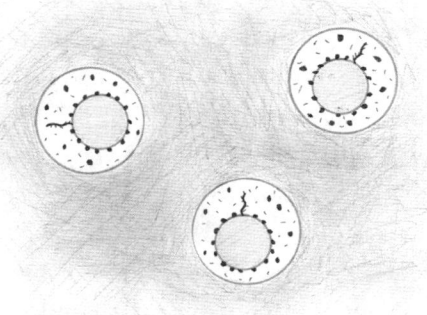

The surface of the beads is decorated with primers to initiate DNA polymerase's DNA synthesis. The watery solution contains DNA polymerase and nucleotides, as well as primers that enable replication of the single strand's complement. Each droplet therefore holds all the ingredients to generate vast numbers of copies, around a million typically, of the starting fragment. After the replication is done, one collects the droplets and adds soap or alcohol to lower the surface tension that keeps each drop isolated in oil (chapter 11). The water coalesces and flows over a plate dotted with tiny wells just larger than a single bead, so that each well is home to just one DNA-coated sphere. At this point the

DNA is double-stranded, with one of the strands held by the surface-attached primers. Melting the double-stranded DNA and washing away the unbound strands leaves an array of tiny globes, each covered with its own forest of identical single-stranded DNA molecules.

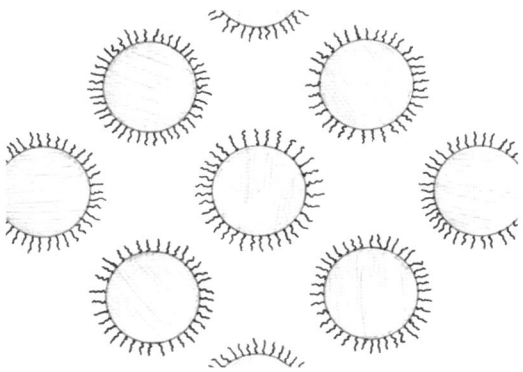

Now pyrosequencing can begin, watching for flashes of light in each well as the million-fold-replicated DNA grows, a million times brighter than would be the case for a single DNA molecule. If there are repeating letters in the DNA sequence, the pulses will be brighter, twice as bright for two As in a row, for example, as for one. The researchers, or more accurately their machines, tabulate the sequence of colors and intensities, and thereby read the DNA.

This elaborate scheme actually works, and is robust enough to have been the first next-generation sequencing method to be commercialized, by a company called 454 Life Sciences in 2005. For about $500,000, you could buy a machine to perform pyrosequencing and output the nucleotide letters making up the DNA input. At half a million dollars, you wouldn't buy one for your living room, but it was well within the grasp of moderately sized research institutes. (For scale, the median house price in the United States in 2005 was about $240,000.) Pyrosequencing machines are no longer sold, however, superseded by other sequencing technologies that are no less amazing.

A similar method to pyrosequencing makes use of another piece that's left over from the stitching of a nucleotide onto a DNA strand. This piece is so small and seemingly insignificant that its detection is especially astounding. It's a single proton. All the everyday matter surrounding us is made up of protons, neutrons, and electrons. The lightest element, hydrogen, is simply one proton, positively charged, and one electron, negatively charged, bound together. A lone proton isn't even that. Detecting it quickly and robustly is made possible by a very nonbiological technology: the transistor.

Transistors make up the circuitry of cell phones, computers, and countless other electronic devices. In every transistor, the flow of electrical current from one point to another is governed by what happens at a third point, a bit like boat traffic along a river being governed by a drawbridge operator. In so-called field effect transistors, the controlling factor is an electric field, for example, from a proton in proximity to the transistor surface.

As in pyrosequencing, beads coated with cloned DNA fragments settle into individual wells. The array of wells sits atop a semiconductor chip, constructed so that each well has a field effect transistor at its base. Rather than pulses of light, one detects pulses of electrical current. This technology was commercialized by Ion Torrent, a company founded by prolific biotechnologist Jonathan Rothberg, who previously headed 454 Life Sciences. Ion Torrent introduced its Personal Genome Machine in 2010. It fit on a tabletop and cost just $50,000, less than twice the average price of a new car sold that year.

The dominant next-generation sequencing technology, however, made use of clever chemical modification of DNA rather than sophisticated detection of nucleotide addition. Recall that in Sanger sequencing the growth of a DNA strand is terminated with an unnatural nucleotide. By the late 1990s, more flexible methods were in hand: reversibly terminating, reversibly fluorescent nucleotides. One at a time, the researcher adds to replicated DNA fragments a modified A, C, G, or T, each with a different color. Washing away the free nucleotides reveals the color corresponding to the one that was glued by DNA polymerase onto the next position of the growing DNA. The fluorescent part of the molecule and the part that blocks DNA polymerase activity are both linked to the normal nucleotide part by a strand of atoms that can be chemically cleaved. Flowing in the cleaving agents leaves regular, unadorned DNA, ready for the next nucleotide addition; the process repeats. Typically, the process is performed using glass slides dotted with patches of replicated DNA fragments, with cameras capturing the flashes of color appearing and disappearing at each spot on the slide.

This technology is commonly known as *Illumina sequencing*, after the company that acquired its original developer. From them, one can buy instruments at prices from about $100,000 to millions of dollars, depending on parameters like how many nucleotide bases are read per run. As with all sequencing technologies, it takes time and money to prepare the DNA sample and assemble the chip with wells, the slide with patches of DNA clones, or some other platform. Apart from purchasing the sequencing machine, the cost to sequence a billion DNA bases ranges from about $5 to $150 for Illumina sequencing, in contrast to about $10,000 for semiconductor sequencing, contributing in large part to Illumina's appeal. I comment more on price shortly, but it's worth noting that all these numbers are far smaller than the $3 billion it cost to read three billion bases in the initial Human Genome Project!

SINGLE WORDS FROM SINGLE MOLECULES

It's harder to identify the boundary between second- and third-generation sequencing methods than that between first- and second-. The invention of new techniques doesn't come in bursts separated by periods of rest, but rather by continuous activity along multiple overlapping fronts. Nonetheless, a convenient and roughly chronologically correct distinction is that third-generation methods involve the sequencing of *single* DNA molecules, without the need for replicas. In an approach commercialized by a company called Pacific Biosciences, for example, individual DNA polymerase molecules are anchored in tiny pits carved into a metal film, the optical properties of which dictate that fluorescent nucleotides are visible only while being stitched onto the growing DNA molecule.

The most remarkable new DNA sequencing technology doesn't involve growing DNA, colorful nucleotides, or even DNA polymerase. We could ignore everything we've learned over the past few pages about methods in which sequencing is linked to synthesis, and return to the naive picture of simply proceeding base by base along an already formed DNA strand and recording what's there. As noted, one can't use light to distinguish bases. One can, however, use electricity.

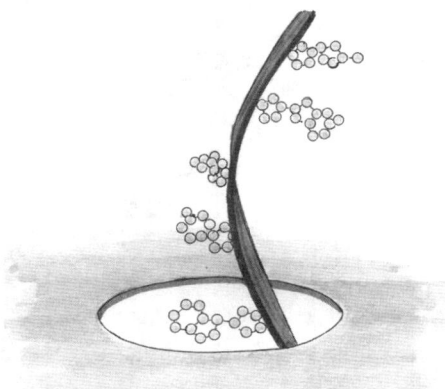

Imagine a pore in a membrane that separates two liquid reservoirs. The watery solution on each side contains ions, which are electrically charged atoms or molecules. A voltage applied across the membrane causes ions to flow through the pore, and one can measure this electrical current. If the pore is partially blocked, the current will be smaller; the magnitude of the current reflects the accessibility of the pore to ions. Now imagine a single strand of DNA threading the pore. Each base, A, C, G, or T, has a different number of atoms and a different physical size, and therefore allows a different current flow if it is situated in the pore. To turn this into a sequencing method, one needs to drag the DNA through the hole. In early setups, the voltage difference was also the driver of DNA motion; as we've seen, DNA itself is electrically charged. This works poorly, however, mainly because the transport is too fast to allow precise current measurements for each base passing by. The solution to this problem? Proteins. There are many different proteins that ratchet themselves along a DNA strand as part of their normal function—DNA polymerase, for example. Fixing one of these proteins to the pore (green in the illustration below), the now stationary protein's ratcheting equipment feeds DNA through the

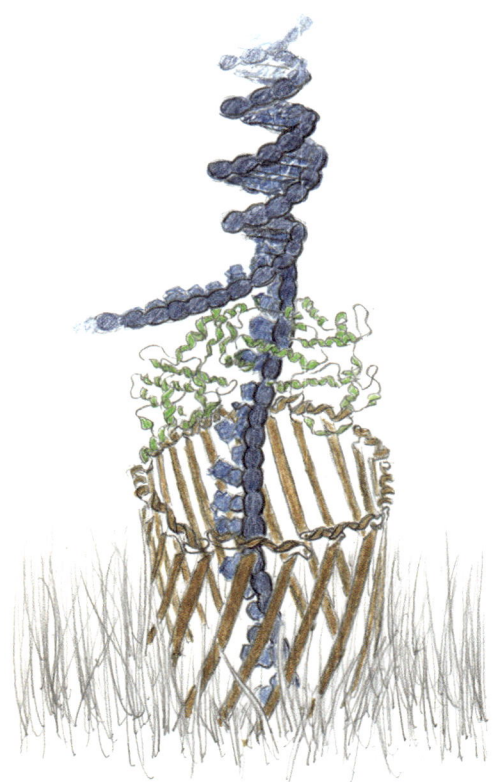

opening. We saw in chapter 2 that many proteins are natural nanometer-scale machines. Here that machinery is harnessed to perform work in an otherwise inaccessible space. Proteins can also form the pores themselves, employing molecules like those we saw in chapter 2 that fold into shapes with channels running through them. Once again, we see the manifestations of self-assembly.

This *nanopore sequencing* scheme is conceptually simple and has been hypothesized since the 1980s as potentially viable. It took a lot of work to actually realize it, however. Researchers first demonstrated the reliability of pore insertion into artificial barriers, not with the aim of DNA sequencing but simply to understand pore structure and develop platforms for working with membranes and proteins together. Early studies of electric-field-driven DNA motion through pores helped researchers understand the physics underlying transport in constrained spaces. Even with the principles in place by the early 2000s, many casual observers, myself included, found it hard to imagine that nanopore sequencing would become a robust, reliable technology. One must put together pores, DNA-ratcheting proteins, few-nanometer-thick membranes, and electrodes. Then one must perform exquisitely sensitive current measurements while ensuring that the pores don't clog due to contaminants or damaged proteins. This daunting engineering challenge was taken up by Oxford Nanopore, a company focused, as you'd guess from the name, on nanopore-based detection of molecules. Its first instrument was released to researchers in 2014 as a pilot program to evaluate its efficacy, and a few different machines are now available commercially. The smallest is thumb-sized, like a USB memory stick, and in fact plugs into a computer's USB port. A major part of its appeal is portability together with low cost. In 2014, for example, scientists sequenced Ebola virus DNA from 14 human patients in Guinea. They mapped the viral genome within two days after collecting samples, highlighting the potential for rapid sequencing in the field that may enable fast identification of the origin of outbreaks or the discovery of potentially harmful variants as they arise.

The drawback of nanopore sequencing is relatively low accuracy; a few percent of bases will be read incorrectly. Illumina sequencing, in contrast, misidentifies around 0.1%. Whether this is a concern or not depends on the application. For a detailed map of a large genome or prenatal tests for small but significant mutations, minimizing errors is crucial. For surveillance of potentially harmful microbes, or surveys of the membership of bacterial communities, lower accuracy may be a fine trade-off to gain speed and convenience.

THE EVER-CHEAPER GENOME

Advances in DNA sequencing have been stunning, and the future will likely bring even more. I've briefly mentioned the price of these technologies, but now let's look more closely at the cost of sequencing—how much a researcher would pay to determine the specific series of nucleotides making up a DNA sample. A good measure is the cost of sequencing per human genome, in other words, the cost of sequencing three billion bases. Recall the $3 billion cost of the first human genome, corresponding to $1 per base. Even by 2006 this had fallen over 99% to about $13 million. The graph of cost over time, reproduced here, is amazing.

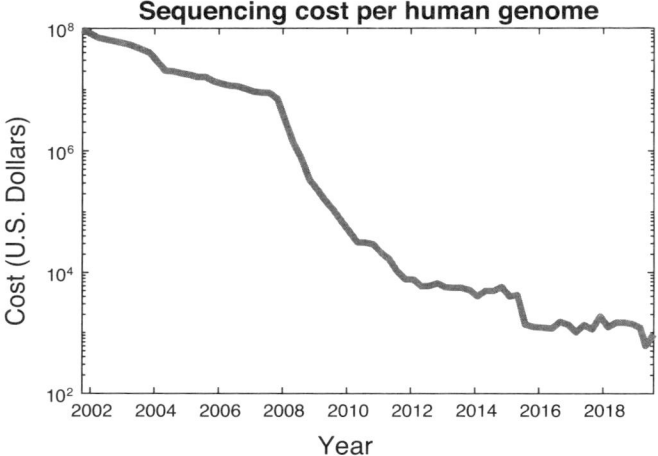

The vertical axis is logarithmic, with divisions being powers of 10; the horizontal axis is not. The decrease in sequencing costs has been strong and sustained. The plunge around 2007, a thousandfold drop in half a decade, marks the transition to next-generation sequencing methods. The time required for sequencing has plummeted along with the cost, from several years for the Human Genome Project to a few days for contemporary instruments.

Perhaps the closest comparably rapid technological advance has been our ability to make transistors. The first transistors were bulky, finicky, and millimeters in size. Now we routinely place billions of transistors on centimeter-square chips, sold for as little as a few dollars apiece. In 1965, Gordon Moore, later CEO of Intel, noted that the number of transistors on a chip doubled every year (or every two years in a revised estimate), an observation that came to be known as Moore's law. Moore's law served as both a description and a manifesto of electronic technology. DNA sequencing, at least as measured by the cost per nucleotide, has advanced at an even more rapid rate. Fittingly, the first human genome to be sequenced using Ion Torrent's Personal Genome Machine was that of Gordon Moore. The drivers of the sequencing revolution are numerous: human ingenuity, the economic incentives of commercialization, public funding for basic research as well as specific programs targeting sequencing innovations, and of course the accumulated understanding of biology, chemistry, and physics that we've managed to amass.

WHEN IS A DNA SEQUENCER NOT A DNA SEQUENCER?

There's a lot of information we can glean from DNA sequences: differences in the genetic code between species, between individuals, and between cancerous and normal cells, for example. Nonetheless, it may turn out that the biggest impact of DNA sequencing technologies lies not in sequencing genes or genomes at all, but rather in using these tools to reveal the inner lives of cells.

As an example, let's consider RNA. As noted in part I, genes are read by the machinery of RNA polymerase, which transcribes a DNA sequence into an RNA sequence that is then translated into a particular amino acid chain. Ignoring egg and sperm cells and certain immune cells, all the cells in your body have the same genome; sequencing every cell's DNA would be redundant. Knowing what RNA molecules are present in each cell, however, would be interesting—we'd see what genes were on or off as cells developed into blood cells or neurons, or as cells responded to changes in diet or stress. For this we would need to sequence RNA molecules—that is, read their sequence of nucleotides—keeping track of which RNA came from each cell of interest. By now you might guess at aspects of the methodology: harnessing physical forces and properties, with some assistance from biological tools that nature has already developed.

Rather than jellyfish or fireflies, this time our natural aides are viruses. The "central dogma" of biology is that DNA encodes RNA, which encodes proteins. It was a shock, therefore, when the labs of David

Baltimore and Howard Temin in 1970 each independently discovered that certain viruses are capable of the opposite work flow, transcribing their RNA genome into DNA that is inserted into the genomes of their hosts. The protein machine the viruses use is called, appropriately, *reverse transcriptase*, and like DNA polymerase and other such tools, we can co-opt it.

Extracting RNA from a cell, one can add reverse transcriptase and free nucleotides to generate DNA that is complementary to the RNA strands. If the RNA reads "CAGUUGGA," for example, the complement in DNA is "GTCAACCT"—recall that U in RNA is the analog of T in DNA (chapter 3). Subsequently sequencing the DNA reveals its code, and therefore the RNA's code as well. To resolve the RNA library within individual cells, researchers employ a suite of methods similar to those we've already seen, for example, encapsulating single cells along with beads and the requisite biochemical ingredients in vinaigrettes of water and oil. Each RNA molecule is transcribed into DNA, the DNA is sequenced, and we're presented with a readout of what genes were "on" in each cell.

Though single-cell RNA sequencing is a destructive measurement—cells don't survive being burst open—one can apply it to similar cells collected at different times during some process, or with and without some treatment, to reconstruct cellular trajectories. For example, researchers have studied immune cells from organisms exposed, or not, to some pathogenic stimulus to see how gene expression changes in response to provocation. As a second example, RNA from embryos of zebrafish or mice captured at different stages postfertilization reveals the flow of gene expression patterns that guide cells into their roles during animal development.

RNA sequencing is one of many technologies enabled by DNA sequencing. Researchers can now also assess which segments of DNA are wrapped around histones, which have methyl group tags attached, which have transcription factors bound to them, and more. We asked at the top of this section, "When is a DNA sequencer not a DNA sequencer?" The answer? When it's an RNA sequencer, or a DNA packaging mapper, or a guide to gene regulation, or any other such machine.

Living things routinely handle the information in DNA, copying it when cells reproduce and transcribing and translating its sequences into RNA and proteins. The outputs of these processes depend on the sequences of As, Cs, Gs, and Ts, so the processes themselves are, in a sense, reading DNA molecules. For about four billion years, these methods for reading DNA were all that existed. Now we've invented radically new tools—fast, cheap, and almost magical in their efficacy—that place in our sights the information encoded inside every organism. This amazing technological transformation occurred because we took seriously the tangible, physical characteristics of biological molecules, building interfaces between them and other manifestations of our technologies. We turn next to what we can infer from the information encoded in DNA.

14

Genetic Combinations

The information encoded in all sorts of organisms, including ourselves, now lies before us thanks to our marvelous ability to read DNA. What can it tell us? We asked this question in part I and focused on the nature of genes and gene regulation. It is often tempting to think it straightforward to connect a creature's genetic code with its traits: we merely assess what variations of a gene correspond to variations in the characteristic of interest. Even from our discussions in part I, we know that reality isn't so simple; it isn't just the genes in a genome that dictate biological activity, but also the regulatory circuitry encoded in DNA that switches the transcription of genes on and off. Now we'll see that nature's complexity is even greater than we may have thought; many of the traits and diseases we care about are influenced by the DNA at thousands of sites in the genome, forming a web of connections that is daunting to make sense of.

The theme of randomness introduced earlier comes to our rescue, giving us conceptual and practical tools for handling genetic information. These tools are so effective that we can often get by with maps of the genome that are much sparser, and cheaper to acquire, than the whole genome sequences introduced in the previous chapter. Understanding the meaning of randomness and predictability is crucial for making sense of technologies that already have a large impact on our world, intersecting with topics in health, industry, and ethics in ways that we will touch upon.

WHERE'S THE TALLNESS GENE?

There are many characteristics that fall under the umbrella "genetic," meaning that they're at least influenced, if not wholly determined, by one's sequence of As, Cs, Gs, and Ts—the code we get from our parents. In some cases, including several involving debilitating disorders, there's a simple mapping between what the body does and which span of DNA is ultimately responsible, with just a single gene involved. Cystic fibrosis provides a good example.

We all have mucus lining our lungs, making up the liquid coating we encountered in chapter 11. Cells secrete mucus and drive it along the airways, toward the mouth, sweeping away fluid as well as dirt, pollen, bacteria, and other particles we may have inhaled. The mucus in people suffering from cystic fibrosis is exceptionally viscous, stagnating in the lungs rather than flowing, which leaves the lungs susceptible to bacterial infection. All this is the fault of a single gene that encodes a single protein, the cystic fibrosis transmembrane conductance regulator, which sits in cellular membranes and facilitates the flow of ions into or out of the cells. In cystic fibrosis patients, a mutation in the cystic fibrosis transmembrane conductance regulator gene alters the structure of the protein. As a result, the concentrations of ions on each side of the membrane aren't what they should be, causing water to be sucked out of the mucus, the viscosity of the mucus to increase, and the patient to suffer.

At the other extreme from cystic fibrosis are traits like height. There are nongenetic influences on height—nutrition, for example, plays a major role—but the genetic code you carry from the moment of your conception strongly influences how tall you'll be when you're fully grown. There is not, however, a height gene. Rather, there are tens of thousands of sites in the human genome at which the identity of that site—A, C, G, or T—somehow influences height. These sites aren't all located at genes; many are parts of sequences that regulate gene expression or DNA packing (chapter 3), pulling the strings that guide the genes themselves.

The case of height is much more typical than that of cystic fibrosis, at least for the complex traits and diseases that occupy much of our attention these days. There's no single gene that determines your risk of colon cancer or even your hair color, but rather a polyphony of pieces of the genome. This shouldn't surprise us. After all, with only 20,000 protein-coding genes in the human genome and a complexity far greater than could be described by 20,000 protein-based instructions, it would be astonishingly unlikely that one-to-one mappings between gene and trait would be the norm. The protein encoded by a gene can play roles in many different traits; conversely, a trait can be controlled by the coordinated activity of many different genes. Most important of all, as we saw in chapters 3 and 4, the 99% of the genome that doesn't consist of genes has a major influence on how genes work, regulating their activation and deactivation.

Let's look more closely at height, both because it's familiar and because it's an excellent example of what genetics can and can't tell us. Height is influenced both by an individual's environment and by genes, and these two driving forces—nurture and nature—are not mutually exclusive.

On average, people were shorter in the past. The average Frenchman born in 1800 was 5 feet, 5 inches tall (164 cm); if born in 1980, he was almost 5 feet, 10 inches (176.5 cm) tall. A typical Japanese woman born in 1900 was 4 feet 8 inches tall (143 cm); her descendant born in 1980 was 6 inches (15 cm) taller. The same patterns have been repeated worldwide, especially as countries develop modern economies. These century-or-two spans haven't seen mysterious plagues that wiped out short people, nor striking mutations of the human genome. Rather, the increases in height are driven primarily by nutrition. Our modern Frenchman has about twice the calories per day at his disposal as his turn-of-the-nineteenth-century counterpart. Calories aren't everything, but the abundance of dietary energy correlates with an abundance of nutrients that together allow the human body to develop to its full potential. Height is influenced by other nongenetic factors as well, such as childhood diseases and environmental pollutants, exposure to which has decreased markedly for vast swathes of humanity over the past few hundred years.

Now, however, let's consider the population *within* a modern, industrialized country. The heights of its adults, even separated by sex, aren't all the same. What's more, we all know that taller parents tend to have taller children. Children tend to grow to heights more similar to their biological parents than to random adults or to adoptive parents. Genetics matters. How much? Which regions of the genome are responsible?

We saw in the last chapter exquisite tools for reading DNA. For rare traits, subtle variations, or comprehensive characterizations, it's useful to sequence entire genomes. Height and many other characteristics, however, turn out to be robust enough that we can use simpler methods. Your genome and mine are over 99% identical, and so we can focus our attention on the regions where we differ. Consider a rare site of disagreement, a location in the genome where most people have an A nucleotide, for example, but a sizable fraction of people have a C nucleotide instead. Such sites of relatively common differences are called *single nucleotide polymorphisms*, abbreviated SNPs and evocatively pronounced "snips." There are a few million common SNPs in the human genome, where "common" means that at least 1% of the population has the rarer nucleotide. A few million is a large number, but it's much smaller than the three billion nucleotides of the full human genome, allowing us to find these SNPs fairly easily. For example, one can coat microscopic beads with short, single-stranded DNA whose sequence includes the complement of the dominant form of an SNP, and then assess whether a test subject's DNA, fragmented and replicated, binds to the beads. Binding indicates that the test sequence matches the SNP; failure to bind indicates a mismatch. I'm glossing over the details and there are several different technologies available, but they're all echoes of the elegant approaches we examined in the last chapter that harness fluorescent nucleotides, DNA polymerase, mass-produced glass slides dotted with millions of beads each with millions of DNA clones, and so on. The end result is that for well under $100 per test, less than many people pay for a pair of shoes, one can determine the ensemble of SNPs that characterize a genome and that therefore characterize much of the genetic variation represented in that individual.

It would be reasonable to guess that knowing one's SNPs would provide little insight, because these sites are only a small fraction of the genome and genomes are complicated. Initially, this seemed to be the case. The first studies mapping SNPs to height, reported in 2008, uncovered about 40 genetic variants that together did a detectable but feeble job of accounting for how tall the people studied were. Even in 2008, it was clear that examining more people would be crucial. The reason is not biology per se, but rather the relationship between randomness and predictability.

As in chapter 6, let's think about flipping coins. Imagine 10 tosses of a fair coin. On average, we'd expect the coin to land heads five times and tails five times, but it wouldn't shock us if we found six heads and four tails. In fact, the probability of this outcome is just 83% that of getting five heads. If you toss the coin 1000 times, 500 heads and 500 tails would be the most probable outcome, and the greater number of flips smooths over the variation so that finding 600 heads and 400 tails becomes far less likely, in fact about *one billion* times less probable than 500 heads. Suppose you suspect that your coin is counterfeit and unbalanced, with a greater than 50% chance of landing heads. If you flip the coin 10 times, you wouldn't be very sensitive to this imbalance—if you found six heads, you wouldn't necessarily conclude that the coin has a 60% chance of landing heads. If 1000 flips gives you 600 heads, however, you'd have a strong indication that your coin is unfair. More precisely, our sensitivity to biased coins scales as the square root of the number of coin flips. This square root may remind you of the statistical properties of our random walker in chapter 6. The connection is more than superficial; the dependencies arise for very similar mathematical reasons.

Returning to the genome, our SNPs are like the coins, and the challenge is to figure out how "fair" or "unfair" each one is—that is, how much each SNP contributes to a trait being different than its average, expected value. An SNP at which the rare variant is equally likely to be found in a tall individual as a short one would be analogous to our fair coin, equally likely to contribute heads or tails. An "unfair" SNP, in contrast, gives a nudge toward being taller or shorter than the average person, like an unfair coin gives a nudge toward the total number of heads being greater or smaller than 50%. The nudges may be very small. The analog to tossing lots of coins is examining lots of people's genomes. To assess the bias of any particular SNP, we can look at the correlation between a person's DNA at that SNP position and the person's height, assessed for as many people as we can. The more genomes we examine, the more sensitive we are to identifying height-related SNPs.

We're now firmly entrenched in the era of large-scale genome studies. A group led by Michigan State University's Stephen Hsu, a physicist by training, examined data from nearly half a million people collected as part of the United Kingdom's Biobank project, detecting from their statistical signatures any SNPs that correlated with the subjects' heights. The researchers found far more than the 40 SNPs of the 2008 study—in fact, about 20,000. Such analyses are challenging and it's easy to be fooled by false patterns. There are mathematical tests to evaluate the reliability of one's assessments, but the best test is to see whether the SNPs one has identified as being correlated with height in the genomes under study work as predictors of height in a different set of people. In other words, one examines most, but not all, of the Biobank dataset, discovering, for example, that SNP number 312 corresponds, on average, to a 0.05 cm increase in height relative to the average; SNP number 3092 corresponds to −0.02 cm; SNP number 4512 corresponds to +0.08 cm; and so on. Next, in the part of the dataset so far left unexamined, one looks at each person's SNPs and adds up the inferred effects, predicting what their height should be. One then assesses how well the prediction matches their actual height. Hsu and colleagues did this in their 2018 paper, finding that most of the measured heights were within an inch (3 cm) of the SNP-based predictions. To get a better sense of what this sort of accuracy looks like, let's make some graphs.

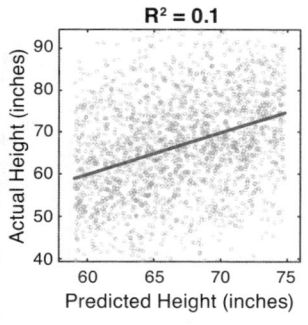
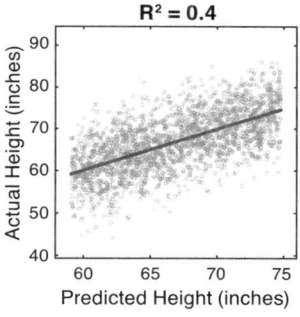
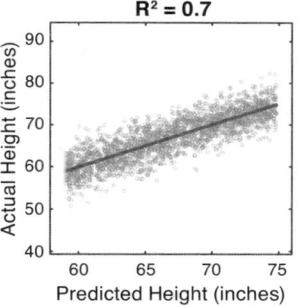

Each graph shows a cloud of hypothetical data points, the predicted heights measured on the horizontal axis and the actual heights on the vertical, with each dot corresponding to one person. The actual and predicted heights are correlated in all three plots. Moreover, the line that best fits the data points is nearly identical in each graph. How well that "best-fit" line actually describes the data, however, is markedly different in the three cases. On the left, the cloud is very diffuse; in the middle, the data points are better localized near the line; on the right, the measured values quite tightly cluster around the prediction. This variation can be quantified by a statistical property called the *coefficient of variation*, often denoted by the symbol R^2. For an intuitive description of R^2, first imagine measuring how scattered the points would be around a flat, horizontal line going through the middle of the graph. (If you know some statistics, imagine the variance of the measured data points.) Next suppose we measure how scattered the points are around our best-fit line. This amount of variation is smaller—it's what's left over after accounting for the relationship revealed by the line. The ratio of this second variance to the first is a number between zero and one, which is smaller the more tightly the points cling to the best-fit line. One minus this ratio is the variance "explained" by the linear relationship; that's R^2. In the leftmost graph, the diffuse cloud, R^2 is 0.1, meaning that the best-fit relationship between prediction and measurement accounts for 10% of the scatter of the data points. In the rightmost graph, R^2 is 0.7; 70% of the scatter is accounted for.

In the SNP-based height analysis of Hsu and colleagues, R^2 is about 0.42, similar to the middle graph—not a perfectly tight spread but not a formless cloud either, and good enough to correspond to the precision of an inch mentioned earlier. An inch may not seem so impressive, but it's more accurate, it turns out, than the prediction you'd make about someone's height by knowing how tall their parents were. And, of course, the SNP-based assessment doesn't require any information about identity or ancestry—just DNA and a cheap test. As Hsu points out, the debris from a crime scene now suffices to tell you how tall a completely unknown person is, and it will reveal a host of other physical characteristics as well.

How good could the R^2 of height get? From studying family members of different levels of relatedness, including identical twins (whose genomes are exactly the same), geneticists have long known that the heritability of height is about 80%. In other words, heredity explains about 80% of the variation between individuals. Is the gap between 0.4 and 0.8 due to the DNA that's inaccessible to SNP studies, or does it arise from more mysterious biological mechanisms? In 2019, Australian geneticist Peter Visscher and colleagues examined whole genome sequences from over 20,000 people and found that the information encoded by DNA does, in fact, explain 80% of the variation in human height. At least for modern Europeans, the vagaries of diet, disease, and exercise give rise to 20% of the spread in the heights we see.

BUILD A BETTER CHICKEN . . .

Of course, all this applies to much more than humans. Rather than the heights of our neighbors, we could just as well ask about the genetic contributions to the spots on a leopard, the petals of a rose, or the mass of an amoeba. Controlling variation in the traits of organisms is central to agriculture. The human population of our planet doubled from two to four billion between 1930 and 1970, and has doubled again since then. That this stunning increase hasn't been met by widespread starvation is due to a remarkable range of innovations. A central feature of the Green Revolution of the 1950s and 1960s, for example, was the selective breeding of new strains of wheat and rice. American agronomist Norman Borlaug, working in Mexico in the mid-twentieth century, bred wheat varieties with large, seed-bearing heads. These plants, however, had a tendency to fall over—as we saw in chapter 10, it's hard to be big. Crossing the plants with dwarf strains—mutants from Japan— gave robust, high-yield wheat. With these and similar advances, Borlaug is considered to have saved a *billion* human lives.

We want shorter wheat but larger chickens. Contemporary chickens raised for consumption in North America weigh four times as much as their ancestors of the 1950s, even if raised on the same feed. (To get a sense of how dramatic this increase has been, picture a world in which the average person weighs 700 pounds (320 kg).) There is variety in chicken body types, some amount of which is due to genetics, and the hulking modern fowl are the result of consistently selecting the larger chickens as those to breed. In humans, by the way, the human genome study of Visscher and colleagues noted above assigns to DNA sequence about 40% of the variation in body mass index, a measure of relative body size and mass.

Nowadays, rather than relying solely on easily identifiable traits, one can use SNPs to help choose which animals or plants to mate. Selecting a large hen and a large rooster, for example, may give large chicks, but what one would really like is for the size-enhancing variants in the mother's genome to be different from the father's, so that the children are more likely to possess two distinct sets of genetic nudges toward being large—two different unfair coins in the chick's pocket. Collecting SNP data, or "SNP genotyping," is therefore increasingly common. In 2019, the US dairy database, for example, contained genotypes from three million cows, up from two million just two years before. Dozens of crops from wheat to tomatoes to sunflowers are covered by SNP tools and databases.

. . . AND A BETTER WATERMELON

As we saw for human height, and as holds for a host of disorders that we'll come to shortly, SNPs provide predictive power despite their sparsity across the genome. This is surprising, and the minimal commentary I gave earlier that perhaps the single nucleotide variation lies at a gene or a regulatory region is perhaps unsatisfying. There's another reason SNPs convey so much meaning that recalls a key point from the first chapter: DNA is a physical object, a long, chain-like molecule. The DNA we received from each of our own two parents is combined in the cells that seed our offspring. In the cell that creates a sperm or egg cell, a strand from one parent and a strand from the other are close together. Protein machines can exchange segments of the strands, cutting and pasting to mix the two genomes, resulting in final strands that are each mixtures of genetic information from the two parents. The randomness of where and whether exchanges happen ensures that each sperm or egg cell is, for all practical purposes, unique, one of a vast number of possible splices. The exchanged DNA segments may include genes, regulatory regions, and SNPs. The closer together any regions are on the parental genome, the more likely they'll travel together in these DNA exchanges and, when sperm and egg meet, be passed on to the resulting child. An SNP, therefore, isn't just an SNP; it can be a marker of a larger expanse of DNA surrounding it that more directly shapes the organism. Sequencing these single nucleotide variants gives us a proxy for more detailed genetic information.

Watermelons illustrate the consequences of physical proximity between genes. Ten thousand years ago, there were no watermelons. The ancestor of the modern plant was a desert vine whose small fruits held pale yellow, bitter flesh. They readily retained water, however—a valuable trait. The ancient Egyptians farmed and domesticated the gourd, selecting generation after generation the seeds that came from the least bitter fruits, generating the sweetness that we continue to love. One of the genes most responsible for taste, whose variants in DNA sequence result in varying degrees of bitterness or sweetness in the fruit, sits close to a gene that encodes a protein that influences color. Selecting for the sweeter variants, therefore, brought redder flesh along for the ride. Continued selective breeding for the next few millennia, focusing on size, taste, color, rind thickness, and other traits, has given us the intensely sweet, red watermelon we know today.

GENES, DISEASE, AND RISK

Let's return to humans and to traits more important than height or even sweetness. Genetics plays a major role in many maladies of current concern, like cardiovascular diseases and cancers—the two leading causes of death globally—as well as diabetes, Alzheimer's disease, and numerous mental disorders. In some cases, there are single genes that have an oversized role in susceptibility. Altered sequences in genes named BRCA1 and BRCA2 (short for breast cancer genes 1 and 2) are found in about 5% of breast cancer patients, and women with these variants have a roughly fivefold greater risk of developing breast cancer at some point in their lives compared to the general population. It's more often the case, however, that there aren't just a few genes at play. Rather, as is the case for height, a host of genetic variants each makes a small contribution. SNP-based studies have revealed a web of connections between an individual's genome and disease that promises to be further illuminated by whole genome sequencing. As with height, these connections are probabilistic: we can't say with certainty that a person will develop coronary artery disease or not, nor will we likely ever be able to, given that nongenetic factors, such as diet and exercise, are also important. However, we can acknowledge the spread in possible outcomes and speak in terms of variation for groups of people, as in our height graphs above, or risks for specific individuals.

Given one's DNA sequence, how much greater or lower is one's probability of developing type 2 diabetes, for example, compared to the average person? In 2018, a group led by Sekar Kathiresan at Massachusetts General Hospital and Harvard Medical School showed that, for a host of diseases, an SNP-based score derived from the same UK Biobank cohort noted earlier can robustly identify high risks; for example, a fivefold increase in the likelihood of developing coronary artery disease. A factor of five is comparable in magnitude to rarer single-gene variants that are already relevant to clinical screening and diagnosis, like BRCA mutations. The researchers proposed that "it is time to contemplate the inclusion of polygenic risk prediction in clinical care." They note, however, that the data so far used in their study and others are based primarily on people of European ancestry, and therefore the predictions derived from them are less accurate for members of other ethnic groups. To similarly treat a broader range of people, important for both individual health and social equity, we need more large-scale genomic studies across a wide range of ethnicities.

Of course, an elevated risk uncovered by a genetic test is just that—a risk and not a certainty, a probability connected to the development of some disorder. Its utility lies in the hope that the identification of risk will suggest changes in lifestyle or will direct the application of diagnostic or preventive tools specifically to those likely to need them. For various cancers, for example, early detection improves the odds of successful treatment, but the detection methods themselves can carry harm. Rather than indiscriminate testing on everyone, using genetic scores to identify candidates for whom the risks from disease outweigh the risks from detection would be, on average, beneficial to everyone.

PICK A GENOME

If you find yourself with a genome that puts you at high risk for various problems, you might ask yourself, Could it have been different? Your genome is a random composite of your parents' and grandparents' genomes. The die was cast long ago—too late for you to do anything about it. The genome of your future children, however, has yet to be fixed, raising the question of whether you could guide its DNA sequence. In the next chapter, we'll see how to rewrite DNA sequences to be whatever we like. Here we look at the more limited, though easier, method of selecting one particular genome from the set of possibilities. This selection is, to use a loaded term, unnatural. It's based on another unnatural procedure that caused shock and outrage upon its introduction, but that is now widely accepted and appreciated: in vitro fertilization.

In 1969, Robert Edwards, Barry Bavister, and Patrick Steptoe announced that they had successfully fertilized human egg cells with human sperm cells in vitro—meaning outside of a human body, or any other body. (*In vitro* literally means "in glass" and is the standard term for experiments performed in an artificial environment, like a petri dish or a test tube.) The second half of the two-sentence abstract of their paper drily notes that "there may be certain clinical and scientific uses" for the procedure. The obvious use is the treatment of infertility, as the authors were well aware. Steptoe, in fact, was a gynecologist who worked closely with women suffering from fertility problems. It took nearly another 10 years of work, led mainly by Edwards, Steptoe, and a nurse named Jean Purdy, until the steps of egg cell gathering, external fertilization, and implantation into the mother all fell into place. The first "test tube baby," Louise Brown, was born in 1978 to an infertile couple, making headlines around the world. The intervening decade saw a great deal of discussion, filled with both excitement and trepidation, concerning the new intersection of technology and conception. A 1969 *Life* magazine cover featured an illustrated fetus next to a photograph of a mother and child, with text posing questions about what would happen as a result of the "new methods of human reproduction," including "Will children and parents still love each other?" As if to answer, a poll of Americans in the same issue reported that only about 60% of respondents (55% of men, 61% of women) believed that a child born through in vitro fertilization "would feel love" for his or her family. (It doesn't state how many naturally conceived children the survey group believed love their families.) Only about a third approved of in vitro fertilization methods for infertile parents. By 1978, however, according to a Gallup poll also conducted in the United States, 60% had a favorable view of in vitro fertilization, and just over half reported that they would be willing to try it if they were infertile. These days mention of technologically assisted reproduction hardly raises an eyebrow. There have been in total about eight million babies born worldwide through in vitro fertilization as of 2018. In the United States, about 2% of births each year are "test tube babies"; Denmark has the highest rate, about 9%. As far as I know, no one considers the resulting people as anything less (or more) than human.

It may strike us as silly that it could have ever been otherwise, but issues of creation and reproduction tap into deeply held and often unstated notions of identity and humanity. Our views can change, however, as we examine them critically. We've been aided in this case by the harmlessness that decades of in vitro fertilization methods have demonstrated and also, I suspect, by a growing realization of the physical nature of biology. From the perspective of their structure, their function, or the information they encode, it doesn't matter where strands of DNA happen to be brought together—in vitro or in utero, in a petri dish or in a body.

As described so far, there's no connection between in vitro fertilization and knowledge of genetic traits. In practice, in vitro fertilization involves the fertilization of multiple eggs, typically around a dozen. The collection, handling, and implantation of eggs can all fail in various ways, so one works with several to minimize the odds of ending up with zero viable candidate embryos. These fertilized eggs have different genomes thanks to the shuffling of DNA in sperm and egg cells described earlier. What if we could peek at these nascent genomes? Could we avoid choosing an embryo with a high risk for a debilitating disease, or select a particular embryo with traits we want for the child? This peeking is, in fact, possible.

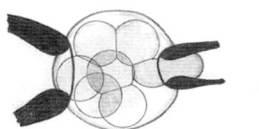 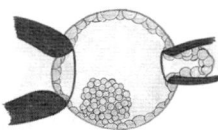

At three days postfertilization, each of us consisted of just six to ten cells. If one of these cells were removed, we'd be fine, even though one cell is a sizable fraction of the whole. Again, we see the wonders of self-assembly that we encountered in chapter 7—the remaining cells and their progeny fill in the gaps and pick up the cues corresponding to their locations, not their ancestry, and develop into the proper ensemble of cells characteristic of the organism. One can perform this extraction with fine pipettes under a microscope, as shown, holding the embryo with one pipette and applying suction with another to gently remove a single cell. Alternatively, one could wait a few more days, until there are around a hundred cells that have differentiated into two distinct groups: a cluster that will develop into the fetus and a shell that will eventually form the placenta. Extraction of 10 or so cells from the perimeter leaves the progenitors of the fetus completely undisturbed and still reveals the child's genome. Compared to the single-cell extraction, there's less time available between this procedure and the implantation of the embryo in the mother, so the genomic analysis must be fast. As we've seen, however, techniques for reading DNA are increasingly rapid. With either method, one gathers cells that can report on the genetic makeup of their embryonic cohort.

If there is more than one viable embryo to choose from for implantation, as is frequently the case, we could in principle use the information gathered from the embryonic biopsy and DNA characterization to choose among them, rather than relying on random chance. This is, in fact, already widely done to screen for embryos that have a high risk of single-gene-derived disorders, such as cystic fibrosis. To be concrete, imagine three embryos, one of whose genomes contains the transporter gene mutation that leads to cystic fibrosis; the other two do not. All three are equally unnatural from the perspective of being in vitro fertilized, or equally natural from the perspective of arising from the normally accessible combinations of parental genomes. There isn't any genetic editing involved; no DNA sequence is changed. There is, however, an element of choice—the deliberate selection of one genome over another.

There are some disorders that are even easier to diagnose. Genomes aren't single, continuous strands of DNA, but are divided into pieces called *chromosomes,* as noted in chapter 3. Humans have 46 (23 pairs), but errors in cell division sometimes lead to a missing or an extra chromosome. This is usually fatal—the fetus doesn't survive—but not always. Down syndrome, for example, arises from having three of chromosome number 21 rather than a pair. The extra genetic material leads to an overabundance of the proteins encoded by genes on chromosome number 21, which manifests itself in a wide range of neurological and physical symptoms. Detecting an extra or a missing chromosome is easy. In fact, it doesn't require in vitro fertilization or removing any cells from an embryo. In a normal pregnancy, one can sample the amniotic fluid surrounding the fetus, which contains more than enough sloughed-off fetal cells to assess their genetic architecture.

While peeking at embryonic genomes in vitro can, and does, allow for selection of embryos for which genetic signatures of particular traits are clear (for example, single-gene determinants or alternations in chromosome number), would it work for more complex traits? In principle, yes. In practice, the answer is more nuanced.

As we've seen, the genetic component of many disorders is determined by thousands of genes, ensembles that are being revealed by present-day mapping studies. We can identify the rare, high-risk genomes, let's say the 1 per 200 that correspond to a fivefold greater risk of contracting coronary artery disease, a level already considered actionable for single-gene factors. These extremes of risk, therefore, are within the range of present-day embryo selection, where their result is not a course of diagnosis or treatment but rather preventing occurrence entirely. Again, given three embryos, one that has a high probability of eventually contracting a disease while the other two do not, we could select one of the latter for implantation. Note, however, that in the vast majority of cases none of the three embryos would show a high risk; it's the rare occurrences that we'd occasionally find and might select against. Neither the numbers nor the method are hypothetical. The prevalence and risk factor of coronary artery disease are taken directly from the 2018 SNP-based study noted earlier.

One might think that if selecting against harmful traits is feasible, selecting for desirable traits would work just as well. This is not the case. The symmetry is broken by the nature of randomness. Let's again consider height, though the same argument holds for hair color, intelligence, or any other characteristic with a large hereditary component governed by a large number of sites on the genome. A set of three embryos samples three of the vast number of genetic rearrangements that could have occurred. It is three instances of flipping 10,000 coins, representing the roughly 10,000 genetic determinants of height. It's possible that you'd find in one of the three instances 10,000 heads, the exceptional child a foot taller than average, but it is astronomically unlikely. It's vastly more probable that you'd find 4987 or 4672 or 5115 heads. It's true that among these we could select the one with the most height-promoting variants, the 5115-head case in our coin analogy, but the net effect would be a tiny fraction of the coordinated pull of all 10,000 variants, and would, moreover, likely be swamped by the nongenetic variability also at play. Screening to reject an extreme genome in the unlikely event that it turns up is very different than hoping that a few random genomes will happen to be extreme.

More precise analyses reach similar conclusions. A 2019 study led by Shai Carmi at the Hebrew University in Jerusalem found that, with current technologies and data, one might expect embryo selection to give a height advantage of about 1 inch and an IQ advantage of 2.5 points, overall a "limited utility."

THE IMPLICATIONS OF CHOICE

Influencing the traits that will make up a future child is both awe-inspiring and frightening, and it's not surprising that the notion of "designer babies" has captivated the public. There are deep ethical issues that intersect embryo selection, a full discussion of which could fill multiple books. My goal here is to describe the science that should underlie each of our assessments of how to apply, or not to apply, new biotechnologies, and not to fully survey the field of biological ethics. Still, I'd be remiss if I didn't at least outline some of the key concerns.

The notion of shaping future people has a long and dispiriting history, most notoriously manifested in the genocidal philosophy of the Nazis but also applied in nominally free democracies in the early twentieth century. People deemed "weak-minded," for example, could be forcibly sterilized in several US states, a procedure declared legal nationwide by the Supreme Court in 1927. About 30,000 forced sterilizations had been performed in the United States by 1939. Several other countries had similar policies. The promoters of the eugenics movement were optimistic that these activities would broadly benefit society and aid human well-being. Supreme Court Justice Oliver Wendell Holmes wrote, "We have seen more than once that the public welfare may call upon the best citizens for their lives. It would be strange if it could not call upon those who already sap the strength of the State for these lesser sacrifices, often not felt to be such by those concerned. . . . The principle that sustains compulsory vaccination is broad enough to cover cutting the Fallopian tubes." Whatever the justifications its proponents invoked, however, the implementation of forced sterilization tended to target the poor and others on the margins of society, conveniently applying flimsy or capricious assessments of moral or intellectual character to quash the rights of those deemed unworthy. The Nazis stretched notions of genetic culling to even more horrific extremes. These historical examples should inform our present-day decisions, because of their similarities but also because of their differences.

One key difference is that, at least so far, modern tools of embryo selection are driven by individuals rather than the state, with prospective parents making decisions about their reproduction, subject to legal constraints. This brings up, however, issues of access—whether embryo selection technologies will disproportionately serve the wealthy. This is a concern with all health technologies, including in vitro fertilization, which can be ameliorated by decreasing costs and ensuring widespread medical coverage. Embryo selection also intersects more general issues regarding the relationships between parents, children, and society. Most people consider it acceptable, for example, for parents to buy their children athletic training, music lessons, or tutoring. Whether nudges obtained before birth are similar or categorically different from other such benefits that better-resourced parents can confer on their children isn't obvious and calls for informed deliberation.

Even if embryo selection is an individual choice, it can alter the overall makeup of the human population simply by diminishing the propagation of genetic variants to future generations. Of course, we already influence the makeup of the gene pool by choosing our partners, and the signatures of "assortative mating," in which people of similar educational or socioeconomic backgrounds pair up, are evident in genetic analyses of populations. In a sense, embryo selection makes this more deliberate. If, extrapolating wildly, no children at all carrying the BRCA1 or BRCA2 mutations that lead to high breast cancer risks are born, these gene variants won't be part of the pool that makes up subsequent generations. The potential benefits are clear, but there may be downsides as well. It's conceivable, though unlikely, that these variants could be beneficial in some unforeseen context, perhaps some future plague against which the aberrant BRCA1 or BRCA2 gives resistance. More plausibly, the absence of the traits we select against may be regrettable in itself. No one will mind if breast cancer disappears from the earth, but what about Down syndrome, which despite its difficulties is compatible with a joyful and fulfilling life? Many works of brilliance and beauty have been created by people who suffered from depression, a trait that is partly hereditary. We may lose some of the richness of humanity by eliminating disorders like this, reducing genetic diversity in favor of some narrow conception of normalcy. How "normal" is defined by society, and who gets to decide, further complicates the issue. On the other hand, no one would argue that we should deliberately increase the number of people affected by Down syndrome or depression, leading to the question of why the status quo is optimal. Finally, even if we all agree that preserving human genetic variation is important, the notion that reproductive autonomy is subservient to the public good brings back echoes of last century's eugenics. These are exceptionally difficult issues to think about. I encourage the reader not to look away from them, unsettling as they may be, as their already sizable relevance to our society will only grow in the future.

The individualistic nature of embryo selection highlights the importance of education. As we've seen, concepts of variability, uncertainty, randomness, and the very nature of genes and genetic traits are central to knowing what can and can't be done with modern tools. Misunderstanding any of these can lead to false hopes and misplaced expectations that may end up on the shoulders of future generations. Education is crucial. This is perhaps the least controversial statement in all of biological ethics, but how and when to implement education isn't obvious. My own belief is that an understanding of cells, genes, development, and technology should be part of the tool kit carried by every educated adult and not merely the subject of discussion at a fertility clinic.

All the concerns outlined above are important, interesting, and challenging. None, however, involve genetic engineering; none involve changes to DNA itself. The mere knowledge of the possible and the actual physical rearrangements of genomes, coupled with tools to handle cells and read DNA, gives us the means to reshape life in ways that would have been inconceivable a generation ago. The ability to rewrite DNA opens further possibilities. In the next chapter, we see how this is done.

15

How We Write DNA

We routinely reshape organisms. We reset broken bones, tie saplings to posts, and feed antibiotics to livestock to make them grow larger. These actions affect proteins, cells, and tissues and influence whether genes are turned on or off, but they don't change the sequence of As, Cs, Gs, and Ts that make up the organism's genome. Now, however, we have the tools to make such alterations and rewrite DNA, revising the instruction set for life's components directly.

Ever since the first chapter, we've seen that, for DNA and all the other materials that make up living things, biological function and physical form are inseparable. When examining the astounding tools with which to read DNA sequences, I noted that their invention required taking seriously the notion of DNA as a physical object, developing techniques to cut, grow, move, and monitor it that take into account its material characteristics as well as overarching themes related to self-assembly and randomness. This perspective applies to gene editing as well, as we seek to modify a strand of DNA. Here, however, we manipulate this molecule inside living cells, rather than in a machine, calling for a different set of approaches. In this chapter, we see how to edit DNA using a revolutionary twenty-first-century technology known as *CRISPR/Cas9*. To understand why this tool is so stunning, we first look at the methods that preceded it, remarkable in themselves.

HIJACKING BACTERIA

The heroes of the gene editing saga are bacteria, whose abilities have amazed us several times already in past chapters. There are many species of bacteria that are easy to grow in vast quantities, such as the lab stalwart *E. coli*. A warm bucket of nutrient-rich broth can sustain trillions of *E. coli*, each growing, dividing, and making lots of proteins. The proteins are those encoded in its genome—proteins that digest sugars, propel the cells through liquid, build membranes, and more. What if we could give *E. coli* a human gene, with which it would create a human protein, such as insulin? The bacterium would become a small, living, nearly infinitely replicable pharmaceutical factory.

The insulin example isn't hypothetical. People with type 1 diabetes don't produce insulin, leading to a debilitating and potentially fatal inability to regulate blood sugar levels. Since the 1920s, diabetics had been treated with insulin from pigs and cattle, the purification of which is difficult and expensive. In the 1970s, even after decades of technical refinements, preparing one pound of insulin required piles of pancreases from more than 20,000 animals. Moreover, the nonhuman material often induced allergic reactions in patients. Like the sonic hedgehog proteins we saw in chapter 7, the amino acid sequences of the animal insulins are very similar to the human insulin sequence—similar enough to work—but they aren't exactly the same, and these slight differences along with other substances carried by the animal extractions can trigger the immune system to mount a vigorous response. For all these reasons, the idea of mass-producing human insulin was tantalizing. For all of our species' history until the 1970s, however, humans were the only large-scale source of human proteins. That changed when we learned how to move genes between species, starting with the transformation of bacteria.

I'll describe in some detail the process of turning bacteria into machines that serve our ends, because the philosophy behind it is so different from our approach to nonbiological engineering. In a *Calvin and Hobbes* comic strip, the six-year-old Calvin asks his father, "How do they know the load limit on bridges, Dad?" The reply: "They drive bigger and bigger trucks over the bridge until it breaks. Then they weigh the last truck and rebuild the bridge." Calvin is stunned, and we're amused, since this is ridiculous. Similarly, no one would build a car by randomly connecting engines, axles, wheels, and other components in different arrangements and searching for the one-in-a-million that turned out to be a functioning vehicle. This is, however, a sensible approach to bioengineering, and it's made possible by the ability of living materials to self-assemble and by the modular, reproducible components with which they build themselves.

Bacteria possess powerful tools for manipulating DNA, including proteins called *restriction enzymes* that cut the double helix. The cutting isn't haphazard; each restriction enzyme recognizes a specific DNA sequence, typically a sequence present in the genome of a virus that can infect the bacterium. The restriction enzymes, therefore, serve as a defense against viral invasion, part of an age-old struggle to which we return later in the chapter. Most restriction enzymes leave the cut ends of the DNA "sticky"—that is, in fact, the technical term—ready to adhere to other cut ends. The stickiness isn't from some artificial adhesive but is a consequence of the shape of the cut: one of the single strands of the double-stranded DNA overhangs the other. The overhanging sequence, therefore, can bind to another cut DNA strand, as long as the two overhangs are complementary—As matching Ts and Cs matching Gs.

Another idiosyncrasy of bacteria is a fondness for taking up DNA from their surroundings, potentially scavenging useful traits from now deceased neighbors. Loops of DNA much smaller than the full genome are particularly amenable to being transported in or out of bacterial cells. Many bacteria contain such loops, called *plasmids*.

This repertoire of tools lets us move a human gene into a bacterium. First, one acquires the segment of DNA corresponding to the gene of interest—human insulin, for example. This can be cut out of an existing genome or synthesized from scratch—tractable even in the 1970s for a small gene like insulin. Recall that, thanks to the polymerase chain reaction, another technology enabled by microorganisms (chapter 1), even a tiny amount of DNA can be duplicated to give millions of identical copies. Restriction enzyme cleavage sites flanking the gene sequence of interest provide a multitude of sticky DNA fragments. Meanwhile, the researcher also collects plasmids from bacteria, grown in large numbers, and similarly make cuts in these. The vials are mixed together, and the target gene fragments and the opened plasmids meander through their watery surroundings, encountering each other through the randomness of Brownian motion. Some stick to each other to give new loops, composed of the original DNA plus the target gene (the latter indicated by curved bars in the illustration).

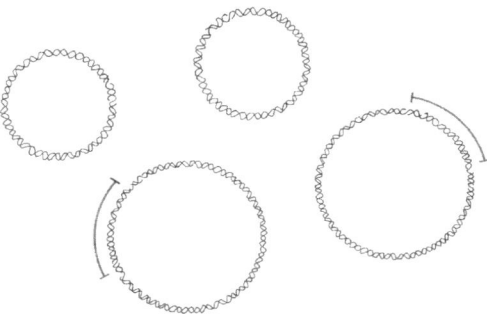

Some loops may close without grabbing the target gene fragment; these duds won't matter, as we'll see shortly. At least some of the loops will have incorporated the sequence we want. (If you look carefully, you'll find gaps drawn in the DNA backbone where the sticky ends meet; bacterial proteins repair these gaps.)

The next task is to get bacteria to swallow the plasmids. Uptake is rare, but can be enhanced by pulses of heat or electricity, both of which open transient pores in the bacterial membrane. Still, only a tiny fraction of the bacteria will find themselves in possession of an engineered circle of DNA. One can, however, winnow the crop to keep only these microbes, killing all the rest. The researcher selects or designs the plasmid to have other genes present, for example, genes for proteins that confer resistance to antibiotics. Exposing the bacteria to antibiotics, therefore, leaves alive only those that have taken up the plasmid. It's fine if these are few in number; bathed in broth, billions more grow. Of course, some of these surviving bacteria will have plasmids that closed without capturing our target gene, but straightforward techniques for determining the size or the sequence of the loop let the researcher identify and reject these cases, and again grow billions of the correct bacteria.

At the end of this process of cutting, gluing, and sifting, one is left with an organism that didn't exist before: a bacterium encapsulating and expressing a foreign gene. Working together, the labs of Stanley Cohen at Stanford University and Herbert Boyer at the University of California at San Francisco announced the first successful bacterial transformation in 1973. The potential soon became clear: microbes could be turned into microscopic assembly lines, churning out materials they wouldn't naturally make. Because insulin was an especially appealing target, several labs raced to genetically engineer microbes to produce it. Walter Gilbert's group at Harvard University—the same Walter Gilbert who pioneered DNA sequencing (chapter 13)—used DNA purified from humans to provide the insulin gene, an approach subject to stringent regulations on its handling. A general wariness about moving genes between species led to a moratorium on all such work in Cambridge, Massachusetts, the town in which Harvard is situated, following lively city council meetings. As one of the councillors said later, "I tried to understand the science, but I decided I couldn't make a legitimate assessment of the risk. When I realized I couldn't decide to vote for or against a moratorium on scientific grounds, I shifted to the political." He voted for the moratorium. Meanwhile, in a makeshift lab a few miles south of San Francisco, a small biotech start-up used chemically synthesized DNA for the insulin gene, avoiding the regulatory and social drama associated with human-derived molecules. The arrangements of atoms into As, Cs, Gs, and Ts are exactly identical, regardless of what process made them—a molecule is a molecule is a molecule—but law and public opinion don't necessarily recognize this. The company, Genentech, was founded by Boyer and venture capitalist Robert Swanson, and in August of 1978 its vats of microbes created the first-ever bacterially produced human insulin proteins. The small firm partnered with pharmaceutical giant Eli Lilly to tackle clinical tests, manufacturing, and administrative approval. In 1983, its insulin was brought to market, revolutionizing the treatment of diabetes and becoming the first human therapeutic drug produced by genetic engineering.

The bacterial production of insulin opened the floodgates to biologically produced pharmaceuticals, designed to treat an ever-expanding range of diseases and making up a market currently worth around $250 billion. These days insulin itself is mostly produced from engineered yeast rather than bacteria, as are many other drugs. Yeast are also single-celled microorganisms but are eukaryotes like us. We share with yeast various machineries for modifying proteins that bacteria lack. Insulin is initially formed in your pancreas as a single amino acid chain (chapter 2) that is subsequently cut into three pieces, two of them linked by new chemical bonds and one discarded. Bacteria can't perform these chemical steps, so the Genentech researchers engineered the two final fragments as separate genes whose resulting proteins joined themselves together. Yeast allow a more direct and productive approach.

SPLICING NICER RICE

It took further insights and inventions to enable the insertion of genes into eukaryotes. Few eukaryotes have plasmids, and eukaryotic genes generally need to be inserted into chromosomes, becoming part of the overall genome, for cells to express them. (This insertion can be relevant to bacterial engineering as well: for really robust modification of bacteria, one wants the gene of interest to be integrated into the main bacterial genome, not into plasmids that might be lost as cells divide.) We've had methods for modifying eukaryotic genomes for the past few decades, but until recently they've been inefficient, imprecise, and labor intensive. CRISPR/Cas9 has changed all this, so I won't say much about other methods except to note the general tactics and to give an example of why one might go to the trouble of using them.

If we think of a eukaryotic genome as a large library whose stacks are closed to us, like the Library of Congress or a private collection, our task is to somehow sneak a new book of our own onto the shelves. We could try leaving our book in a public part of the building, but it's unlikely that a librarian would decide to shelve it; eukaryotic cells don't take up and integrate random scraps of DNA. We're more likely to succeed if the library is renovating or relocating, and amid the shuffle our book is grouped with the others. Newly fertilized egg cells, in the interval before the parents' DNA comes together, provide such an opportunity. DNA carefully injected into the nascent embryo can find itself inserted into the genome. When first developed, this technique had a low success rate and allowed only insertion at random locations in the genome, potentially disrupting existing genes; but refinements have made it more robust and have even enabled some degree of targeting. The approach is a standard one for generating transgenic mice, for example, with fluorescent protein genes that serve as reporters of cellular activities (chapter 2).

Another approach is to recruit someone else who can more easily enter the library—perhaps a skilled burglar. We often employ viruses. Viruses infiltrate cells and replicate their genomes, either as stand-alone pieces or integrated into their host's genome. Being small and efficient, viruses don't easily lend themselves to the addition of new cargo, but it can be done, with the modified viruses delivering genes to cells in a variety of different organisms. Viral insertion of DNA has advantages over microinjection in that the virus itself, rather than a lab technician with a fine needle, takes care of getting material into the cell, and the cell type needn't be restricted to newly fertilized eggs. It suffers similar disadvantages, though, of not being as reliable as we'd like and being quite random regarding where in the genome the delivered gene is inserted. If we're making transgenic mice, we don't mind if our success rate is far from perfect. Again we winnow, studying only those in which the transformation has worked. If we want to create a human therapy, however, our standards for robustness and precision are much higher.

Bacteria, even those that can enter other cells and cause harm, can't in general alter eukaryotic genomes. There are rare exceptions, however. The soil microbe *Agrobacterium tumefaciens*, when the opportunity arises, infects plants. The microorganism snips a certain piece of its own DNA and injects it into a plant cell, along with proteins that direct the package to the nucleus and induce the plant to use its DNA repair machinery to integrate the bacterial DNA into its genome. The eventual consequence is the development of tumors that encourage bacterial growth. Scientists have engineered *Agrobacterium tumefaciens* in which the tumor-causing genes are replaced by whatever genes we like, turning the bacterium into a potent, and harmless, gene delivery tool for plants.

One of the most interesting applications of *Agrobacterium*-mediated gene delivery is the engineering of rice to fight vitamin A deficiency. Inadequate vitamin A causes blindness in 250,000 to 500,000 children annually—it's the largest cause of preventable blindness in children. About half of these kids die within a year of becoming blind, tragically highlighting the overall importance of the vitamin to health. Our bodies produce vitamin A from various precursors, most notably beta-carotene, which contributes to the orange color of foods like carrots and sweet potatoes. Beta-carotene is not, however, found in rice, a cheap, abundant staple food in many regions where vitamin A deficiency is prevalent. Rice plants are, in fact, naturally capable of making beta-carotene—they do so in leaves, where the molecule plays a role in photosynthesis. They don't, however, express the relevant genes in the starchy grains we eat. Researchers led by Ingo Potrykus of the Swiss Federal Institute of Technology and Peter Beyer of the University of Freiburg, Germany, therefore developed "golden rice," adding to the rice genome two genes and their promoters, derived from the daffodil and a bacterium, that lead to beta-carotene synthesis in the edible part of rice. After years of work, the success of the project was announced in 2000. Further improvements developed in conjunction with the biotechnology company Syngenta increased the beta-carotene levels by over a factor of 20; Syngenta subsequently donated all patents, technologies, and transgenic seeds associated with golden rice to the public.

Clinical studies showed that the strikingly yellow-orange rice is safe and effective, delivering beta-carotene that the human body converts into vitamin A. The American Society for Nutrition noted that "Golden Rice could probably supply 50% of the Recommended Dietary Allowance (RDA) of vitamin A from a very modest amount—perhaps a cup—

of rice" per day. Despite this, adoption of golden rice has been exceptionally slow; the plants are anathemas to various groups opposed to genetically modified organisms. Especially in recent years, the complaints are not about the plant per se, whose utility is hard to deny, but rather that its use opens the door to other modified foods, or that it would be better if vitamin A were supplied by providing the poor with a generally more nutrient-rich diet. Still, there is some motion. Bangladesh, 21% of whose children suffer from vitamin A deficiency, became in 2019 the first country to approve the planting of golden rice seeds. In the Philippines, the fraction of vitamin A–deficient children between six months and five years of age increased from 15% to 20% between 2008 and 2013, numbers that correspond to blindness and death for thousands of children. In 2019, the Philippines acknowledged the safety of golden rice, setting the stage for its planting. Twenty years after the invention of golden rice, debates and controversies surrounding it and other genetically modified crops continue, informed in some cases by an understanding of the underlying science, in some cases by complex issues of economics and commerce, and in many cases by amorphous perceptions and opinions. More recent technologies offer even more scope for discussion.

CRISPR AND THE GENE EDITING REVOLUTION

As we've seen, it's been possible for decades to alter genomes in all sorts of organisms. The methods I've sketched are difficult and inelegant, though. It takes trial and error to get a creature to take up a new gene, and where in its genome the gene ends up is either random, with random consequences for its expression, or crudely controlled at the cost of even more tedious engineering. For a bacterium or a mouse into which we'd like to insert an insulin gene or a fluorescent reporter, it is perhaps acceptable to keep trying until we get the outcome we want, but this strategy has its limitations, especially if we dream of curing genetic diseases in humans. What we'd really like is a way to edit genomes simply and precisely, identifying specific stretches of DNA and either replacing them with our own designs or cleanly excising them.

This ability is now in our grasp. I focus on the CRISPR/Cas9 system, but as is often the case with technology, more than one revolutionary method emerged at around the same time. Here the alternatives, *zinc finger nucleases* and *transcription activator-like effector nucleases* (TALENs), aren't quite as fast, cheap, and easy to use as CRISPR/Cas9; I mention them both for completeness and to suggest that the atmosphere of the early twenty-first century was full of the seeds of genome editing, ready to sprout.

CRISPR/Cas9 is an ancient technique, or a very new one. It was discovered, or it was invented. Each of these alternatives is correct depending on one's perspective. We first meet the primordial practitioners of genome manipulation before turning to their modern, human-driven manifestations.

Nature, to borrow Tennyson's phrase, was "red in tooth and claw" long before teeth and claws existed. Even microorganisms battle one another. Bacteria stab and poison their fellows, amoebas consume bacteria, viruses infect cells and vandalize their genomes. In addition to weapons, all these creatures have developed defenses. Until recently, it was thought that bacterial defenses act solely in the present, detecting cues as they occur without any memory of past insults. We now know, however, that bacteria do remember. Like our own immune systems, they keep a record of prior antagonists to enable rapid responses if they're encountered again. The discovery of this bacterial immune system is in itself emblematic of the modern biotechnological age: it was made possible by DNA sequencing and computers.

In 1987, a Japanese team studying a gene in the workhorse bacterium *E. coli* noted in passing "an unusual structure," a sequence of 29 nucleotides that repeated five times, with each repetition separated by 32 nucleotides. What its purpose might be, and whether it exists in other genomes, were mysteries. The observation attracted little attention; life is full of oddities, mostly of no consequence.

In the 1990s, Francisco Mojica and colleagues in Spain discovered similar repeating DNA patterns in the genomes of several archaea. Archaea are single-celled creatures without nuclei or organelles (chapter 5). While superficially resembling bacteria, they form a distinct branch of the tree of life; like us, they are separated from bacteria by a few billion years of evolution. Mojica came across the Japanese group's paper and realized that the presence of a similar genetic structure in such different organisms was a clue that the pattern plays an important role in the lives of microbes. In the late 1990s, as increasing numbers of bacterial genomes were sequenced and computational tools allowed one to pore over them at a desktop computer, Mojica and colleagues discovered many more instances of spaced, repeating DNA in bacteria and archaea. Their persistence and the puzzle they uncovered didn't deliver fame and fortune, however; Mojica's lab struggled for years to get funding, with the lack of relevance of the repeating sequences being a persistent critique. The genomes continued to pile up, however. In 2002, Mojica and Ruud Jansen of Utrecht University coined the term *CRISPR* for the genomic entities: clustered regularly interspaced short palindromic repeats. Researchers also noticed that a handful of genes were often found in the neighborhood of the spaced repetitions; these were named *Cas* (CRISPR-associated) genes. Though CRISPR now had a name, its function was still a mystery.

Again, insights came from genomes and computers. Three papers published in 2005, one from Mojica's group, one from Christine Pourcel and colleagues in France, and one from Alexander Bolotin and colleagues, also in France, announced that the DNA sequences of spacers matched nonbacterial and nonarchaeal DNA sequences, especially those of viruses. Since time immemorial, viruses have infected cells, and the viral signature suggested that CRISPR may be part of some sort of previously unrealized defense strategy. Though attention would slowly build, none of these papers were particularly well appreciated at their advent. Mojica's, the first to be written, was repeatedly rejected from prestigious journals before eventually finding a more modest home. Nonetheless, the essence of CRISPR was beginning to be unraveled. Why keep a carefully curated arrangement of pieces of DNA from viruses that infect you? So that you can use these mug shots to recognize and deactivate the viruses, should they invade again.

As a bacterial immune system, CRISPR/Cas works by making use of the complementarity of DNA and RNA (chapter 3), as well as protein machines that can cleave DNA. First, fragments of DNA encountered within the cell—likely from viral intruders and not the bacteria's or archaea's own DNA—are cut and inserted into the genome in between the repeating units by the proteins Cas1 and Cas2. The spacers, therefore, delineate a library of past viral encounters. Next, the cells take the seemingly foolhardy step of transcribing sections of the repeats and the spacers into RNA, called crRNA (for CRISPR RNA). Recall that transcription (chapter 3) is typically the first step to making proteins—potentially a dangerous activity if viral genes are the starting point. The RNA of the repeat segment, however, is in part complementary to RNA transcribed from another region in the CRISPR neighborhood of the bacterial or archaeal genome, called tracrRNA. The crRNA and tracrRNA together form an RNA double helix, with the viral-derived segment dangling unbound (left in the illustration). The CRISPR-associated protein, Cas9, recognizes and binds this RNA assembly (middle).

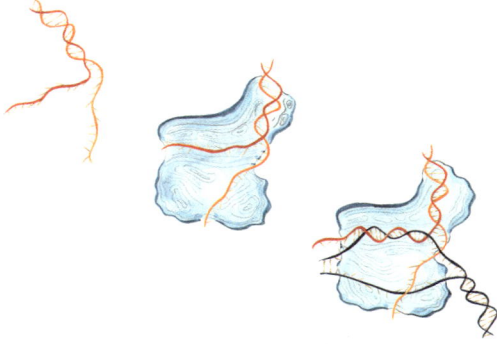

The RNA-Cas9 complex meanders around the cell thanks to Brownian motion. If it encounters a specific, short, and fairly common DNA sequence among the molecules it collides with, it attaches; this same DNA motif formed part of Cas1 and Cas2's identification of where to excise DNA. Cas9 destabilizes the bonds between the two strands of the DNA it has latched onto, essentially melting the double helix (chapter 1) in a small neighborhood. If the single-stranded RNA that Cas9 has been carrying is complementary to the opened-up DNA, which is only the case if the DNA is of the same virus from which the crRNA was derived, the two strands form a DNA-RNA hybrid. The peculiar but stable double-helical construction lies nestled in a groove in Cas9, lined with positive electrical charges that hold the highly negative DNA and RNA in place (right; DNA in dark blue).

Cas9 also contains two sections capable of slicing the DNA backbone. As the protein grips the DNA-RNA helix, these cleavers change their orientation, each situating themselves in contact with one of the viral DNA strands, the one that's joined with the RNA and the one left over, its former partner. I've illustrated the most dramatic shape change, in which one of the cleavers, depicted in green, rotates nearly 180 degrees. Cas9 makes the cuts and the viral DNA is crippled, with whatever genes the segment contains rendered nonfunctional. Cas9 wanders off, its job complete.

 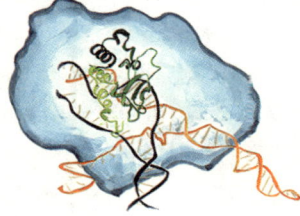

There's variety among CRISPR sequences and Cas proteins that I won't go into. The basic mechanism is the same for all the variants, and I focus on the relatively simple and elegant Cas9 protein, which has become the workhorse of most biotechnological applications.

As a bacterial immune system, CRISPR/Cas9 is impressive, combining memory and defense. At a deeper level, it reveals that Nature, billions of years ago, solved one of the grand challenges of genetic engineering: finding one particular DNA sequence among billions of possibilities, and then altering it. Here the finding involves viral DNA libraries, and the alteration is simply destruction by cleavage. Several researchers realized, however, that Nature's solutions could be tweaked to be far more generally applicable, and even constructive.

In 2012, the teams of Jennifer Doudna at the University of California at Berkeley and Emmanuelle Charpentier at the University of Umeå in Sweden, working together, published an enormously influential paper elegantly demonstrating CRISPR/Cas9's power as a DNA editor outside of bacteria or archaea. Charpentier's research group discovered tracrRNA and had deciphered many aspects of Cas9's activity. Doudna and coworkers were experts in RNA, and their wide range of research targets included the complex landscape of CRISPR-associated proteins. In their natural context, tracrRNA and virus-derived crRNA together direct Cas9. Doudna and Charpentier realized that replacing the viral sequence with whatever sequence you want programs a new destination into Cas9, and furthermore that the tracrRNA and crRNA can be replaced by a single, appropriately designed piece of RNA that folds back on itself, referred to as a single-guide RNA. Therefore, using just one easy-to-synthesize RNA strand and one easy-to-produce protein, Cas9, should enable simple, general targeting of specific DNA sequences. To cut DNA at the location of a particular nucleotide sequence, the recipe is as follows: make single-guide RNA molecules that contain that sequence (with the Ts replaced by Us, as is always the case for RNA (chapter 3)), add Cas9, mix into your system of interest, and you're done. The two research groups showed that this works in a test tube, outside the familiar confines of a bacterial cell. Reporting the results in a 2012 paper, the authors noted that the system is "efficient, versatile, and programmable," adding that it "could offer considerable potential for gene-targeting and genome-editing applications."

Doudna and Charpentier's 2012 paper, submitted to the prestigious journal *Science* on June 8 and published online on June 28, elicited a torrent of attention; it was clear that it demonstrated a breakthrough with far-reaching potential. The two researchers each received a $3 million Breakthrough Prize in Life Sciences in 2015, an award funded by Facebook's Mark Zuckerberg, Google's Sergey Brin, and other tech titans, awarded at a celebrity-laden Silicon Valley gala.

Returning to 2012, another research group, led by Virginijus Šikšnys at Vilnius University, Lithuania, submitted a paper on April 6 also describing RNA-guided DNA cleavage in a test tube by CRISPR/Cas9, again with a general version of the tracrRNA/crRNA template, though using two RNA fragments rather than a single-guide RNA, and similarly noting that the findings "pave the way for engineering of universal programmable RNA-guided DNA" manipulation. The manuscript was brusquely rejected; the authors sent it to a different journal, where it was published in late September, three months after Doudna and Charpentier's paper. In science, the acclaim often goes to those first across the finish line, and Šikšnys has been referred to as "the forgotten man of CRISPR." However, in 2018, he shared with Doudna and Charpentier the $1 million Kavli Prize for Nanoscience. It was certain that *someone* would win a Nobel Prize for CRISPR; the buzzing questions were "when?" and "who?" These were answered in 2020, as the Nobel Prize in Chemistry went to Doudna and Charpentier.

After the pioneering papers of 2012, many more appeared, from many research groups. In 2013, for example, the labs of Feng Zhang at the Broad Institute of MIT and Harvard University and George Church at Harvard Medical School demonstrated CRISPR/Cas9 in mouse- and human-derived cells. Popular accounts mushroomed as well; by 2019, there had been over 100 articles mentioning CRISPR in the *New York Times* alone. There is much more to the history of the study of CRISPR/Cas9, including patent fights and dramas that intersect broader issues of how the modern scientific enterprise works. My goal is to focus on the science and its uses, and so to the science we return.

CRISPR/Cas9 is purely destructive in our depiction so far, slicing DNA at the desired spot. This destruction can be used to deactivate a gene. Cells contain mechanisms to sense broken DNA and repair it, necessary for dealing with ever-present damage from chemical byproducts of metabolism or physical dangers such as ultraviolet light. Repair is performed by proteins that join the ends of DNA fragments. The joining is error-prone, however, altering the sequence at the seam or adding or deleting nucleotides. These mistakes likely lead to a nonfunctional protein.

Repair can more constructively enable the insertion of new genes into the genome. We can load the cell with fragments of DNA that encode whatever gene we want. Repair proteins aren't fussy—they'll glue together any ends they find—so in some cases one of our fragments is inserted at the break, joined to the genome at each end. You might be thinking that "some cases" is masking a lot of uncertainty, and you're right: the probabilistic nature of the microscopic world manifests itself in many ways. Cut by Cas9, the free ends wander by Brownian motion. The proteins that sense and repair DNA damage wander as well. Whether they find a free end of the genome and a free end of the introduced DNA before finding two free ends of the genome is up to the vagaries of their random walks, though we can bias the outcome by adding lots of copies of our new gene to make it more likely to find one. As all this is happening, Cas9 is still present—it may again find its target DNA sequence and again cut it, repeating the process over and over. There's some probability of getting the final result we want, but it's by no means a certainty. As with the older methods, we need to screen lots of failures to find successes. Still, if it works, the method puts the new gene exactly where we want it in the genome.

Though barely a decade has elapsed since the advent of programmable CRISPR/Cas9 systems, researchers have already invented much better gene insertion schemes. A powerful method called *prime editing*, developed in 2019 by the group of David Liu at the Broad Institute of MIT and Harvard University, paints with a broad palette of cellular machineries. I won't go into the details, but the essence is to fuse Cas9

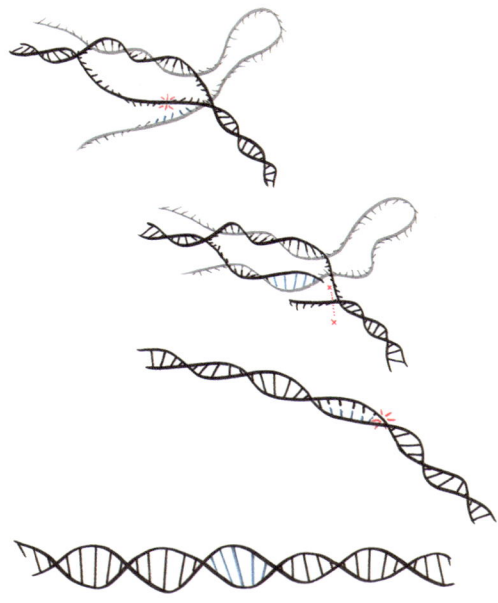

and reverse transcriptase—the protein we met in chapter 13 that makes DNA from RNA—and include as part of the guide RNA the desired insertion sequence. The chimera of Cas9 and reverse transcriptase recognizes and cuts DNA and generates new DNA at the cut. The illustration gives a flavor of the multistep process; DNA is black, RNA is gray, the insertion sequence is blue; the protein is omitted.

Liu's group demonstrated prime editing in human- and mouse-derived cells and noted that 89% of the genetic variants that are associated with human diseases are the types that can be targeted by the technique. (The remaining 11% include diseases associated with multiple copies of genes, for example, that simple rewriting won't fix.)

This elegant tailoring of DNA, making use of the needles and threads of proteins and nucleotides, is a stunning manifestation of the physicality of biological molecules. DNA is a code carrying genetic information, and to alter this code, we develop tools to quite viscerally grab, cut, paste, and glue. The precision is remarkable even from a nonbiological perspective. State-of-the-art integrated circuit chips in your computer, for example, have features as fine as about 10 nanometers. A single DNA nucleotide is about a third of a nanometer in length, and can be controllably altered by prime editing and other present-day methods.

As we've learned, simply having or not having a gene isn't the sole determinant of cellular activity. Regulation is crucial—controlling whether a gene's nucleotide code is actually read to give rise to a protein (chapter 4). Already, researchers have figured out how to use CRISPR/Cas9 to program regulation. Here again, its guide RNA serves as a navigator to specific sites in the genome. Rather than a standard Cas9, however, one uses modified variants that can't cut DNA. To turn genes off, the deactivated Cas9 simply sits on the DNA, blocking transcription by RNA polymerase like a normal repressor. The switch I described in chapter 9 that stops bacteria from swimming uses this technique. One can also turn genes on, for example, by linking a deactivated Cas9 to activators that recruit RNA polymerase. Thanks to CRISPR, we have exquisite access not only to genomes but to genetic circuits.

ON TOMATOES, T-CELLS, AND THERAPIES

Plants, the targets of genetic manipulations for millennia, can be more easily and elegantly transformed by CRISPR-based methods than by earlier techniques. The ancestor of the tomato, still existent in the wild, produces small, pea-sized fruits. Its domestication gives the larger and more nutritious tomato with which we're familiar. Selective breeding has picked particular genetic variants that we like, but at the cost of genetic diversity and tolerance to stress relative to the wild cousin. In 2018, researchers used CRISPR/Cas9 to insert just four of the genes from domesticated tomatoes into the wild plant's genome, generating fruit with three times the normal weight. The precision of CRISPR not only preserves the target genome but avoids the linkage-induced dragging along of potentially unwanted genes that is a by-product of conventional breeding, as we saw for watermelons.

The organism we care most about, of course, is us. With amazing speed, therapies and treatments based on gene editing are being developed and deployed. In addition to being useful, these techniques highlight some of the subtleties of using gene editing in practice. For all the cutting and pasting of CRISPR/Cas9 to actually happen, these molecular machines need to get inside the cells we want to modify. The most effective approach is to use viruses that are engineered to carry DNA encoding Cas9 and the requisite RNA sequences. The viruses mindlessly recognize, latch onto, and enter their target cells.

Exposing only certain cells to the viruses can be very challenging, however. One approach is to take the cells of interest out of the body, edit them, and put them back in. In 2015, researchers at the University of Pennsylvania piped blood cells and circulating immune cells out of and into the bodies of patients infected with the human immunodeficiency virus (HIV), which attacks T-cells. Years earlier, scientists noticed that a mutation in a gene called CCR5, which encodes a T-cell membrane protein, can confer protection against HIV. The Pennsylvania group extracted T-cells, disrupted the CCR5 gene, and reintroduced the modified immune cells to the patients' bloodstream. Improvements in health were minimal, but the method showed itself to be safe, spurring several groups to continue to develop the treatment. Another pioneering application in 2015 also involved immune cells, in this case removed from a donor, edited to withstand anticancer drugs, and placed into a one-year-old girl who suffered from leukemia and was unresponsive to all other treatments. Typically, donated immune cells trigger strong and potentially lethal responses in the recipient, unless perfectly matched. Here the editing also disabled the genes that drive such immune responses. The child's body adopted the modified cells, and her leukemia went into remission.

Neither of these gene editing examples used CRISPR/Cas9, but rather the slightly earlier-developed tools noted at the start of the chapter, zinc finger nucleases (for HIV) and TALENs (for leukemia). Now a flurry of CRISPR/Cas9–based therapeutics are being developed. Concurrently, treatments are moving into the human body itself, accessing tissues and organs that aren't amenable to out-of-body experiences.

In early 2020, a person with a rare genetic mutation that causes blindness became the first human to receive direct delivery of CRISPR/Cas9, via viruses injected into the eye. The mutation is in a gene that encodes a protein necessary to construct the towers containing light-sensitive proteins in certain cells of the retina. To ensure that gene editing occurs in the appropriate cells, the Cas9-encoding DNA includes a promoter sequence specific to transcription factors expressed in those retinal cells alone. If the clinical trial, yet to conclude, is successful, it will mark a milestone in our ability to treat blindness, and one achieved not through complex surgeries or electrical implants but by reconfiguring a person's own instructions for self-assembly.

CRISPR AND ANTI-CRISPR

Understanding the role of the biophysical principles of self-assembly and regulation gives us insight into how gene editing works. I've noted in passing another theme, that of randomness, in the context of Cas9 finding specific stretches of DNA. Randomness plays a larger role, however, and accounting for or controlling it is a challenge for gene editing applications. I've said that Cas9, via the guide RNA it cradles, binds to a specific DNA sequence complementary to the guide sequence. This is true, but it is an oversimplification. Like all molecular binding, the affinity is greatest for a perfect match but isn't exactly zero for an imperfect match. There is always some probability for Cas9 to target the wrong DNA. Whether this probability matters or not depends on its magnitude and on timescales. Suppose the chance of off-target Cas9 binding is 1 in 1000, and for whatever our goal may be, 1 in 1000 is an acceptable error rate. (It gives us, for example, only a few duds in a retina full of repaired cells.) Suppose further, however, that Cas9 is continuously present in the cells, expressed from the DNA we've delivered. Perhaps over 1 week the errors are 1 in 1000, but over 1000 weeks (20 years) it is close to certain that an off-target edit occurs in each cell. The actual numbers depend on the details of the binding and the mathematics of chance. In some cases, the benefits outweigh the risks; in some, they don't. We could push the odds dramatically in our favor if we could turn Cas9 off when its intended task is complete rather than letting it roam, active, forever.

In the incessant competition between organisms, it seems that every ploy has a counterploy, every weapon is met by a defense, and every defense has a weapon that evolves to subvert it. It perhaps shouldn't surprise us, then, that if CRISPR exists so does anti-CRISPR. Viruses have developed tools to disable the machinery of the bacterial immune system. Anti-CRISPRs were discovered around 2012 as graduate student Joe Bondy-Denomy, in the lab of Alan Davidson at the University of Toronto, noticed that some viruses managed to mount a successful invasion of bacteria that possessed CRISPR sequences against them. Earlier invaders, it turned out, had inserted into the bacterial genome genes for proteins that thwart Cas. There are a multitude of different anti-CRISPRs—over 50 have been discovered so far—targeting every imaginable component of the process sketched above. Some anti-CRISPRs prevent Cas proteins from loading guide RNA. Some cut the guide RNA. Some block the binding of DNA by Cas proteins. Some work by mechanisms yet to be deciphered. In every case, though, biophysical interactions are central. In blocking DNA binding, for example, some anti-CRISPR proteins mimic the electrical charge profile of DNA, mirroring the strength and spatial arrangement of its negatively charged surface to snuggle into the Cas binding groove, rendering it unavailable for its intended DNA. Reminiscent, perhaps, of our applying our knowledge of DNA's electrical properties to move it through gels (chapter 13), viruses make use of these properties for their own ends.

Almost immediately after their discovery, researchers realized that anti-CRISPRs could be crucial additions to the gene editing tool kit. In 2017, Jennifer Doudna and others, including Bondy-Denomy (by this time head of his own lab at the University of California at San Francisco), showed in human cells that delivering an anti-CRISPR after Cas9 effectively shuts down further gene editing and hence reduces unwanted genetic alterations.

In the past few chapters, we've looked at DNA from a broader physical perspective than we applied in part I. Understanding the nature of this molecule and how it interfaces with other materials, whether proteins like polymerases and Cas9 or inorganic structures like the semiconductors of chapter 13, has given us the ability to read and write in the language of genes. In the next and final chapter, I say more about what we can do with this ability, focusing especially on the editing of human genomes and the reshaping of ecosystems, both of which bring issues of ethics to the fore. I also comment on the type of understanding that a biophysical perspective gives us, and its relevance to questions both practical and profound.

16

Designing the Future

For as long as we've been human, we've sought principles that make sense of the complexity of life. It was widely believed in medieval Europe and the Islamic world, for example, that every species on land has its counterpart in the sea, that Creation had provided "the horse and the sea-horse, the dog and the dog-fish, the snake and the eel," as T. H. White noted in an introduction to a twelfth-century bestiary. This belief has faded—it provides neither descriptive accuracy nor predictive power—but we can sympathize with its hope that some simple symmetry governs living things. Biology often seems a cacophony, beautiful but dizzying. Evolution provides one unifying scaffold, but it primarily illuminates the processes that mold forms over many life spans rather than explaining the connections between form and function.

Over the past several decades, however, we've grown to understand the principles underlying life—physical underpinnings involving self-assembly, randomness, networks of regulatory interactions, and scaling relationships that guide the activities of all creatures. These frameworks connect the microscopic and macroscopic worlds, linking the structure and dynamics of molecules to the workings of cells, tissues, and whole organisms. We could extend our descriptions to still larger scales: communities of species and whole ecosystems are also subject to similar rules. The mathematics of population growth and the vagaries of foraging, for example, tend toward chaotic dynamics with random-looking features that mirror the randomness of their microscopic counterparts. Competition and cooperation lead to self-organizing networks of interactions that govern population sizes. Areas and abundances, as we briefly noted in chapter 12, may obey general scaling relationships. Our focus has mostly been on the scales from molecules to individuals, however, which provide more than enough mystery to keep us occupied.

Understanding biophysical principles brings with it not only a deeper appreciation of life but the ability to reshape it. We've seen several examples of how insights into physical mechanisms intersect issues of health and disease, from the behaviors of pathogenic microbes to the mechanics of organs to risk prediction based on reading DNA. In the next few sections, we look briefly at contemporary examples of modifications to ourselves and our environment that even more dramatically bring to the fore ethical and social concerns. While not central to our aim of describing the workings of life, we'd be remiss to ignore these issues. Moreover, I claim, the biophysical perspective we've developed can help us grapple with difficult questions at the intersection of science and ethics, clarifying the possibilities and impossibilities of technologies that are often popularly described in vague or sensational terms.

Our set of biotechnological tools continues to expand along with its applications. It is tempting to conclude with a survey of possible future feats—resurrected mastodons and mammoths, an end to animal-derived meat, cures for every disease, engineered plagues, dystopias or utopias of designer babies. The list is limitless, however, and the future is notoriously hard to predict. Instead, we touch on broader themes that, I hope, will be useful regardless of where the future takes us. We end by asking more deeply what our biophysical understanding reveals about the wonders of life, addressing along the way the question of what "understanding" means.

WHAT DOES IT MEAN TO EDIT AN EMBRYO?

The most dramatic new tools are those that edit genomes. We examined in the last chapter therapies related to CRISPR/Cas9 that promise relief from debilitating diseases beyond the reach of conventional treatments. We also saw in chapter 14 that reading the DNA of embryos enables prediction of future traits. It takes little imagination to combine the two and edit embryonic genomes. Delivering CRISPR/Cas9 to the single cell of a fertilized egg transforms the targeted gene, with the change faithfully copied upon each round of cell division, reaching not only each cell of the fully grown organism but also each of its children, each of their children, and so on. Editing embryos was successfully demonstrated in mice and zebrafish in 2013 and in monkeys in 2014. The latter especially was heralded as a means of generating mimics of human disorders in a similar animal, thereby aiding the development of therapies. It should surprise no one after the preceding chapters that this technique will also work in human embryos—the basic machineries of life are nearly identical in humans as in other animals, to a degree that would have been incomprehensible to our medieval ancestors.

Whether human embryo editing *should* be done, however, is a very different issue. The current consensus is no, at least for embryos that would be implanted to seed viable pregnancies. In fact, such procedures are illegal in dozens of countries, including the United States. The caution arises from the same sorts of ethical concerns highlighted in chapter 14, amplified by the ability of CRISPR/Cas9 to make truly novel changes to genomes rather than just rearranging existing variations. Moreover, solutions to complications such as off-target edits are, as we saw in the last chapter, currently under development. While very promising, few would describe the story as complete. An overwhelming number of scientists, ethicists, and policy makers therefore believe that the time for human embryonic genome editing has not come. Not everyone agrees, however.

In November 2018, Chinese researcher He Jiankui stunned the world with the announcement that his team had used CRISPR/Cas9 to produce the first ever gene-edited babies, a pair of twin girls. Condemnations from around the globe were strong and swift. Chinese authorities quickly shut down He's lab and declared his work "extremely abominable in nature." One can imagine making a case for rushing ahead with human embryo editing if there's imminent danger, perhaps the certainty of a lethal genetic defect, but this was not the case here. The alteration was the disruption of a gene we've seen already, CCR5, whose absence lowers the likelihood of contracting HIV. The choice is puzzling. The girls' father is HIV-positive, but transmission from a father to a child is exceptionally unlikely. The stated aim was to lower the odds of infection during the children's lives, but there are plenty of standard means of effectively preventing HIV. What's more, CCR5 isn't a useless gene. There's good evidence that it helps people resist the impact of other viruses, such as influenza and West Nile virus. Its deletion, therefore, comes with a cost. Even if our target were better justified and the potential benefits outweighed the risks, we should still only proceed if the people involved, namely the parents, are educated about the meaning and consequences of the action. The latter underlies the important ethical principle of informed consent, and it was violated in the He debacle. He's information for the parents was minimal and misleading, and documentation for ethical review procedures was found to be forged. In 2019, He was sentenced to three years in prison.

One could imagine an alternative scenario, in which the first direct alteration of a human embryonic genome occurred after transparent deliberation, and in which the target was an obviously harmful genetic mutation, like the ones that cause muscular dystrophy, cystic fibrosis, or a host of other severe diseases. Such was not the case, however, and despite its technical success, the CCR5 episode will likely muddy the waters of CRISPR-based biotechnologies. The intersection between science and human drama is often messy.

More relevant to our aims here, these events highlight the importance of education. What does "informed consent" mean in the era of genome sequencing and gene editing? At least some awareness of the nature of genes, regulation, probability, and variation, and how all these together orchestrate health and disease, are essential for being an informed participant in any cutting-edge biotechnological therapy. In the absence of understanding, it is all too easy to imagine hopes raised to unrealistic heights, automatic rejection of any genetic technologies, or simple acquiescence to the desires of doctors or the state. Fostering education will be at least as big a challenge as the technical tasks of building organs on a chip or rewriting immune cell genomes. There are precedents, however. The germ theory of disease, for example, has permeated our awareness, with nearly all of humanity possessing a basic mechanistic understanding that microbes exist, can replicate, can infect other creatures, and can cause disease. As noted in chapter 14, almost no one looks in horror at in vitro fertilization; again, a mechanistic understanding of conception has become commonplace. The same awareness can, and should, occur for more recent insights into the workings of life.

THE UNNATURAL WORLD

Modifying plants is contentious. It is unnatural. We've looked at rice and tomatoes; let's now consider carrots, specifically the "baby carrots" you may have seen at a grocery store. They're small, smooth, perhaps even cute, and they account for about 70% of carrot sales in the United States. One might think that they're young carrots, as the name implies; they are not. One might think they are a miniature variety, as ponies are to horses; they are not. Rather, they are full-size, craggy, ugly carrots whittled by machines into small cones. The shaved-off bulk is waste to be discarded. To again use a loaded term, baby carrots are unnatural. If one were to bioengineer an actual miniature carrot, controlling its shape and form, designing cuteness into its genome, it would be unnatural as well. One can wonder which form of unnaturalness would be more palatable, and why.

The existence of baby carrots leads to a more general statement: our world is much less pristine than many people believe. Consider nitrogen, a necessary component of many of the molecules central to life, including DNA and proteins. Nitrogen is an abundant element; as a gas, it makes up about 78% of the atmosphere. This form is inaccessible to animals and plants; none can extract nitrogen from the air. Some bacteria can, however—especially in symbioses with certain plants, they create the nitrogen-containing chemicals that serve as precursors for the molecules with which the plants build themselves and that fertilize the soil after their death, providing nitrogen for other plants and for the animals that eat them. The process works slowly, and insufficient nitrogen is often the limiting factor for the productivity of agricultural land. In the early twentieth century, German chemists Fritz Haber and Carl Bosch developed an artificial process for extracting nitrogen from the air, allowing the industrial production of nitrogen-containing fertilizers. The impact on food production, and civilization, has been enormous. The Haber-Bosch process is the largest single factor underlying humanity's population explosion, and it is estimated that without it, half of the people on the planet would not exist. (The world's population is currently just under 8 billion; in 1900, it was 1.6 billion.) A staggering 50% of the nitrogen atoms in your body, incorporated into DNA strands, amino acid chains, and other materials, are derived from the Haber-Bosch process rather than the natural route of bacterial activity.

The imprint of unnatural processes extends beyond the atoms in our bodies. About 40% of the earth's land is used for agriculture; the percentage is even higher if we exclude barren expanses, like the Sahara and the Antarctic, from the denominator. The atmosphere as well is being reshaped: For the past 10,000 years, its concentration of carbon dioxide ranged between 250 and 280 parts per million. Now, it is over 400, the increase being a consequence of fossil fuel combustion. Through the well-understood physics of thermal radiation, rising carbon dioxide levels lead to an overall increase in the temperature of the planet.

We live, therefore, on a planet in which human activity isn't a small perturbation but is instead a major factor in global systems. Many of us believe that nature is worth preserving, that even aside from its utility there is an intrinsic value to having an awe-inspiring variety of species and landscapes. Preserving nature is, I would argue, paradoxically ill-served by dialing back our unnatural activities. To put it bluntly, that ship has sailed. Eliminating chemical fertilizers and pesticides, for example, with no new technologies brought in to replace them would lead to a roughly fourfold increase in the amount of land required for agriculture. That extra land doesn't exist. The human population continues to grow, and pressures on wilderness and wildlife grow alongside. Bioengineering crops to achieve greater efficiency, however, could enable a reduction in our agricultural footprint, which is hard to imagine occurring by any other means.

ECOSYSTEM ENGINEERING

If accelerating the human manipulation of the living world seems risky, that's because it is. History provides clear instances of ecosystem tinkering gone awry. The story of cane toads provides an example. A century ago, sugar cane crops in Australia were being attacked by beetles. In other parts of the world, cane toads consume beetles and other pests. Why not bring these voracious amphibians, native to Central and South America, down under? And so in the 1930s, cane toads were deliberately introduced to Australia to protect the sugar cane crop. The plan failed catastrophically. The toads gleefully multiplied, with a population now numbering around 200 million and covering a range of about 200,000 square miles (500,000 square kilometers). They ignored the cane beetles, instead gorging on native insects and frogs, bird eggs, and more. Being poisonous, the toads killed many potential predators that might have kept them in check, as well as domestic pets. Rather than being a benefit, the cane toads became a major pest, sending ripples through the continent's habitats. It is still unclear how to deal with them.

Not all plans succeed, but one can hope that we may learn from the past, design better tests and experiments, and consider different tools. In the previous chapter, we looked at CRISPR/Cas9 applied to cells and individual organisms. Here, as our final biotechnological illustration, we see how CRISPR/Cas9 can modify, and even eliminate, an entire species via a construction known as a *gene drive*. Whether or not to use this is a difficult question, as we'll see shortly.

Imagine a trait conferred by a single gene—the common variant, for example, leads to gray insects and a particular mutation in this gene gives black insects. (Or, foreshadowing an application described below, the mutation leads to sterility in mosquitoes if present in both copies of the genome.) Suppose just one individual has the gray mutation. When it mates with a nonmutant, the randomness of recombination gives a 50% chance of the mutation being passed on in the sperm or eggs of the following generation, and unless there's some advantage to being black, the mutation is unlikely to spread through the population.

(I've illustrated this for pairs that always have two offspring, in a schematic that ignores the randomness of inheritance.)

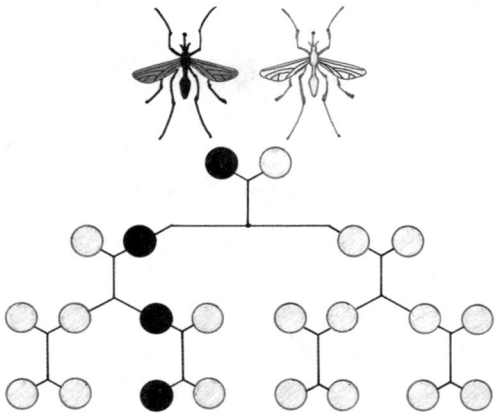

Now suppose instead that, along with the black mutation, this individual's genome has been engineered to contain CRISPR/Cas9, with its target, encoded by the CRISPR sequence, being the "normal" (gray) form of the gene. If the modified DNA encounters an unmodified genome—provided by the unedited mate—Cas9 breaks the gray pigment gene. As we've seen, cells repair broken DNA, and they often use the counterpart provided by the other half of a chromosome pair as a template. The template, containing the black mutation and the CRISPR/Cas9 sequence, is now present in *both* strands. All the future offspring

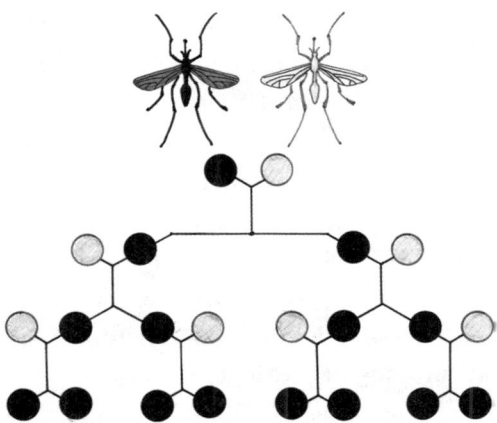

will be black, and will carry the CRISPR/Cas9 machinery that ensures continued edits upon mating with unmodified individuals. The black insects, therefore, will spread through the population.

Why would anyone want to do this? One compelling reason is to eliminate malaria. Each year over 400,000 people die from malaria and more than 200 million are infected, suffering fever, fatigue, and headaches even if they eventually recover. The disease is caused by a parasite transmitted by the female *Anopheles* mosquito, which delivers it to and from human hosts as it drinks their blood. Mosquitoes spread other serious diseases as well, such as dengue fever, West Nile fever, and Zika fever. A variety of government agencies, nonprofit organizations, and public health groups are intensely discussing gene drive targeting of malaria-transmitting *Anopheles* mosquitoes, either by engineering parasite resistance into the insects or by eliminating these mosquitoes entirely. The latter can be done by spreading a mutation that induces maleness, for example, leading to an overwhelmingly male population that can't reproduce, or by disrupting genes related to egg production and therefore spreading sterility.

Though laboratory studies of gene drives in mosquitoes have unfurled as expected, the modified insects have not been released in the wild. However, other genetically altered mosquitoes, modified by older methods that don't propagate through a population, have been set loose. Over a billion male mosquitoes with a genetic variation lethal to their offspring have been set free in Brazil, Malaysia, and the Cayman Islands since 2009. Oxitec, the biotechnology company that developed the insects, notes that their deployment has been successful, causing, for example, an 80% decline in the mosquito population at its Grand Cayman test site and a 90% decline in Jacobina, Brazil. In 2021, the insects were released in the Florida Keys, a region that has seen a surge in cases of dengue fever and Zika. The approval of the Florida plan, the result of a 4–1 vote by the local Mosquito Control District Board of Commissioners in August 2020, followed contentious hearings on whether to adopt the new approach; the standard control method is aerial spraying of pesticides, thought to kill 30%–50% of the mosquitoes.

Intriguingly, Oxitec's modified mosquitoes highlight a biophysical mechanism we've encountered: the lethal variation is not in a gene that encodes some biochemically specific activity, but in a transcription factor (chapter 4), altering the creatures' regulatory circuitry. It seems that the mutant mosquitoes produce an activator that overstimulates its own production, giving a feedback loop of ever-increasing but pointless protein production that jams the insects' cellular machinery. In the lab, males, females, and their progeny can be kept alive by a drug that inhibits the regulatory circuit; this is absent in the wild. Young, rapidly growing mosquitoes are highly sensitive to the regulatory imbalance and die. There are similarities and differences between this approach and that of a gene drive. For example, both involve changes in the genome: the former requires continued introduction of modified individuals, and the latter propagates changes through generations. Understanding these contrasts is important to making informed decisions about how to tackle pressing public health problems like insect-borne diseases.

Returning to gene drives: Their potential deployment in the wild brings trepidation. The elimination of a species could harm its predators or the broader food webs of which it's part, in addition to diminishing the diversity of animal life on the planet. For malaria-linked mosquitoes, the case is stronger than for most applications: the insects aren't a primary food source for other creatures, and even if we erase a few mosquito species, over 3000 other mosquito species remain. And, of course, on the other side of the ethical coin is a vast amount of human suffering. There are still more issues to consider, however. For example, who decides to deploy a gene drive? For mosquitoes and malaria, there is widespread agreement that actions in Africa, home to over 90% of cases, should be decided in Africa. But who in Africa? Mosquitoes don't recognize national borders, and a gene drive propagates to wherever its targets disperse.

These decisions intersect issues of conservation as well. Many fragile ecosystems are threatened by invasive species. For example, rats on the Galápagos Islands plague local wildlife, devouring the eggs of birds and turtles. Would a gene drive for rats be a better method of protecting native species than the rat-killing poisons currently used? Similar questions apply to other regions and other species, including agricultural crops and their pests.

Using gene drives and analogous technologies brings to mind all our past failures of ecosystem engineering, as it should. One worries that we will simply see new versions of the cane toad saga. We can learn from our mistakes, though, and also take heart that our current technologies are more precise and focused than those of the past. Additional enhancements that offer safeguards are already under development, such as DNA-encoded mechanisms to block or even reverse gene drives in a population. Surgery used to be crude and often lethal due to exposure to infection; now it is routinely successful thanks to better insights and methods. It is not unreasonable, I claim, to work toward similarly effective ends for larger-scale interventions. This may seem unduly optimistic, but, as noted above, the alternatives to embracing modern methods may be even less palatable.

There's much more we could write about the intersection of biotechnology and society. New developments and implementations, exhilarating as well as risky, will continue to arrive not only because of the depth of our insights into the living world but because the tools we're creating are cheap and easy to use. I see these days advertisements offering to sequence my entire genome for $300, a far cry from the $3 billion of 30 years ago or even the $10,000 of 10 years ago. Machines for performing PCR, purifying proteins, and growing cells are more accessible than ever. In fact, communities of amateur enthusiasts share biochemical recipes and designs for 3-D-printed equipment with which to explore biotechnological crafts from their garages. Members of the Open Insulin Project, for example, aim to develop "freely available, open organisms for insulin production" that will liberate diabetics from commercial insulin suppliers.

One might be aghast that biotechnological paraphernalia is so readily available, or delighted at its democratization, but regardless of our response, the ubiquity of such tools is a reality. The technologies that form the basis of futuristic applications exist now, and they make possible such widespread and uncontroversial applications, like diagnosing disease and monitoring microbial contaminants, that it is hard to imagine we will stop using them or stop thinking of new targets at which to direct them. I've often used the word *we* when writing, glossing over who exactly "we" refers to. In the context of biotechnology, for better or worse, "we" can be anyone. Any reasonably advanced country, or even institution, is capable of using the tools we've encountered in this book. We'll see how these instruments are put to use in the years to come. Once again, I claim, education is crucial to ensuring that the decisions we make are the best they can be, informed by an understanding of how the technologies at our disposal work and what they can and cannot do. Education, of course, has benefits beyond the practical. We end by stepping back to the deeper, and more inspirational, lessons drawn from looking into the workings of life.

UNDERSTANDING UNDERSTANDING

The most important theme of this book is that themes exist. The living world isn't just a collection of anatomical structures and biochemical reactions. There are principles, motifs, and mechanisms that unify all of its components and their activities. This statement shouldn't come as a surprise after the preceding 15 chapters. I've avoided until now, however, the question of whether this unity matters, either practically or aesthetically. This question underlies a tension present in much of contemporary science that I illustrate with a nonbiological example before we return to the living world.

There wasn't a singular experience that drove me to study physics, but one event I remember vividly occurred in my high school physics class, an activity in which we released a ball from a ramp onto a table, along which it rolled until plummeting off the table's edge to fall into a plastic cup on the floor. Our task: to predict beforehand where to place the cup to catch the ball. It may seem dull on paper, but to see that the ball did, in fact, land exactly where universal laws of motion say it should was amazing to me. We needn't memorize facts about balls, ramps, or cups, or even care about them; instead, there are general themes that guide us. Searching for and applying broad principles is central to physics and underpins the biophysical approach that I've championed throughout this book.

There is, however, another way we could have made our prediction. From prior experiments, we could have tabulated a vast list of outcomes. For a range of ramp heights, tilt angles, ball masses, ball materials, table heights, air temperatures, and whatever other parameters might possibly be relevant, we could have released balls and recorded where they landed. Then, for the conditions of our new roll, we could look up the outcome from our giant database. This, to use the current jargon, is the "big data" approach. There's nothing wrong with it, and it's often very successful. We've seen this already, in the correlations observed between genomic features and traits such as human height (chapter 14). These correlations enable predictions, but we have no understanding of why the correlations exist.

The contrast between understanding basic principles and cataloging information arises with increasing frequency throughout science, as our tools for generating and processing data become more powerful. Sometimes, as in the applications of Brownian motion (chapter 6), the basic principles are so profound and well understood that it would be absurd to ignore them. Sometimes, as in the genomics example above, looking for correlations is the only possible approach, at least for now. Sometimes it is unclear what to do. In chapter 4, we looked at the biological circuits that regulate genes, composed of repressor and promoter proteins that orchestrate gene activity. To some researchers, identifying basic circuit motifs—feedback loops, oscillators, clocks, and so on—is the path to progress. To others, this is irrelevant; plugging in the details of all the interactions between all the components of a regulatory network, even if they number in the hundreds or thousands, is the more useful strategy. It isn't obvious whether the former, more biophysical approach or the latter, more phenomenological approach is better. They aren't mutually exclusive; pursuing both in parallel may be the best plan, especially as the detailed models might help uncover core concepts that are yet unknown. Still, the tension between these philosophies is real, and it colors debates about funding and research directions.

My own bias is toward illuminating basic, minimal sets of principles. After all, a giant database of falling ball trajectories would enable prediction of future experiments on falling balls, but it's the deeper understanding of Newton's laws of motion that gave us the trajectories that sent astronauts to the moon. Which approach is the more practical, therefore, may depend on the timescale: immediate application to some problem at hand, or to problems of the future that are still unimagined. Aside from the aims of practicality, there's also the deeply human appeal of understanding. *Understanding* is hard to define. Still, extracting from complexity simple yet powerful explanations has compelled us for as long we've been human, powering myths as well as science.

TWO ENDINGS TO OUR STORY

There is a way one may have expected this book would end. We began with the ingredients of life, molecules and machineries that generate the inner dynamics of cells. We then examined communities of cells, as in organs and embryos, and then some of the principles that guide larger-scale shape and form. Finally, we reconnected with the subcellular world, learning how to read and write the molecular code of the genome. One might think, therefore, that the loop is closed and that we could conclude by describing the procedure for designing organisms at will—perhaps nutritious, drought-resistant plants, or animals resurrected from human-induced annihilation—by crafting and executing the code that gives creatures of whatever body and behavior we want. For example, we would sketch how to transform a tiny dik-dik antelope into a massive wildebeest, poring over the clues in its DNA, nudging the nucleotide letters in the appropriate genes to spin new amino acid chains that fold into new loops and sheets, guiding the expression of other genes and forming the substance of bones and muscle, eyes and lungs, all designed to accommodate the pull of the earth, the inhalation of air, and the demands of the environment.

We do not, however, know how to design such a transformation, and this is not how the book ends. Our ignorance isn't a sign of failure;

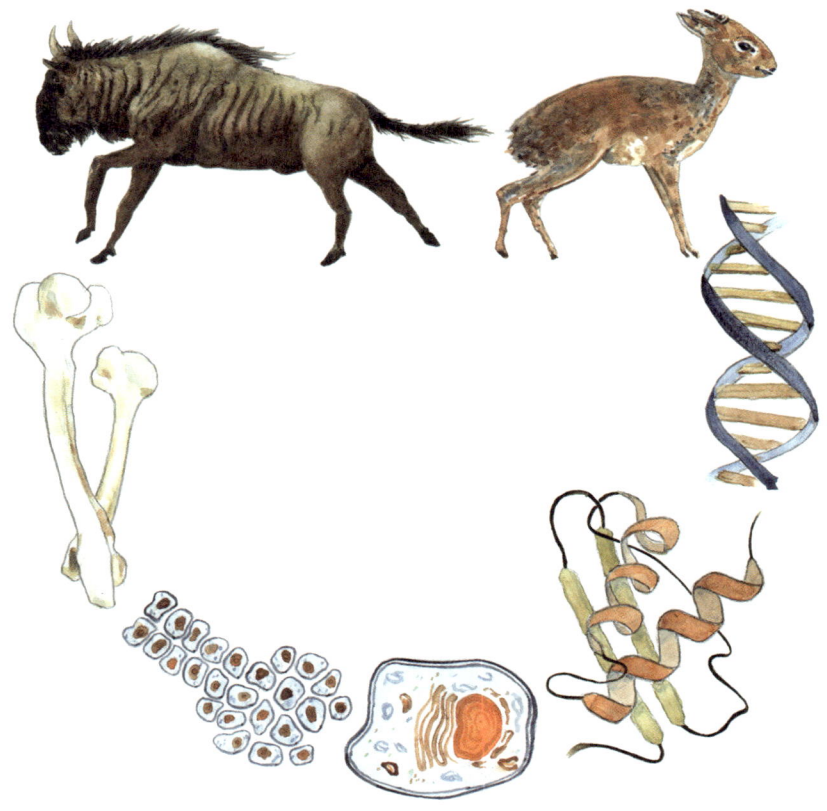

rather, it tells us that the story we've been exploring is far from finished. We know the basic pieces and principles that make up life. Our grasp of the detailed contexts in which they are manifested, however, is small, but growing. It is as if, having recently learned to read, a whole library awaits us. It is not certain that we will be able to make sense of it all. It may be that, though the general themes are clear, the minutiae of genetic interactions, chemical reactions, and microscopic forces that give rise to bamboo rather than beech, that separate the horse and the hippo, and that set the precise ratio of femur width to length will be too difficult to entirely comprehend or may not be worth the effort. This remains to be seen. Even if so, it would not diminish the accomplishment of understanding the architecture of life, nor dampen our enthusiasm for further exploration.

The fundamentally human urge to know, to seek connections among phenomena and to feel joy when we find such connections, provides the most powerful motivation for studying nature. Especially in the past few chapters, we've looked a lot at technological applications of our modern understanding of life. Technology and its intersection with health, disease, and society are important, and to many people they're the most important facet of science. I would study biophysics, however, and I would have written this book even if these applications didn't exist. The living world is full of an awe-inspiring variety of forms and activities. In our own backyards and urban parks, squirrels leap among branches, sunlight glints off the diaphanous wings of dragonflies, and trees pull carbon from the air to build towers tens of feet tall. In more exotic locales, lions stalk the savannas, dolphins play among waves, and fish miles deep in ocean trenches generate their own light. We can see all of this and be struck by its beauty. We can, in addition, now appreciate the miles of DNA packed inside the lion, the dance of neurotransmitters as it decides to pounce, the march of motor proteins along the fibers of its cells—layers of phenomena as remarkable as those that are more obvious. What's more, we can contemplate a unity among all creatures, including ourselves, beyond the dreams of our ancestors. Whether on land or sea, newborn or grown, microscopic or gigantic, every organism is made of the same molecular building blocks, a handful of molecules that encode information and build themselves into three-dimensional shapes. Every organism is shaped with the physical forces of the universe; the bones of both the lion and the antelope she stalks are governed by the constraints of gravity, and the dolphin could only be a dolphin in a large liquid home. Every organism not only tolerates the microscopic chaos of jostling molecules but transforms the chaos into computation, regulating the activity of genes, proteins, cells, and organs in response to stimuli from within and without. Beyond superficial appearances, as wonderful as they are, is a profound and elegant framework that makes life work and that we now begin to appreciate.

Acknowledgments

The idea of writing this book dates back a decade. Around the same time, I created a course called "The Physics of Life," with the aims of conveying the wonders of biophysics to non-science-major undergraduates and using biophysics as a tool to foster scientific literacy. This book is much broader than the course and is structured quite differently, but nonetheless I'd like to thank the course's many students not only for their frequent enthusiasm but also for their moments of polite but stone-faced indifference. As a scientist, one tends to find fascinating almost everything even tangentially connected to one's field; it can be shocking to learn that some topics are, to most people, quite dull. The spur to constantly try different tactics to enliven certain subjects, or to drop them entirely, was invaluable. I'd also like to thank the University of Oregon's Science Literacy Program, especially Elly Vandegrift, Michael Raymer, and Judith Eisen, for creating a program that encourages thinking deeply about science communication.

I'm grateful to Sergio, Miguel, Pablo, Jesse, Chilo, and everyone else who works at Espresso Roma, along with its owners, Miguel and Maria Cortez, for great coffee and a wonderful environment in which to think and write.

David Rabuka and Phil Nelson read drafts of every chapter and generously provided extensive, enthusiastic, and insightful comments. I am delighted to be able to thank them in print. Phil furthermore helped launch this project, not only with words of encouragement but also by serving as matchmaker with my editor-to-be, Jessica Yao of Princeton University Press. To Jessica, warm thanks not only for championing this book but also for an abundance of perceptive suggestions. Sincere thanks also go to Ingrid Gnerlich, the project's enthusiastic second editor.

Finally, I am deeply grateful to my wife, Julie, and to the kids, Kiran and Suryan, who are always wonderful and who put up with the many hours I've spent distracted by reworking sentences and drawing DNA.

References

GENERAL

Standard biology textbooks cover many well-established facts that are especially relevant to part I. See, for example:
- B. Alberts, A. D. Johnson, J. Lewis, D. Morgan, M. Raff, K. Roberts, P. Walter, *Molecular Biology of the Cell* (W. W. Norton, New York, sixth edition, 2014). The fourth edition is freely available: http://www.ncbi.nlm.nih.gov/bookshelf/br.fcgi?book=mboc4.

Excellent textbooks on biophysics, aimed at advanced undergraduates or graduate students, include
- P. Nelson, *Biological Physics: Energy, Information, Life* (Chiliagon Science, Philadelphia, student edition, 2020).
- R. Phillips, J. Kondev, J. Theriot, H. Garcia, *Physical Biology of the Cell* (Garland Science, London, second edition, 2012).
- W. Bialek, *Biophysics: Searching for Principles* (Princeton University Press, Princeton, NJ, 2012).

INTRODUCTION

p. 1 "Ancient Indian texts applied a variety of classifiers"
- B. K. Smith, Classifying animals and humans in ancient India. *Man* **26**, 527–548 (1991).

p. 6 "arrangements of adjoining cells in all sorts of tissues resemble the arrangements of soap bubbles"
- D. W. Thompson, *On Growth and Form* (Cambridge University Press, Cambridge, UK, 1942).

p. 6 "Takashi Hayashi ... and Richard Carthew ... looked at the cluster of photoreceptor cells situated in each of a fruit fly's compound eyes"
- T. Hayashi, R. W. Carthew, Surface mechanics mediate pattern formation in the developing retina. *Nature* **431**, 647–652 (2004).

p. 7 "assessing the mechanical stiffness of the neighboring tissue"
- E. Puklin-Faucher, M. P. Sheetz, The mechanical integrin cycle. *Journal of Cell Science* **122**, 179–186 (2009).

- A. del Rio, R. Perez-Jimenez, R. Liu, et al., Stretching single talin rod molecules activates vinculin binding. *Science* **323**, 638–641 (2009).

p. 9 "In the twentieth century alone, for example, more than 300 million people died of smallpox"
- D. A. Henderson, The eradication of smallpox—an overview of the past, present, and future. *Vaccine* **29**, D7–D9 (2011).

p. 11 "At the end of *On the Origin of Species*, Darwin writes"
- C. Darwin, *On the Origin of Species by Means of Natural Selection, or, the Preservation of Favoured Races in the Struggle for Life* (J. Murray, London, 1859), 490.

CHAPTER 1. DNA: A CODE AND A CORD

p. 15 "A beige gelatinous slab speckled with bacterial colonies hangs in the National Portrait Gallery"
- D. Nelkin and M. S. Lindee, *The DNA Mystique: The Gene as a Cultural Icon* (W. H. Freeman, 1995).

p. 15 "it carries the actual instructions that led to the creation of John"
- BBC News, "Gallery puts DNA in the frame," September 19, 2001. http://news.bbc.co.uk/2/hi/entertainment/1550864.stm.

p. 20 "James Watson and Francis Crick figured out the structure of double-stranded DNA." Much has been written about this history, and the roles of Watson, Crick, Rosalind Franklin, and others. See, for example:
- J. D. Watson, *The Double Helix: A Personal Account of the Discovery of the Structure of DNA* (Atheneum Press, New York, 1968).
- A. Sayre, *Rosalind Franklin and DNA* (W. W. Norton, New York, 2000).
- M. Cobb, Sexism in science: Did Watson and Crick really steal Rosalind Franklin's data? *The Guardian*, June 23, 2015. http://www.theguardian.com/science/2015/jun/23/sexism-in-science-did-watson-and-crick-really-steal-rosalind-franklins-data.
- S. Mukherjee, *The Gene: An Intimate History* (Scribner, New York, 2016).

p. 23 On phase transitions in general:
- N. Goldenfeld, *Lectures on Phase Transitions and the Renormalization Group* (Addison-Wesley, Reading, MA, 1972).

pp. 23–24 On the physics of DNA melting:
- M. Peyrard, Biophysics: Melting the double helix. *Nat. Phys.* **2**, 13–14 (2006).
- M. Peyrard, Nonlinear dynamics and statistical physics of DNA. *Nonlinearity* **17**, R1–R40 (2004).

p. 24 "In 1983, the recipe that combines nucleotides, polymerases, temperature, and DNA came to scientist Kary Mullis." A first-person account of the discovery of the polymerase chain reaction:
- K. B. Mullis, The unusual origin of the polymerase chain reaction. *Sci. Am.* **262**, 56–65 (1990).

CHAPTER 2. PROTEINS: MOLECULAR ORIGAMI

p. 30 On the history of X-ray crystallography:
- J. C. Brooks-Bartlett, E. F. Garman, The Nobel science: One hundred years of crystallography. *Interdisciplinary Science Reviews* **40**, 244–264 (2015).
- M. Jaskolski, Z. Dauter, A. Wlodawer, A brief history of macromolecular crystallography, illustrated by a family tree and its Nobel fruits. *FEBS Journal* **281**, 3985–4009 (2014).

pp. 30–31 "Kendrew's team struggled with myoglobin from porpoises, penguins, seals, and other creatures"; and the quote from Kendrew, "the most marvelous . . . gigantic crystals"
- S. de Chadarevian, John Kendrew and myoglobin: Protein structure determination in the 1950s. *Protein Science* **27**, 1136–1143 (2018).

p. 32 On green fluorescent protein and other fluorescent proteins:
- R. N. Day, M. W. Davidson, The fluorescent protein palette: Tools for cellular imaging. *Chem. Soc. Rev.* **38**, 2887–2921 (2009).
- E. A. Rodriguez, R. E. Campbell, J. Y. Lin, et al., The growing and glowing toolbox of fluorescent and photoactive proteins. *Trends in Biochemical Sciences* **42**, 111–129 (2017).
- N. C. Shaner, The mFruit collection of monomeric fluorescent proteins. *Clin. Chem.* **59**, 440–441 (2013).

p. 33 The drawing of green fluorescent protein is based on structure 1EMA in the Protein Data Bank:
- https://www.rcsb.org/structure/1EMA.
- M. Ormö, A. B. Cubitt, K. Kallio, et al., Crystal structure of the *Aequorea victoria* green fluorescent protein. *Science* **273**, 1392–1395 (1996).

p. 33 On the structure of potassium channels:
- Q. Kuang, P. Purhonen, H. Hebert, Structure of potassium channels. *Cell Mol. Life Sci.* **72**, 3677–3693 (2015).

p. 33 The drawing of a potassium channel is based on structure 1BL8 in the Protein Data Bank:
- https://www.rcsb.org/structure/1BL8.

- D. A. Doyle, J. Morais Cabral, R. A. Pfuetzner, et al., The structure of the potassium channel: Molecular basis of K+ conduction and selectivity. *Science* **280**, 69–77 (1998).

p. 33 The drawing of kinesin is based on structure 1N6M in the Protein Data Bank:
- https://www.rcsb.org/structure/1N6M.
- M. Yun, C. E. Bronner, C.-G. Park, et al., Rotation of the stalk/neck and one head in a new crystal structure of the kinesin motor protein, Ncd. *EMBO Journal* **22**, 5382–5389 (2003).

p. 34 The drawing of the glucocorticoid receptor is based on structure 1R4O in the Protein Data Bank:
- https://www.rcsb.org/structure/1R4O.
- B. F. Luisi, W. X. Xu, Z. Otwinowski, et al., Crystallographic analysis of the interaction of the glucocorticoid receptor with DNA. *Nature* **352**, 497–505 (1991).

p. 36 "a group of other proteins called *chaperones* comes to their aid"
- F. U. Hartl, M. Hayer-Hartl, Molecular chaperones in the cytosol: From nascent chain to folded protein. *Science* **295**, 1852–1858 (2002).

pp. 39–40 On kuru:
- M. P. Alpers. The epidemiology of kuru: Monitoring the epidemic from its peak to its end. *Philosophical Transactions of the Royal Society London, B: Biological Sciences* **363**, 3707–3713 (2008).
- J. T. Whitfield, W. H. Pako, J. Collinge, M. P. Alpers, Mortuary rites of the South Fore and kuru. *Philosophical Transactions of the Royal Society London, B: Biological Sciences* **363**, 3721–3724 (2008).
- R. E. Bichell, When people ate people, a strange disease emerged. NPR.org, September 6, 2016. https://www.npr.org/sections/thesalt/2016/09/06/482952588/when-people-ate-people-a-strange-disease-emerged.

p. 40 On prions and their discovery:
- S. B. Pruisner, The prion diseases. *Scientific American* **272**, 48–57 (1995).
- NobelPrize.org, Nobel Media AB press release (1997). https://www.nobelprize.org/prizes/medicine/1997/press-release/.

p. 40 On meat-and-bone meal:
- C. Ducrot, M. Paul, D. Calavas, BSE risk and the use of meat and bone meal in the feed industry: Perspectives in the context of relaxing control measures. *Natures Sciences Societes* **21**, 3–12 (2013).

pp. 41–42 On predicting protein structures:
- F. M. Richards, The protein folding problem. *Scientific American* **264**, 54–63 (1992).
- K. A. Dill, J. L. MacCallum, The protein-folding problem, 50 years on. *Science* **338**, 1042–1046 (2012).

p. 42 "to commission bespoke supercomputers devoted to the biophysical challenge of protein folding"
- A. Elliot, David E. Shaw's supercomputer is uncovering secrets of human biology. Columbia Engineering, April 7, 2017. https://engineering.columbia.edu/news/engineering-icons-david-shaw.

p. 42 "the 'folding@home' program that runs in the background of volunteers' computers"
- Folding@home—fighting disease with a world wide distributed super computer. https://foldingathome.org/.
- K. Greene, 2002, Folding@home takes to the lab. *Science*, October 21, 2002. https://www.sciencemag.org/news/2002/10/foldinghome-takes-lab.

p. 42 "a free protein folding game"
- S. Cooper, F. Khatib, A. Treuille, et al., Predicting protein structures with a multiplayer online game. *Nature* **466**, 756–760 (2010).

p. 42 "DeepMind, a company affiliated with Google"
- R. F. Service, "The game has changed." AI triumphs at protein folding. *Science* **370**, 1144–1145 (2020).
- E. Callaway, "It will change everything": DeepMind's AI makes gigantic leap in solving protein structures. *Nature* **588**, 203–204 (2020).

CHAPTER 3. GENES AND THE MECHANICS OF DNA

p. 46 "Differences of even a single three-nucleotide group . . . can lead to subtle but measurable shifts in color perception . . . several forms of color blindness"
- S. S. Deeb, The molecular basis of variation in human color vision. *Clin. Genet.* **67**, 369–377 (2005).
- MedlinePlus, Color vision deficiency. https://medlineplus.gov/genetics/condition/color-vision-deficiency/(reviewed January 1, 2015; updated August 18, 2020).

p. 47 "An RNA called 'Growth arrest-specific 5'" and other aspects of non-coding RNA:
- T. Kino, D. E. Hurt, T. Ichijo, et al., Noncoding RNA gas5 is a growth arrest- and starvation-associated repressor of the glucocorticoid receptor. *Sci. Signal* **3**, ra8 (2010).
- M. Guttman, J. L. Rinn, Modular regulatory principles of large non-coding RNAs. *Nature* **482**, 339–346 (2012).

pp. 47–48 "The bacteria that cause tuberculosis and cholera each have about 4000 genes in their genome"

- *Kyoto Encyclopedia of Genes and Genomes* (KEGG). https://www.genome.jp/kegg/; for *Mycobacterium tuberculosis*, https://www.genome.jp/kegg-bin/show_organism?org=mtu; for *Vibrio cholerae*, http://www.genome.jp/kegg-bin/show_organism?org=vch (accessed January 14, 2021).

p. 48 "The genome of the *Lactobacillus delbrueckii* subspecies commonly employed to turn milk into yogurt has about 2000 protein-coding genes"
- National Center for Biotechnology Information. https://www.ncbi.nlm.nih.gov/genome/?term=Lactobacillus%20delbrueckii[Organism] (accessed January 14, 2021).

p. 48 "The human genome contains about 20,000 protein-coding genes." For this, and the number of noncoding genes, see
- I. Ezkurdia, D. Juan, J. M. Rodriguez, et al., Multiple evidence strands suggest that there may be as few as 19 000 human protein-coding genes. *Hum. Mol. Genet.* **23**, 5866–5878 (2014).
- C. Willyard, New human gene tally reignites debate. *Nature* **558**, 354–355 (2018).

pp. 48–49 On genomes of various organisms:
- D. M. Church, L. Goodstadt, L. W. Hillier, et al., Lineage-specific biology revealed by a finished genome assembly of the mouse. *PLOS Biology* **7**, e1000112 (2009).
- C. M. Wade, E. Giulotto, S. Sigurdsson, et al., Genome sequence, comparative analysis, and population genetics of the domestic horse. *Science* **326**, 865–867 (2009).
- X. Zhan, S. Pan, J. Wang, et al., Peregrine and saker falcon genome sequences provide insights into evolution of a predatory lifestyle. *Nature Genetics* **45**, 563–566 (2013).
- R. A. Ohm, J. F. de Jong, L. G. Lugones, et al., Genome sequence of the model mushroom Schizophyllum commune. *Nature Biotechnology* **28**, 957–963 (2010).
- J. K. Colbourne, M. E. Pfrender, D. Gilbert, et al., The ecoresponsive genome of *Daphnia pulex*. *Science* **331**, 555–561 (2011).
- Rice Annotation Project database (RAP-DB): 2008 update. *Nucleic Acids Res.* **36**, D1028–D1033 (2008).
- Gramene database. http://ensembl.gramene.org/Zea_mays/Info/Annotation/.

p. 49 On noncoding genes in rice and maize:
- H. Wang, Q.-W. Niu, H.-W. Wu, et al., Analysis of non-coding transcriptome in rice and maize uncovers roles of conserved lncRNAs associated with agriculture traits. *Plant J.* **84**, 404–416 (2015).

- L. Li, S. R. Eichten, R. Shimizu, et al., Genome-wide discovery and characterization of maize long non-coding RNAs. *Genome Biology* **15**, R40 (2014).

pp. 49–50 On genome sizes:
- The BioNumbers database, http://bionumbers.hms.harvard.edu/search.aspx, contains information on genome sizes, including those of the bread mold *Neurospora crassa* and the soil-dwelling amoeba *Dictyostelium discoideum*; search "number of genes" or the species name.

pp. 54–56 On DNA packaging, histones, and chromatin:
- H. Schiessel, The physics of chromatin. *J. Phys. Condens. Matter* **15**, R699–R774 (2003).
- D. J. Tremethick, Higher-order structures of chromatin: The elusive 30 nm fiber. *Cell* **128**, 651–654 (2007).

p. 54 The drawing of DNA wrapped around a histone complex is based on structure 1AOI in the Protein Data Bank:
- https://www.rcsb.org/structure/1AOI.
- K. Luger, A. W. Mader, R. K. Richmond, et al., *Nature* **389**, 251–260 (1997).

p. 55 "researchers at the Salk Institute . . . developed a method to stain DNA in intact nuclei, decorating them with metal atoms that are readily visible in an electron microscope"
- H. D. Ou, S. Phan, T. J. Deerinck, C. C. O'Shea, ChromEMT: Visualizing 3D chromatin structure and compaction in interphase and mitotic cells. *Science* **357**, eaag0025 (2017).

p. 56 On DNA packaging and disease:
- C. Mirabella, B. M. Foster, T. Bartke, Chromatin deregulation in disease. *Chromosoma* **125**, 75–93 (2016).
- A. DeLaurier, Y. Nakamura, I. Braasch, Histone deacetylase-4 is required during early cranial neural crest development for generation of the zebrafish palatal skeleton. *BMC Developmental Biology* **12**, 16 (2012).

p. 56 "the determinants of what DNA regions are wrapped around histones"
- E. Segal, Y. Fondufe-Mittendorf, L. Chen, et al., A genomic code for nucleosome positioning. *Nature* **442**, 772–778 (2006).
- F. G. Brunet, B. Audit, G. Drillon, et al., Evidence for DNA sequence encoding of an accessible nucleosomal array across vertebrates. *Biophysical Journal* **114**, 2308–2316 (2018).

pp. 57–59 On the packaging of DNA in viruses:
- A. Evilevitch, L. Lavelle, C. M. Knobler, et al., Osmotic pressure inhibition of DNA ejection from phage. *Proc. Natl. Acad. Sci.* **100**, 9292–9295 (2003).

- W. M. Gelbart and C. M. Knobler, Virology: Pressurized viruses. *Science* **323**, 1682–1683 (2009).

CHAPTER 4. THE CHOREOGRAPHY OF GENES

p. 63 The illustration of the lac repressor bound to DNA is based on structures 1EFA and 1TLF in the Protein Data Bank and a composite illustration by David Goodsell:
- https://www.rcsb.org/structure/1EFA.
- https://www.rcsb.org/structure/TLF.
- D. Goodsell, http://pdb101.rcsb.org/motm/39.
- A. M. Friedman, T. O. Fischmann, T. A. Steitz, Crystal structure of lac repressor core tetramer and its implications for DNA looping, *Science* **268**, 1721–1727 (1995).
- C. E. Bell, M. Lewis, A closer view of the conformation of the lac repressor bound to operator. *Nat. Struct. Biol.* **7**, 209–214 (2000).

p. 63 "The protein must therefore loop the DNA into a tight circle"
- R. Schleif, DNA Looping. *Annual Review of Biochemistry.* **61**, 199–223 (1992).

p. 63 "The looped DNA interferes with the normal binding of RNA polymerase"
- Z. Vörös, Y. Yan, D. T. Kovari, et al., Proteins mediating DNA loops effectively block transcription. *Protein Sci.* **26**, 1427–1438 (2017).
- N. A. Becker, J. P. Peters, L. J. Maher, T. A. Lionberger, Mechanism of promoter repression by lac repressor-DNA loops. *Nucleic Acids Res.* **41**, 156–166 (2013).

p. 64 "This phenomenon was discovered in the 1940s by Jacques Monod." For more of the history, see
- M. Morange, *A History of Molecular Biology* (Harvard University Press, Cambridge, MA, 2000).

p. 65 "the human genome contains over 1500 genes for transcription factors"
- S. A. Lambert, A. Jolma, L. F. Campitelli, et al., The human transcription factors. *Cell* **172**, 650–665 (2018).

p. 65 "a transcription factor bound to a segment of DNA can influence expression of a gene that would be distant if the DNA were laid out in a straight line"
- S. Schoenfelder, P. Fraser, Long-range enhancer–promoter contacts in gene expression control. *Nature Reviews Genetics* **20**, 437–455 (2019).

p. 66 "a paper from 2001 by Heidi Scrable and colleagues at the University of Virginia"
- C. A. Cronin, W. Gluba, H. Scrable, The lac operator-repressor system is functional in the mouse. *Genes Dev.* **15**, 1506–1517 (2001).

p. 68 "As Monod himself dramatically and presciently noted"
- H. C. Friedmann, From "butyribacterium" to "E. coli": An essay on unity in biochemistry. *Perspect. Biol. Med.* **47**, 47–66 (2004).

pp. 68–73 On memories, clocks, and other genetic circuits:
- P. C. Nelson, *Physical Models of Living Systems* (W. H. Freeman, 2015).
- U. Alon, *An Introduction to Systems Biology: Design Principles of Biological Circuits* (CRC Press, Boca Raton, FL, 2007).

pp. 70–72 On the circadian rhythm and its cellular clock:
- S. A. Brown, E. Kowalska, R. Dallmann, (Re)inventing the circadian feedback loop. *Dev. Cell* **22**, 477–487 (2012).
- E. S. Maywood, L. Drynan, J. E. Chesham, et al., Analysis of core circadian feedback loop in suprachiasmatic nucleus of mCry1-luc transgenic reporter mouse. *Proc. Natl. Acad. Sci.* **110**, 9547–9552 (2013).
- J. P. Pett, A. Korenčič, F. Wesener, et al., Feedback loops of the mammalian circadian clock constitute repressilator. *PLOS Comput. Biol.* **12**, e1005266 (2016).
- D. R. Weaver, The suprachiasmatic nucleus: A 25-year retrospective. *Journal of Biological Rhythms* (2016).

p. 72 On engineering the repressilator into cells:
- M. B. Elowitz, S. Leibler, A synthetic oscillatory network of transcriptional regulators. *Nature* **403**, 335–338 (2000).

p. 72 "a variety of other precise and tunable oscillators have been engineered into cells." See, for example:
- J. Stricker, S. Cookson, M. R. Bennett, et al., A fast, robust and tunable synthetic gene oscillator. *Nature* **456**, 516–519 (2008).

p. 73 On histone modification:
- M. Lawrence, S. Daujat, R. Schneider, Lateral thinking: How histone modifications regulate gene expression. *Trends in Genetics* **32**, 42–56 (2016).

p. 73 "This modification of histones is particularly important during the early development of an embryo"
- L. Ho, G. R. Crabtree, Chromatin remodelling during development. *Nature* **463**, 474–484 (2010).

p. 74 On epigenetics:
- C. D. Allis, T. Jenuwein, The molecular hallmarks of epigenetic control. *Nature Reviews Genetics* **17**, 487–500 (2016).
- Bošković, O. J. Rando, Transgenerational epigenetic inheritance. *Annual Review of Genetics* **52**, 21–41 (2018).
- B. T. Heijmans, E. W. Tobi, A. D. Stein, et al., Persistent epigenetic differences associated with prenatal exposure to famine in humans. *PNAS* **105**, 17046–17049 (2008).

p. 74 On studies of the 1944–45 Dutch famine survivors:
- R. C. Painter, C. Osmond, P. Gluckman, et al., Transgenerational effects of prenatal exposure to the Dutch famine on neonatal adiposity and health in later life. *BJOG: An International Journal of Obstetrics & Gynaecology* **115**, 1243–1249 (2008).
- M.V.E. Veenendaal, R. C. Painter, S. de Rooij, et al., Transgenerational effects of prenatal exposure to the 1944–45 Dutch famine. *BJOG: An International Journal of Obstetrics & Gynaecology* **120**, 548–554 (2013).

CHAPTER 5. MEMBRANES: A LIQUID SKIN

p. 76 "These membrane-associated proteins account for over a third of the human genome." An even larger number are peripherally associated with the membrane:
- L. Dobson, I. Reményi, G. E. Tusnády, The human transmembrane proteome. *Biol. Direct* **10**, 31 (2015).
- M. S. Almén, K. J. Nordström, R. Fredriksson, H. B. Schiöth, Mapping the human membrane proteome: A majority of the human membrane proteins can be classified according to function and evolutionary origin. *BMC Biology* **7**, 50 (2009).

p. 78 "interactions between two cell types known as T-cells and antigen presenting cells"
- A. Grakoui, S. K. Bromley, C. Sumen, et al., The immunological synapse: A molecular machine controlling T cell activation. *Science* **285**, 221–227 (1999).
- S. K. Bromley, W. R. Burack, K. G. Johnson, et al., The immunological synapse. *Annu. Rev. Immunol.* **19**, 375–396 (2001).

p. 79 "similar synapses form at the contacts between immune cells transmitting the human T-cell leukemia virus as well as the human immunodeficiency virus"
- V. Piguet, Q. Sattentau, Dangerous liaisons at the virological synapse. *J. Clin. Invest.* **114**, 605–610 (2004).

p. 81 "In London at the start of the nineteenth century, 30% of all deaths were due to tuberculosis"
- S. A. Waksman, *The Conquest of Tuberculosis* (University of California Press, Berkeley, 1964).

p. 81 "remained the first or second leading cause of death each year in the United States through the first decade and a half of the twentieth century"
- Centers for Disease Control (USA), Leading causes of death, 1900–1998. https://www.cdc.gov/nchs/data/dvs/lead1900_98.pdf.

p. 81 "Even now, about one million people die of tuberculosis annually"
- World Health Organization, WHO global tuberculosis report, 2017. http://www.who.int/tb/publications/global_report/en/.

p. 81 "We've known for about a century, for example, that *Mycobacterium leprae* and *Mycobacterium tuberculosis* can survive periods of dehydration lasting several months"
- D. C. Twitchell, The vitaility of tubercle bacilli in sputum. *Transactions of the National Association for the Study and Prevention of Tuberculosis, Annual Meeting*, 221–230 (1905).
- M. B. Soparker, The vitailty of tubercle bacilli outside the body. *Indian J. Med. Res.* **4**, 627–650 (1917).
- C. R. Smith, Survival of tubercle bacilli. *American Review of Tuberculosis* **45**, 334–345 (1942).

p. 82 On the use of trehalose for dehydration resistance:
- J. H. Crowe, F. A. Hoekstra, L. M. Crowe, Anhydrobiosis. *Annu. Rev. Physiol.* **54**, 579–599 (1992).

pp. 82–84 My lab's work on trehalose-containing lipids:
- C. W. Harland, D. Rabuka, C. R. Bertozzi, R. Parthasarathy, The *M. tuberculosis* virulence factor trehalose dimycolate imparts desiccation resistance to model mycobacterial membranes. *Biophys. J.* **94**, 4718–4724 (2008).
- C. W. Harland, Z. Botyanszki, D. Rabuka, et al., Synthetic trehalose glycolipids confer desiccation resistance to supported lipid monolayers. *Langmuir* **25**, 5193–5198 (2009).

p. 87 "chemically perturb cells to create 'blebs' . . . . In these, one finds visibly discernible lipid phases"
- T. Baumgart, A. T. Hammond, P. Sengupta, et al., Large-scale fluid/fluid phase separation of proteins and lipids in giant plasma membrane vesicles. *Proc. Natl. Acad. Sci.* **104**, 3165–3170 (2007).

p. 87 "scientists have observed large, visible domains in the membranes that bound an organelle called the *vacuole* in yeast cells"
- S. P. Rayermann, G. E. Rayermann, C. E. Cornell, et al., Hallmarks of reversible separation of living, unperturbed cell membranes into two liquid phases. *Biophys. J.* **113**, 2425–2432 (2017).

p. 87 "the yeast cells seem to use these domains to enable the digestion of stored fats"
- A. Y. Seo, P.-W. Lau, D. Feliciano, et al., AMPK and vacuole-associated Atg14p orchestrate μ-lipophagy for energy production and long-term survival under glucose starvation. *eLife* **6**, e21690 (2017).

p. 88 "This picture of cell membranes as two-dimensional fluids made possible by the self-organized lipid bilayer was cemented in the early 1970s"

- S. J. Singer, G. L. Nicolson, The fluid mosaic model of the structure of cell membranes. *Science* **175**, 720–731 (1972).

CHAPTER 6. PREDICTABLE RANDOMNESS

pp. 91–92 On the history of Brownian motion:
- R. M. Mazo, *Brownian Motion: Fluctuations, Dynamics, and Applications* (Clarendon Press, Oxford, UK, 2002).
- P. Hänggi, F. Marchesoni, 100 years of Brownian motion. *Chaos* **15**, 026101–026105 (2005).

pp. 92–95 A classic on the properties of random walks, and their importance in biology:
- H. C. Berg, *Random Walks in Biology* (Princeton University Press, Princeton, NJ, 1993).

p. 98 "My brain is relatively slow, but its neurons are much more interconnected than the transistors in my laptop's central processing unit"
- L. Luo, Why is the human brain so efficient? *Nautilus*, April 12, 2018. http://nautil.us/issue/59/connections/why-is-the-human-brain-so-efficient.

p. 99 "Finding a specific target like a DNA binding site is even more challenging. The average time required, it turns out, is roughly proportional to the cell size cubed"
- S. Redner, *A Guide to First-Passage Processes* (Cambridge University Press, Cambridge, UK, 2007).

p. 100 On motor proteins and transport in neurons:
- O. Yagensky, T. Kalantary Dehaghi, J. J. E. Chua, The roles of microtubule-based transport at presynaptic nerve terminals. *Front. Synaptic Neurosci.* **8**, 3 (2016).

p. 101 On bacterial motion and foraging:
- E. M. Purcell, "Life at low Reynolds number," *American Journal of Physics* **45**, 3–11 (1977).

CHAPTER 7. ASSEMBLING EMBRYOS

p. 105 "it was widely believed that this single cell held within it a *homunculus*, a miniature but fully formed human"
- C. Pinto-Correia, *The Ovary of Eve* (University of Chicago Press, Chicago, 1998).

p. 106 On Hans Driesch, including his statement that every embryonic cell "carries the totality of all primordia":

- S. F. Gilbert, *Developmental Biology* (Sinauer Associates, Sunderland, MA, sixth edition, 2000).

p. 107 "Christiane Nüsslein-Volhard and Eric Wieschaus discovered several genes that are important determinants of the body plan of the fruit fly"
- C. Nüsslein-Volhard, E. Wieschaus, Mutations affecting segment number and polarity in Drosophila. *Nature* **287**, 795–801 (1980).

p. 107 "naming one 'hedgehog' because mutations in it give rise to spiky fly larvae"
- D. R. Haskett, Hedgehog signaling pathway. *Embryo Project Encyclopedia* (2015). http://embryo.asu.edu/handle/10776/8685.

p. 108 The drawing of the fruit fly (*Drosophila melanogaster*) hedgehog protein is based on structure 2IBG in the Protein Data Bank:
- https://www.rcsb.org/structure/2IBG.
- J. S. McLellan, S. Yao, X. Zheng, et al., Structure of a heparin-dependent complex of hedgehog and ihog. *Proc. Natl. Acad. Sci.* **103**, 17208–17213 (2006).

p. 108 The drawing of the human sonic hedgehog protein is based on structure 3MXW in the Protein Data Bank:
- https://www.rcsb.org/structure/3MXW.
- H. R. Maun, X. Wen, A. Lingel, et al., Hedgehog pathway antagonist 5E1 binds hedgehog at the pseudo-active site. *J. Biol. Chem* **285**, 26570–26580 (2010).

p. 110 "Transplanting tissue from the protein-producing region of one chick wing bud to the low-concentration side of another wing bud"
- M. Towers, J. Signolet, A. Sherman, et al., Insights into bird wing evolution and digit specification from polarizing region fate maps. *Nature Communications* **2**, 426 (2011).

p. 111 "Hedgehog gradients curate . . . the array of suckers on a cuttlefish arm"
- O. A. Tarazona, D. H. Lopez, L. A. Slota, M. J. Cohn, Evolution of limb development in cephalopod mollusks. *eLife* **8**, e43828 (2019).

p. 111 On hedgehog and cancer, see, for example:
- S. Kim, Y. Kim, J. Kong, et al., Epigenetic regulation of mammalian hedgehog signaling to the stroma determines the molecular subtype of bladder cancer. *eLife* **8**, e43024 (2019).

p. 111 "Morphogens were predicted and named by mathematician and computer science pioneer Alan Turing in 1952"
- M. Turing, The chemical basis of morphogenesis. *Philosophical Transactions of the Royal Society London, B: Biological Sciences* **237**, 37–72 (1952).

p. 114 "fluorescent glow provides a precise, quantifiable reporter of the what, where, and when of developmental activity"

- K. M. Forrest, E. R. Gavis, Live imaging of endogenous RNA reveals a diffusion and entrapment mechanism for nanos mRNA localization in *Drosophila. Current Biology* **13**, 1159–1168 (2003).
- T. Lucas, H. Tran, C.A.P. Romero, et al., 3 minutes to precisely measure morphogen concentration. *PLOS Genetics* **14**, e1007676 (2018).

pp. 114–115 "one can write down the equations of Brownian motion and genetic response functions and calculate patterns of protein abundance and gene activity"

- G. R. Ilsley, J. Fisher, R. Apweiler, et al., Cellular resolution models for even skipped regulation in the entire *Drosophila* embryo. *eLife* **2**, e00522 (2013).
- M. D. Petkova, G. Tkačik, W. Bialek, et al., Optimal decoding of cellular identities in a genetic network. *Cell* **176**, 844–855.e15 (2019).

p. 116 "William Bialek and colleagues at Princeton University connected morphogen measurements and information theory"

- J. O. Dubuis, G. Tkačik, E. F. Wieschaus, et al., Positional information, in bits. *Proc. Natl. Acad. Sci.* **110**, 16301–16308 (2013).

p. 117 On hair cells and lateral inhibition:

- M. Eddison, I. L. Roux, J. Lewis, Notch signaling in the development of the inner ear: Lessons from *Drosophila*. *Proc. Natl. Acad. Sci.* **97**, 11692–11699 (2000).

pp. 117–118 "Lateral inhibition . . . was first clearly demonstrated in a developing animal in the mid-1980s"

- C. Q. Doe, C. S. Goodman, Early events in insect neurogenesis: II. The role of cell interactions and cell lineage in the determination of neuronal precursor cells. *Developmental Biology* **111**, 206–219 (1985).

pp. 117–118 On the discovery of lateral inhibition in developmental patterning:

- K. Bussell, Milestone 3 (1937): Inhibit thy neighbour. *Nat. Rev. Neurosci.*, July 1, 2004. https://doi.org/10.1038/nrn1451.

pp. 117–118 On the Notch protein, its cleavage, and its use in cell signaling:

- W. R. Gordon, K. L. Arnett, S. C. Blacklow, The molecular logic of Notch signaling—a structural and biochemical perspective. *Journal of Cell Science* **121**, 3109–3119 (2008).

pp. 118–119 On Notch and lateral inhibition, more generally:

- M. Sjöqvist, E. R. Andersson, Do as I say, Not(ch) as I do: Lateral control of cell fate. *Developmental Biology* **447**, 58–70 (2019).

p. 119 On somite number in various animals and the early development of snakes:

- C. Gomez, E. M. Özbudak, J. Wunderlich, et al., Control of segment number in vertebrate embryos. *Nature* **454**, 335–339 (2008).

p. 120 "Jonathan Cooke and Erik Christopher Zeeman described an elegant biophysical strategy"
- J. Cooke, E. C. Zeeman, A clock and wavefront model for control of the number of repeated structures during animal morphogenesis. *J. Theor. Biol.* **58**, 455–476 (1976).

pp. 120–123 On the somite segmentation clock:
- I. Palmeirim, D. Henrique, D. Ish-Horowicz, O. Pourquié, Avian hairy gene expression identifies a molecular clock linked to vertebrate segmentation and somitogenesis. *Cell* **91**, 639–648 (1997).
- A. C. Oates, L. G. Morelli, S. Ares, Patterning embryos with oscillations: Structure, function and dynamics of the vertebrate segmentation clock. *Development* **139**, 625–639 (2012).

p. 123 "A joke already decades old." See, for example:
- What is the future of developmental biology? *Cell* **170**, 6–7 (2017).

CHAPTER 8. ORGANS BY DESIGN

p. 124 "you'll shed more than a ton of the cells lining your intestine"
- C. S. Potten, R. J. Morris, Epithelial stem cells in vivo. *J. Cell Sci.* **1988**, 45–62 (1988). (The stated value of about 10^{11} cells per day, at a mass of about 10^{-12} kg per cell, corresponds to about 3000 kg per lifetime, or about 6000 pounds.)

p. 125 "Dennis Discher and colleagues at the University of Pennsylvania grew stem cells of a type that can form neurons, muscle progenitors, or bone progenitors on gels of different stiffnesses"
- J. Engler, S. Sen, H. L. Sweeney, D. E. Discher, Matrix elasticity directs stem cell lineage specification. *Cell* **126**, 677–689 (2006).

pp. 125–126 On stem cells and their mechanical environment:
- J. Keung, S. Kumar, D. V. Schaffer, Presentation counts: Microenvironmental regulation of stem cells by biophysical and material cues. *Annual Review of Cell and Developmental Biology* **26**, 533–556 (2010).

p. 126 "channel-forming membrane proteins . . . whose configuration can be controlled by tension applied to the membrane"
- E. S. Haswell, R. Phillips, D. C. Rees, Mechanosensitive channels: What can they do and how do they do it? *Structure* **19**, 1356–1369 (2011).
- R. Peyronnet, D. Tran, T. Girault, J.-M. Frachisse, Mechanosensitive channels: Feeling tension in a world under pressure. *Front. Plant Sci.* **5**, 558 (2014).

pp. 127–128 "The groups . . . examined mice in which an expanding gel had been placed under the skin"

- M. Aragona, A. Sifrim, M. Malfait, et al., Mechanisms of stretch-mediated skin expansion at single-cell resolution. *Nature* **584**, 268–273 (2020).

p. 128 "fluid flow can induce stem cells to transform into the cell types that line blood vessels"
- K. Yamamoto, T. Sokabe, T. Watabe, et al., Fluid shear stress induces differentiation of Flk-1-positive embryonic stem cells into vascular endothelial cells in vitro. *American Journal of Physiology-Heart and Circulatory Physiology* **288**, H1915–H1924 (2005).
- H. Wang, G. M. Riha, S. Yan, et al., Shear stress induces endothelial differentiation from a murine embryonic mesenchymal progenitor cell line. *Arteriosclerosis, Thrombosis, and Vascular Biology* **25**, 1817–1823 (2005).

p. 129 On the history of cell culture, including Ross Harrison's experiments:
- H. Landecker, *Culturing Life: How Cells Became Technologies* (Harvard University Press, Cambridge, MA, 2010).

p. 129 "embryos of various species could be split apart into discrete cells that . . . could coalesce into aggregates that recapitulate some aspects of normal embryonic form." See, for example:
- M. S. Steinberg, Does differential adhesion govern self-assembly processes in histogenesis? Equilibrium configurations and the emergence of a hierarchy among populations of embryonic cells. *J. Exp. Zool.* **173**, 395–433 (1970).
- M. S. Steinberg, M. Takeichi, Experimental specification of cell sorting, tissue spreading, and specific spatial patterning by quantitative differences in cadherin expression. *Proc. Natl. Acad. Sci.* **91**, 206–209 (1994).

pp. 129–131 On the history of organoids:
- M. Simian, M. J. Bissell, Organoids: A historical perspective of thinking in three dimensions. *J. Cell Biol.* **216**, 31–40 (2017).

p. 130–131 On intestinal organoids:
- T. Sato, R. G. Vries, H. J. Snippert, et al., Single Lgr5 stem cells build crypt-villus structures in vitro without a mesenchymal niche. *Nature* **459**, 262–265 (2009).

p. 131 On organoids that form into a structure like the "optic cup":
- M. Eiraku, N. Takata, H. Ishibashi, M. Kawada, E. Sakakura, S. Okuda, K. Sekiguchi, T. Adachi, Y. Sasai, Self-organizing optic-cup morphogenesis in three-dimensional culture. *Nature* **472**, 51–56 (2011).

p. 131 "mouse stem cells could be grown into balls of connected neurons"
- M. Eiraku, K. Watanabe, M. Matsuo-Takasaki, et al., Self-organized formation of polarized cortical tissues from ESCs and its active manipulation by extrinsic signals. *Cell Stem Cell* **3**, 519–532 (2008).

p. 131 "the lab of Juergen Knoblich at the Austrian Academy of Science in Vienna built 'cerebral organoids'"

- M. A. Lancaster, M. Renner, C.-A. Martin, et al., Cerebral organoids model human brain development and microcephaly. *Nature* **501**, 373–379 (2013).

p. 131 "scientists and philosophers are already collaborating to map the ethical issues involved"
- J. Cepelewicz, An ethical future for brain organoids takes shape. *Quanta Magazine*, January 13, 2020. https://www.quantamagazine.org/an-ethical-future-for-brain-organoids-takes-shape-20200123/.

pp. 132–133 On the "lung on a chip":
- D. Huh, B. D. Matthews, A. Mammoto, et al., Reconstituting organ-level lung functions on a chip. *Science* **328**, 1662–8 (2010).

p. 133 On "body-on-a-chip" devices:
- C. W. McAleer, C. J. Long, D. Elbrecht, et al., Multi-organ system for the evaluation of efficacy and off-target toxicity of anticancer therapeutics. *Science Translational Medicine* **11**, eaav1386 (2019).
- C. D. Edington, W. L. K. Chen, E. Geishecker, et al., Interconnected microphysiological systems for quantitative biology and pharmacology studies. *Sci. Rep.* **8**, 1–18 (2018).

CHAPTER 9. THE ECOSYSTEM INSIDE YOU

p. 134 On the number of human and bacterial cells in the human body:
- R. Sender, S. Fuchs, R. Milo, Revised estimates for the number of human and bacteria cells in the body. *PLOS Biology* **14**, e1002533 (2016).

p. 135 "scooped a few hundred liters of water from the Sargasso Sea . . . and discovered a million previously unknown genes from hundreds of novel bacteria"
- J. C. Venter, K. Remington, J. F. Heidelberg, et al., Environmental genome shotgun sequencing of the Sargasso Sea. *Science* **304**, 66–74 (2004).

pp. 137–139 On the gut microbiota in general:
- J. Durack, S. V. Lynch, The gut microbiome: Relationships with disease and opportunities for therapy. *Journal of Experimental Medicine* **216**, 20–40 (2019).
- A. E. Douglas, *Fundamentals of Microbiome Science: How Microbes Shape Animal Biology* (Princeton University Press, Princeton, NJ, 2018).
- D. Haller, ed., *The Gut Microbiome in Health and Disease* (Springer, Cham, Switzerland, 2018).

p. 137 "There's also considerable, but not perfect, overlap between the set of species in you now and the set inside you a few months ago"

- B. H. Schlomann, R. Parthasarathy, Timescales of gut microbiome dynamics. *Current Opinion in Microbiology* **50**, 56–63 (2019).

p. 138 On the gut microbiome and neurological disorders:
- H. Tremlett, K. C. Bauer, S. Appel-Cresswell, et al., The gut microbiome in human neurological disease: A review. *Ann. Neurol.* **81**, 369–382 (2017).
- J. A. Griffiths, S. K. Mazmanian, Emerging evidence linking the gut microbiome to neurologic disorders. *Genome Medicine* **10**, 98 (2018).

p. 139 On fecal microbiota transplantation:
- L. Drew, Microbiota: Reseeding the gut. *Nature* **540**, S109–S112 (2016).
- E. van Nood, A. Vrieze, M. Nieuwdorp, et al., Duodenal infusion of donor feces for recurrent clostridium difficile. *New England Journal of Medicine* **368**, 407–415 (2013).
- R. J. Colman, D. T. Rubin, Fecal microbiota transplantation as therapy for inflammatory bowel disease: A systematic review and meta-analysis. *J. Crohns Colitis* **8**, 1569–1581 (2014).

p. 140 "compared the microbiome of fecal samples with that obtained by more invasive, direct sampling of the intestine"
- N. Zmora, G. Zilberman-Schapira, J. Suez, et al., Personalized gut mucosal colonization resistance to empiric probiotics is associated with unique host and microbiome features. *Cell* **174**, 1388–1405.e21 (2018).

p. 141 "germ-free animals exhibit a wide range of abnormalities"
- J. L. Round, S. K. Mazmanian, The gut microbiota shapes intestinal immune responses during health and disease. *Nat. Rev. Immunol.* **9**, 313–323 (2009).
- L. V. Hooper, D. R. Littman, A. J. Macpherson, Interactions between the microbiota and the immune system. *Science* **336**, 1268–1273 (2012).
- T. A. Jones, K. Guillemin, Racing to stay put: How resident microbiota stimulate intestinal epithelial cell proliferation. *Curr. Pathobiol. Rep.* **6**, 23–28 (2018).

p. 141 "germ-free fish larvae have a paucity of the insulin-producing beta cells of the pancreas"
- J. H. Hill, E. A. Franzosa, C. Huttenhower, K. Guillemin, A conserved bacterial protein induces pancreatic beta cell expansion during zebrafish development. *eLife* **5**, e20145 (2016).

p. 141 "Microbiome-assisted development has been observed for several other organs and tissues, for example, bone growth in infant mice"
- M. Schwarzer, K. Makki, G. Storelli, et al., Lactobacillus plantarum strain maintains growth of infant mice during chronic undernutrition. *Science* **351**, 854–857 (2016).

p. 142 "Links between human obesity and the composition of the gut microbiome"
- M. M. Finucane, T. J. Sharpton, T. J. Laurent, K. S. Pollard, A taxonomic signature of obesity in the microbiome? Getting to the guts of the matter. *PLoS ONE* **9**, e84689 (2014).
- M. A. Sze, P. D. Schloss, Looking for a signal in the noise: Revisiting obesity and the microbiome. *mBio.* **7**, e01018–16 (2016).

p. 144 On the prevalence and treatment of cholera:
- See articles on the websites of the World Health Organization (https://www.who.int/news-room/fact-sheets/detail/cholera) and the US Centers for Disease Control and Prevention (https://www.cdc.gov/cholera/treatment/index.html).

p. 145 On the type VI secretion system in *Vibrio cholerae* and other bacteria:
- B. Russell, S. B. Peterson, J. D. Mougous, Type VI secretion system effectors: Poisons with a purpose. *Nat. Rev. Micro.* **12**, 137–148 (2014).

pp. 145–147 On my lab's experiments on *Vibrio cholerae* and its type VI secretion system:
- S. L. Logan, J. Thomas, J. Yan, R. P. Baker, D. S. Shields, J. B. Xavier, B. K. Hammer, R. Parthasarathy, The *Vibrio cholerae* type VI secretion system can modulate host intestinal mechanics to displace gut bacterial symbionts. *Proc. Natl. Acad. Sci.* **115**, E3779–E3787 (2018).

p. 149 "we saw large immune responses when the fish was colonized by the normal, motile bacteria"
- T. J. Wiles, B. H. Schlomann, E. S. Wall, et al., Swimming motility of a gut bacterial symbiont promotes resistance to intestinal expulsion and enhances inflammation. *PLOS Biology* **18**, e3000661 (2020).

p. 149 "microbes with memories could record intestinal conditions"
- J. W. Kotula, S. J. Kerns, L. A. Shaket, et al., Programmable bacteria detect and record an environmental signal in the mammalian gut. *Proc. Natl. Acad. Sci.* **111**, 4838–4843 (2014).

p. 150 "devices that mimic aspects of the pulsatile flow of the human gut"
- J. Cremer, I. Segota, C. Yang, et al., Effect of flow and peristaltic mixing on bacterial growth in a gut-like channel. *Proc. Natl. Acad. Sci.* **113**, 11414–11419 (2016).

p. 150 "stretchable gut-on-a-chip devices"
- W. Shin, C. D. Hinojosa, D. E. Ingber, H. J. Kim, Human intestinal morphogenesis Controlled by Transepithelial Morphogen Gradient and Flow-Dependent Physical Cues in a microengineered gut-on-a-chip. *iScience* **15**, 391–406 (2019).

p. 151 "Alvaro Sanchez and colleagues . . . put together hundreds of microbial communities derived from samples of soil and plant leaves"

- J. E. Goldford, N. Lu, D. Bajić, et al., Emergent simplicity in microbial community assembly. *Science* **361**, 469–474 (2018).

p. 152 "ecologist Robert May, who in classic and highly influential work in the 1970s"
- R. M. May, *Stability and Complexity in Model Ecosystems* (Princeton University Press, Princeton, NJ, reprint edition, 2001).
- R. M. May, Will a large complex system be stable? *Nature* **238**, 413–414 (1972).

p. 152 "coexistence can emerge beyond the destabilizing point theorized by May"
- W. Cui, R. Marsland, P. Mehta, Diverse communities behave like typical random ecosystems. bioRxiv.org, 596551 (2019). https://doi.org/10.1101/596551.
- R. Marsland III, W. Cui, J. Goldford, et al., Available energy fluxes drive a transition in the diversity, stability, and functional structure of microbial communities. *PLOS Computational Biology* **15**, e1006793 (2019).

p. 152 "a mathematical description of this sort of constrained resource usage finds a surprisingly large set of parameters for which coexistence occurs"
- A. Posfai, T. Taillefumier, N. S. Wingreen, Metabolic trade-offs promote diversity in a model ecosystem. *Phys. Rev. Lett.* **118**, 028103 (2017).

CHAPTER 10. A SENSE OF SCALE

p. 155 "Large organisms don't simply look like blown-up versions of small ones." Several fascinating works expand on the themes of this chapter:
- S. Vogel, *Life's Devices: The Physical World of Animals and Plants* (Princeton University Press, Princeton, NJ, 1988).
- K. Schmidt-Nielsen, *How Animals Work* (Cambridge University Press, Cambridge, UK, 1972).
- J.B.S. Haldane, *On Being the Right Size and Other Essays* (Oxford University Press, Oxford, UK, 1985).

p. 158 "fluid dynamics pioneer Osborne Reynolds... became in 1868 the second ever 'professor of engineering' in England"
- J. J. O'Connor, E. F. Robertson, Osborne Reynolds—biography. *Maths History* (2003). https://mathshistory.st-andrews.ac.uk/Biographies/Reynolds/.

pp. 159–161 On the importance of the Reynolds number in biology:
- S. Vogel, *Life in Moving Fluids: The Physical Biology of Flow* (Princeton University Press, Princeton, NJ, second edition, 1996).

p. 160 "A classic video of this effect by fluid dynamicist G. I. Taylor is available online"

- *Low Reynolds Number Flow* at the National Committee for Fluid Mechanics Films site, http://web.mit.edu/hml/ncfmf.html; also at https://www.youtube.com/watch?v=51-6QCJTAjU (start at 13:40).

p. 166 "a plot of mass versus leg length for several different cockroaches"
- Henry D. Prange, "The scaling and mechanics of arthropod exoskeletons," in *Scale Effects in Animal Locomotion*, ed. T. J. Pedley (Academic Press, London, 1997), 169–171.

p. 167 "a 1963 paper on lung physiology"
- S. M. Tenney, J. E. Remmers, Comparative quantitative morphology of the mammalian lung: Diffusing area. *Nature* **197**, 54–56 (1963).

p. 168–169 "Thomas McMahon and John Tyler Bonner . . . plot the diameter versus the end-to-end length of the humerus, a leg bone, of lots of bovids"
- T. A. McMahon, J. T. Bonner, *On Size and Life* (Scientific American Library, New York, 1983).

p. 171 "an elephant named Tusko . . . lived a mostly miserable life in early twentieth-century circuses"
- M. Cooper, The elephant in the room. *Oregon Quarterly* (Spring 2014), 34–38.
- C. Lynn, The time Tusko the elephant was abandoned at the Oregon State Fair. *Salem Statesman Journal*, August 25, 2017.

p. 171 "If they did, bones would make up about three-quarters of their body mass"
- S. Vogel, *Life's Devices: The Physical World of Animals and Plants* (Princeton University Press, Princeton, NJ, 1988).

p. 172 "Galileo, in fact, wrote about bone scaling in his 1638 book, *Two New Sciences*"
- Galileo Galilei, *Dialogues Concerning Two New Sciences*, trans. Stillman Drake (University of Wisconsin Press, Madison, 1974).

CHAPTER 11. LIFE AT THE SURFACE

pp. 173–175 On respiration in ants and other insects:
- C. Gillott, ed., *Entomology* (Springer, Dordrecht, Netherlands, 2005), 469–486.
- M. J. Klowden, *Physiological Systems in Insects* (Academic Press, Amsterdam, 2010).

p. 175 "we find in the fossil record abundant evidence of giant insects"
- Ker Than, Why giant bugs once roamed the earth. *National Geographic*, August 9, 2011. https://www.nationalgeographic.com/news/2011/8/110808-ancient-insects-bugs-giants-oxygen-animals-science/.

p. 179 On fire ants and their strategies for staying atop water:
- N. J. Mlot, C. A. Tovey, D. L. Hu, Fire ants self-assemble into waterproof rafts to survive floods. *Proc. Natl. Acad. Sci.* **108**, 7669–7673 (2011).

p. 180 "infant respiratory distress syndrome, or IRDS, the leading cause of death among premature infants"
- L. K. Altman, A Kennedy baby's life and death. *New York Times*, July 30, 2013. https://www.nytimes.com/2013/07/30/health/a-kennedy-babys-life-and-death.html.
- D. Schraufnagel, *Breathing in America: Diseases, Progress, and Hope* (American Thoracic Society, New York, first edition, 2010).

pp. 180–181 On surface tension in the lungs and the role of pulmonary surfactant:
- J. A. Clements, Surface tension in the lungs. *Scientific American* **207**, 120–130 (1962).
- J. A. Clements, Lung surfactant: A personal perspective. *Annual Review of Physiology* **59**, 1–21 (1997).
- L. G. Dobbs, Pulmonary surfactant. *Annu. Rev. Med.* **40**, 431–446 (1989).

CHAPTER 12. MYSTERIES OF SIZE AND SHAPE

p. 183 "If you're curious about these examples, I've listed some readings in the references"
- T. A. McMahon, J. T. Bonner, *On Size and Life* (Scientific American Library, New York, 1983).
- M. Gazzola, M. Argentina, L. Mahadevan, Scaling macroscopic aquatic locomotion. *Nat. Phys.* **10**, 758–761 (2014).
- J. Baumgart, B. M. Friedrich, Fluid dynamics: Swimming across scales. *Nat. Phys.* **10**, 711–712 (2014).
- P. Willmer, G. Stone, I. Johnston, *Environmental Physiology of Animals* (Wiley-Blackwell, Malden, MA, 2004).
- R. Bale, M. Hao, A.P.S. Bhalla, N. A. Patankar, Energy efficiency and allometry of movement of swimming and flying animals. *Proc. Natl. Acad. Sci.* **111**, 7517–7521 (2014).
- T. L. Hedrick, B. Cheng, X. Deng, Wingbeat time and the scaling of passive rotational damping in flapping flight. *Science* **324**, 252–255 (2009).

p. 184 "an evocative description from biochemist and writer Nick Lane, 'an elephant-sized pile of mice would consume twenty times more food and oxygen every minute than the elephant does itself'"
- N. Lane, *Power, Sex, Suicide: Mitochondria and the Meaning of Life* (Oxford University Press, Oxford, UK, 2005).

p. 185 "Max Kleiber considered a range of animals from doves to cattle, plotting their metabolic rates and body masses"
- M. Kleiber, Body size and metabolism. *Hilgardia* **6**, 315–353 (1932).

p. 185 "from a 2003 dataset of over 600 mammals"
- C. R. White, R. S. Seymour, Mammalian basal metabolic rate is proportional to body mass$^{2/3}$. *Proc. Natl. Acad. Sci.* **100**, 4046–4049 (2003).

p. 186 "Max Rubner examined the energy consumption of seven dog breeds"
- M. Rubner, Ueber den einfluss der korpergrosse auf stoffund kaftwechsel. *Zeitschrift fur Biologie* **19**, 535–562 (1883).

p. 186 "Oxygen consumption in guinea pigs shows the same behavior"
- T. A. McMahon, J. T. Bonner, *On Size and Life* (Scientific American Library, New York, 1983), 56.

p. 187 "Peter Dodds, Dan Rothman, and Joshua Weitz at the Massachusetts Institute of Technology evaluated several datasets"
- P. S. Dodds, D. H. Rothman, J. S. Weitz, Re-examination of the "3/4-law" of metabolism. *Journal of Theoretical Biology* **209**, 9–27 (2001).

p. 187 "Considering birds alone, Peter Bennett and Paul H. Harvey . . . found a scaling exponent of 0.67"
- P. M. Bennett, P. H. Harvey, Active and resting metabolism in birds: Allometry, phylogeny and ecology. *Journal of Zoology* **213**, 327–344 (1987).

p. 189 "Clouds are not spheres, mountains are not cones, coastlines are not circles"
- Benoit Mandelbrot, *The Fractal Geometry of Nature* (W. H. Freeman, San Francisco, 1982).

p. 189 "West, Brown, and Enquist proposed a creative explanation for metabolic scaling"
- G. B. West, J. H. Brown, B. J. Enquist, A general model for the origin of allometric scaling laws in biology. *Science* **276**, 122–126 (1997).

p. 190 "several researchers found subtle but important flaws in West and colleagues' proof"
- P. S. Dodds, D. H. Rothman, J. S. Weitz, Re-examination of the "3/4-law" of metabolism. *Journal of Theoretical Biology* **209**, 9–27 (2001).
- R. S. Etienne, M.E.F. Apol, H. Olff, Demystifying the West, Brown & Enquist model of the allometry of metabolism. *Functional Ecology* **20**, 394–399 (2006).

pp. 190–191 "Jayanth Banavar . . . Amos Maritan, and Andrea Rinaldo developed a mathematical model that doesn't require self-similarity"
- J. R. Banavar, A. Maritan, A. Rinaldo, Size and form in efficient transportation networks. *Nature* **399**, 130–132 (1999).

p. 192 "This perspective has been nicely described by biochemist and author Nick Lane"

- N. Lane, *Power, Sex, Suicide: Mitochondria and the Meaning of Life* (Oxford University Press, Oxford, UK, 2005). (See chapter 9, "The Power Laws of Biology.")

p. 192 "This perspective . . . was precisely formulated by Peter Hochachka's group"
- C.-A. Darveau, R. K. Suarez, R. D. Andrews, P. W. Hochachka, Allometric cascade as a unifying principle of body mass effects on metabolism. *Nature*. **417**, 166–170 (2002).

p. 192 "Brown, West, and others have argued for a 'metabolic theory of ecology'"
- J. H. Brown, J. F. Gillooly, A. P. Allen, V. M. Savage, G. B. West, Toward a metabolic theory of ecology. *Ecology* **85**, 1771–1789 (2004).

p. 193 "Geoffrey West and others . . . have argued that cities, like living organisms, are governed by the properties of networks"
- L.M.A. Bettencourt, J. Lobo, D. Helbing, et al., Growth, innovation, scaling, and the pace of life in cities. *Proc. Natl. Acad. Sci.* **104**, 7301–7306 (2007).
- L.M.A. Bettencourt, The origins of scaling in cities. *Science* **340**, 1438–1441 (2013).

pp. 193–194 "The data are, however, noisy, and their trends are even more contentious than those of animal metabolism"
- C. R. Shalizi, Scaling and hierarchy in urban economies. arXiv.org, 1102.4101 [physics, stat.] (2011). http://arxiv.org/abs/1102.4101.
- E. Arcaute, E. Hatna, P. Ferguson, H. Youn, A. Johansson, M. Batty, Constructing cities, deconstructing scaling laws. *Journal of the Royal Society Interface* **12**, 20140745 (2015).

p. 194 "flatworms that grow or shrink dramatically depending on the availability of food"
- A. Thommen, S. Werner, O. Frank, et al., Body size-dependent energy storage causes Kleiber's law scaling of the metabolic rate in planarians. *eLife* **8**, e38187 (2019).

CHAPTER 13. HOW WE READ DNA

p. 201 "Ray Wu and Dale Kaiser managed to decipher 12 of the nucleotides of a viral genome"
- R. Wu, A. D. Kaiser, Structure and base sequence in the cohesive ends of bacteriophage lambda DNA. *Journal of Molecular Biology* **35**, 523–537 (1968).

pp. 201–215 On the history of DNA sequencing:
- J. Shendure, S. Balasubramanian, G. M. Church, W. Gilbert, J. Rogers, J. A. Schloss, R. H. Waterston, DNA sequencing at 40: Past, present and future. *Nature* **550**, 345–353 (2017).

- J. M. Heather, B. Chain, The sequence of sequencers: The history of sequencing DNA. *Genomics* **107**, 1–8 (2016).

pp. 203–204 On DNA in electric fields:
- M. Muthukumar, Theory of electrophoretic mobility of polyelectrolyte chains. *Macromolecular Theory and Simulations* **3**, 61–71 (1994).
- J.-L. Viovy, Electrophoresis of DNA and other polyelectrolytes: Physical mechanisms. *Rev. Mod. Phys.* **72**, 813–872 (2000).

pp. 205–207 On the initial sequencing of the human genome:
- T. Carvalho, T. Zhu, The Human genome project (1990–2003). *Embryo Project Encyclopedia* (2014). http://embryo.asu.edu/handle/10776/7829.
- International Human Genome Sequencing Consortium, Initial sequencing and analysis of the human genome. *Nature* **409**, 860–921 (2001).
- J. C. Venter et al., The sequence of the human genome. *Science* **291**, 1304–1351 (2001).

p. 206 On the completion of the Human Genome Project:
- E. Pennisi, Reaching their goal early, sequencing labs celebrate. *Science* **300**, 409 (2003).
- Genome.gov, Human Genome Project FAQ. https://www.genome.gov/human-genome-project/Completion-FAQ.

p. 207 On second-generation DNA sequencing methods:
- E. R. Mardis, Next-generation DNA sequencing methods. *Annual Review of Genomics and Human Genetics* **9**, 387–402 (2008).
- M. L. Metzker, Sequencing technologies—the next generation. *Nature Reviews Genetics* **11**, 31–46 (2010).

pp. 207–210 On pyrosequencing:
- M. Margulies, M. Egholm, W. E. Altman, et al., Genome sequencing in microfabricated high-density picolitre reactors. *Nature* **437**, 376–380 (2005).
- P. Nyren, B. Pettersson, M. Uhlen, Solid phase DNA minisequencing by an enzymatic luminometric inorganic pyrophosphate detection assay. *Analytical Biochemistry* **208**, 171–175 (1993).

p. 210 "For about $500,000, you could buy a machine to perform pyrosequencing"
- K. Davies, *The $1,000 Genome: The Revolution in DNA Sequencing and the New Era of Personalized Medicine* (Free Press, New York, 2010).

p. 211 On DNA sequencing using field effect transistors (known as ion semiconductor sequencing):
- E. Pennisi, Semiconductors inspire new sequencing technologies. *Science* **327**, 1190 (2010).

- J. M. Rothberg, W. Hinz, T. M. Rearick, et al., An integrated semiconductor device enabling non-optical genome sequencing. *Nature* **475**, 348–352 (2011).

p. 211 "Ion Torrent introduced its Personal Genome Machine in 2010"
- M. Herper, Gene machine. *Forbes*, December 30, 2010. https://www.forbes.com/forbes/2011/0117/features-jonathan-rothberg-medicine-tech-gene-machine.html#12c8a7ed2711.

pp. 211–212 On Illumina sequencing:
- YouTube user Draven1983101, Illumina Solexa sequencing (2010). https://www.youtube.com/watch?v=77r5p8IBwJk.

p. 212 On Pacific Biosciences' DNA sequencing technique:
- J. Eid, A. Fehr, J. Gray, et al., Real-time DNA sequencing from single polymerase molecules. *Science* **323**, 133–138 (2009).

p. 213–215 On nanopore sequencing:
- H. Bayley, Nanopore sequencing: From imagination to reality. *Clin. Chem.* **61**, 25–31 (2015).
- D. Deamer, M. Akeson, D. Branton, Three decades of nanopore sequencing. *Nat. Biotechnol.* **34**, 518–524 (2016).

p. 214 The drawing of the nanopore channel protein is based on structure 3X2R in the Protein Data Bank:
- https://www.rcsb.org/structure/3X2R.
- B. Cao, Y. Zhao, Y. Kou, et al., Structure of the nonameric bacterial amyloid secretion channel. *Proc. Natl. Acad. Sci.* **111**, E5439–E5444 (2014).

p. 214 The drawing of the polymerase attached to the nanopore channel protein is based on structure 3BDP in the Protein Data Bank:
- https://www.rcsb.org/structure/3BDP.
- J. R. Kiefer, C. Mao, J. C. Braman, L. S. Beese, Visualizing DNA replication in a catalytically active Bacillus DNA polymerase crystal. *Nature* **391**, 304–307 (1998).

p. 216 "The graph [of genome sequencing costs] is amazing." The data are from a database maintained by the US National Institutes of Health:
- K. A. Wetterstrand, DNA sequencing costs: Data from the NHGRI genome sequencing program (GSP). www.genome.gov/sequencingcostsdata (accessed February 6, 2020).

p. 217 "public funding for basic research as well as specific programs targeting sequencing innovations." See, for example:
- https://www.nih.gov/news-events/news-releases/nhgri-funds-development-third-generation-dna-sequencing-technologies.

p. 218 On the discovery of reverse transcriptase:

- J. M. Coffin, H. Fan, The discovery of reverse transcriptase. *Annual Review of Virology* **3**, 29–51 (2016).

p. 218 On single-cell RNA sequencing:
- G. X. Y. Zheng, J. M. Terry, P. Belgrader, et al., Massively parallel digital transcriptional profiling of single cells. *Nature Communications* **8**, 14049 (2017).
- J. Shendure, E. Lieberman Aiden, The expanding scope of DNA sequencing. *Nat. Biotechnol.* **30**, 1084–1094 (2012).
- D. Kotliar, A. Veres, M. A. Nagy, et al., Identifying gene expression programs of cell-type identity and cellular activity with single-cell RNA-Seq. *eLife* **8**, e43803 (2019).
- D. E. Wagner, C. Weinreb, Z. M. Collins, et al., Single-cell mapping of gene expression landscapes and lineage in the zebrafish embryo. *Science* **360**, 981–987 (2018).

CHAPTER 14. GENETIC COMBINATIONS

p. 222 "The average Frenchman born in 1800 was 5 feet, 5 inches tall" and "Our modern Frenchman has about twice the calories per day at his disposal"
- M. Roser, C. Appel, H. Ritchie, Human height. *Our World in Data* (2013). https://ourworldindata.org/human-height.
- M. Roser, H. Ritchie, Food supply. *Our World in Data* (2013). https://ourworldindata.org/food-supply.

p. 223 On the heritability of height, nutrition, and whole genome studies:
- C.-Q. Lai, How much of human height is genetic and how much is due to nutrition? *Scientific American*, December 11, 2006. https://www.scientificamerican.com/article/how-much-of-human-height/.

pp. 225–227 On predicting height from SNP data:
- L. Lello, S. G. Avery, L. Tellier, et al., Accurate genomic prediction of human height. *Genetics* **210**, 477–497 (2018).

p. 227 "the information encoded by DNA does, in fact, explain 80% of the variation in human height"
- P. Wainschtein, D. P. Jain, L. Yengo, et al., Recovery of trait heritability from whole genome sequence data. bioRxiv.org, 588020 (2019).
- L. Geddes, Genetic study homes in on height's heritability mystery. *Nature* **568**, 444–445 (2019).

p. 227 On Norman Borlaug and wheat:
- L. F. Hesser, *The Man Who Fed the World: Nobel Peace Prize Laureate Norman Borlaug and His Battle to End World Hunger* (Durban House Publishing, Dallas, TX, 2006).

- G. Easterbrook, Forgotten benefactor of humanity. *The Atlantic* (January 1997) https://www.theatlantic.com/magazine/archive/1997/01/forgotten-benefactor-of-humanity/306101/.

p. 227 "Contemporary chickens raised for consumption in North America weigh four times as much as their ancestors of the 1950s"
- M. J. Zuidhof, B. L. Schneider, V. L. Carney, et al., Growth, efficiency, and yield of commercial broilers from 1957, 1978, and 2005. *Poultry Science* **93**, 2970–2982 (2014).

p. 228 "In 2019, the US dairy database, for example, contained genotypes from three million cows"
- Council on Dairy Cattle Breeding, Annual reports. https://www.uscdcb.com/whats-new/reports/ (accessed April 1, 2020).

p. 229 On watermelons:
- M. Strauss, The 5,000-year secret history of the watermelon. *National Geographic*, August 21, 2015. https://www.nationalgeographic.com/news/2015/08/150821-watermelon-fruit-history-agriculture/.
- M. Jayakodi, M. Schreiber, M. Mascher, Sweet genes in melon and watermelon. *Nature Genetics* **51**, 1572–1573 (2019).
- S. Guo, S. Zhao, H. Sun, et al., Resequencing of 414 cultivated and wild watermelon accessions identifies selection for fruit quality traits. *Nature Genetics* **51**, 1616–1623 (2019).

pp. 229–230 On breast cancer risks:
- A. Antoniou, P.D.P. Pharoah, S. Narod, et al., Average risks of breast and ovarian cancer associated with BRCA1 or BRCA2 mutations detected in case series unselected for family history: A combined analysis of 22 studies. *American Journal of Human Genetics* **72**, 1117–1130 (2003).

p. 231 "In 1969, Robert Edwards, Barry Bavister, and Patrick Steptoe announced that they had successfully fertilized human egg cells with human sperm cells in vitro"
- R. G. Edwards, B. D. Bavister, P. C. Steptoe, Early stages of fertilization in vitro of human oocytes matured in vitro. *Nature* **221**, 632–635 (1969).

pp. 231–232 "The intervening decade saw a great deal of discussion." For a much more thorough treatment of the scientific, historical, and social context of in vitro fertilization, see
- P. Ball, *Unnatural: The Heretical Idea of Making People* (Bodley Head, London, first edition, 2011).

p. 232 "A 1969 *Life* magazine cover"
- *Life*, June 13, 1969.

p. 232 "a poll of Americans in the same issue"
- L. Harris, The *Life* poll. *Life* **66** (23), 52–55 (June 13, 1969).

p. 232 "By 1978, however, according to a Gallup poll also conducted in the United States"
- H. M. Kiefer, Gallup brain: The birth of in vitro fertilization. Gallup.com, August 5, 2003. https://news.gallup.com/poll/8983/Gallup-Brain-Birth-Vitro-Fertilization.aspx.

p. 232 "There have been in total about eight million babies born worldwide through *in vitro* fertilization as of 2018"
- European Society of Human Reproduction and Embryology. ScienceDaily (website), July 3, 2018. https://www.sciencedaily.com/releases/2018/07/180703084127.htm.

p. 232 On the fraction of babies born in Denmark through in vitro fertilization:
- Danish Health and Medicines Authority, 2017 assisted reproduction report. https://sundhedsdatastyrelsen.dk/da/tal-og-analyser/analyser-og-rapporter/andre-analyser-og-rapporter/assisteret-reproduktion (accessed October 26, 2020).

p. 233 On embryo biopsy methods:
- D. Cimadomo, A. Capalbo, F. M. Ubaldi, et al., The impact of biopsy on human embryo developmental potential during preimplantation genetic diagnosis. *BioMed Research International* (2016), e7193075.
- H. J. Stern, Preimplantation genetic diagnosis: Prenatal testing for embryos finally achieving its potential. *Journal of Clinical Medicine* **3**, 280–309 (2014).

p. 235 "one might expect embryo selection to give a height advantage of about 1 inch"
- E. Karavani, O. Zuk, D. Zeevi, et al., Screening human embryos for polygenic traits has limited utility. *Cell* **179**, 1424–1435.e8 (2019).

p. 236 "About 30,000 forced sterilizations had been performed ... by 1939"
- M. Wills, When forced sterilization was legal in the U.S. *JSTOR Daily*, August 3, 2017. https://daily.jstor.org/when-forced-sterilization-was-legal-in-the-u-s/.

p. 236 On eugenics, especially in the United States:
- A. DenHoed, The forgotten lessons of the American eugenics movement. *New Yorker,* April 27, 2016. https://www.newyorker.com/books/page-turner/the-forgotten-lessons-of-the-american-eugenics-movement.
- C. Zimmer, *She Has Her Mother's Laugh: The Powers, Perversions, and Potential of Heredity* (Dutton, New York, first edition, 2018).

p. 237 "Signatures of 'assortative mating,' in which people of similar educational or socioeconomic backgrounds pair up, are evident"
- M. R. Robinson, A. Kleinman, M. Graff, et al., Genetic evidence of assortative mating in humans. *Nature Human Behaviour* **1**, 1–13 (2017).

CHAPTER 15. HOW WE WRITE DNA

p. 240 "insulin from pigs and cattle, the purification of which is difficult and expensive"
- D. Wendt, Two tons of pig parts: Making insulin in the 1920s. *O Say Can You See?* (blog for the National Museum of American History), November 1, 2013. https://americanhistory.si.edu/blog/2013/11/two-tons-of-pig-parts-making-insulin-in-the-1920s.html.
- Genentech (website), Cloning insulin, April 7, 2016. https://www.gene.com/stories/cloning-insulin.

pp. 241–243 A simple, illustrated explanation of bacterial transformation can be found at Khan Academy:
- https://www.khanacademy.org/science/biology/biotech-dna-technology/dna-cloning-tutorial/a/bacterial-transformation-selection.

pp. 241–243 More detailed bacterial transformation protocols that give a sense of what these procedures are like in practice can be found at the websites of vendors of supplies and chemicals. See, for example:
- https://www.thermofisher.com/us/en/home/life-science/cloning/cloning-learning-center/invitrogen-school-of-molecular-biology/molecular-cloning/transformation.html.

p. 243 "Working together, the labs of Stanley Cohen at Stanford University and Herbert Boyer at the University of California at San Francisco announced the first successful bacterial transformation"
- S. N. Cohen, A.C.Y. Chang, H. W. Boyer, R. B. Helling, Construction of biologically functional bacterial plasmids in vitro. *Proc. Natl. Acad. Sci.* **70**, 3240–3244 (1973).

pp. 243–244 On Genetech and the bioengineering of insulin-producing bacteria:
- S. S. Hughes, *Genentech: The Beginnings of Biotech* (University of Chicago Press, Chicago, reprint edition, 2013).
- S. Mukherjee, *The Gene: An Intimate History* (Scribner, New York, 2016).

p. 243 "As one of the councillors said later"
- B. J. Culliton, Recombinant DNA: Cambridge City Council votes moratorium. *Science* **193**, 300–301 (1976).

p. 244 "These days insulin itself is mostly produced from engineered yeast rather than bacteria"
- J. Nielsen, Production of biopharmaceutical proteins by yeast. *Bioengineered* **4**, 207–211 (2013).

p. 245 "DNA carefully injected into the nascent embryo can find itself inserted into the genome"

- R. Behringer, M. Gertsenstein, K. Nagy, A. Nagy, *Manipulating the Mouse Embryo: A Laboratory Manual* (Cold Spring Harbor Laboratory Press, Cold Spring Harbor, NY, fourth edition, 2013).

pp. 245–246 On agrobacterium and DNA insertion into plant genomes:
- T. Tzfira, J. Li, B. Lacroix, V. Citovsky, Agrobacterium T-DNA integration: Molecules and models. *Trends in Genetics* **20**, 375–383 (2004).

p. 246 "Inadequate vitamin A causes blindness in 250,000 to 500,000 children annually"
- World Health Organization, Micronutrient deficiencies: Vitamin A deficiency. https://www.who.int/nutrition/topics/vad/en/ (accessed September 15, 2020).

p. 246 "Researchers . . . developed 'golden rice'"
- X. Ye, S. Al-Babili, A. Klöti, et al., Engineering the provitamin A (β-carotene) biosynthetic pathway into (carotenoid-free) rice endosperm. *Science* **287**, 303–305 (2000).

pp. 246–247 Abundant information about golden rice, including details of the genetics and safety tests, is at
- Golden Rice Humanitarian Board, Golden Rice Project. http://www.goldenrice.org/.

p. 246 "Clinical studies showed that the strikingly yellow-orange rice is safe and effective"
- G. Tang, J. Qin, G. G. Dolnikowski, et al., Golden rice is an effective source of vitamin A. *Am. J. Clin. Nutr.* **89**, 1776–1783 (2009).

p. 246 "Golden Rice could probably supply 50% of the Recommended Dietary Allowance (RDA) of vitamin A"
- American Society of Nutrition, Researchers determine that golden rice is an effective source of vitamin A. http://www.goldenrice.org/PDFs/ASNonGR.pdf (May 15, 2009).

p. 247 On the history and struggles of golden rice:
- E. Regis, *Golden Rice: The Imperiled Birth of a GMO Superfood* (Johns Hopkins University Press, Baltimore, 2019).
- E. Regis, The true story of the genetically modified superfood that almost saved millions. *Foreign Policy*, October 17, 2019. https://foreignpolicy.com/2019/10/17/golden-rice-genetically-modified-superfood-almost-saved-millions/.

p. 247 "Bangladesh . . . became in 2019 the first country to approve the planting of golden rice seeds"
- E. Stokstad, After 20 years, golden rice nears approval. *Science* **366**, 934–934 (2019).

p. 247 "the Philippines acknowledged the safety of golden rice, setting the stage for its planting"
- J. Conrow, Philippine agency rules golden rice is safe. Cornell Alliance for Science December 18, 2019. https://allianceforscience.cornell.edu/blog/2019/12/philippine-agency-rules-golden-rice-is-safe/.

p. 248 On zinc finger nucleases and transcription activator-like effector nucleases:
- T. Gaj, C. A. Gersbach, C. F. Barbas, ZFN, TALEN, and CRISPR/Cas-based methods for genome engineering. *Trends in Biotechnology* **31**, 397–405 (2013).

p. 248 "In 1987, a Japanese team"
- Y. Ishino, H. Shinagawa, K. Makino, et al., Nucleotide sequence of the iap gene, responsible for alkaline phosphatase isozyme conversion in *Escherichia coli*, and identification of the gene product. *Journal of Bacteriology* **169**, 5429–5433 (1987).

p. 248 "Francisco Mojica and colleageus in Spain discovered similar repeating DNA patterns . . ."
- F. J. M. Mojica, G. Juez, F. Rodriguez-Valera, Transcription at different salinities of Haloferax mediterranei sequences adjacent to partially modified PstI sites. *Molecular Microbiology*. **9**, 613–621 (1993).

pp. 248–249 On the early history of CRISPR:
- M. Campbell, Francis Mojica: The modest microbiologist who discovered and named CRISPR. Genomics Research from Technology Networks (website), October 14, 2019. https://www.technologynetworks.com/genomics/articles/francis-mojica-the-modest-microbiologist-who-discovered-and-named-crispr-325093.
- P. D. Hsu, E. S. Lander, F. Zhang, Development and applications of CRISPR-Cas9 for genome engineering. *Cell* **157**, 1262–1278 (2014).
- Y. Ishino, M. Krupovic, P. Forterre, History of CRISPR-Cas from encounter with a mysterious repeated sequence to genome editing technology. *Journal of Bacteriology* **200**, e00580-17 (2018). doi:10.1128/JB.00580-17.
- R. Barrangou, J. van der Oost, eds., *CRISPR-Cas Systems: RNA-Mediated Adaptive Immunity in Bacteria and Archaea* (Springer-Verlag, Berlin, 2013).
- C. R. Fernández, Francis Mojica, the Spanish scientist who discovered CRISPR. https://www.labiotech.eu/interviews/francis-mojica-crispr-interview/ (April 8, 2019).
- E. S. Lander, The heroes of CRISPR. *Cell* **164**, 18–28 (2016). [This paper is rather controversial. See, for example: T. Vence, "Heroes of CRISPR" disputed, *Scientist Magazine,* January 19, 2016, https://www.the-scientist.com/news-opinion/heroes-of-crispr-disputed-34188; and M. Morange,

Why Eric Lander's controversial paper "The Heroes of CRISPR" is not solid historical research, *American Scientist*, February 17, 2016, https://www.americanscientist.org/blog/macroscope/why-eric-lander%E2%80%99s-controversial-paper-%E2%80%9Cthe-heroes-of-crispr%E2%80%9D-is-not-solid-historical.]

pp. 249–251 On CRISPR and the bacterial immune system:
- H. Ledford, Five big mysteries about CRISPR's origins. *Nature News* **541**, 280 (2017).
- R. Sorek, C. M. Lawrence, B. Wiedenheft, CRISPR-mediated adaptive immune systems in bacteria and archaea. *Annual Review of Biochemistry* **82**, 237–266 (2013).

pp. 250–251 On the mechanisms of Cas9's target DNA recognition and cleavage:
- G. Palermo, C. G. Ricci, J. A. McCammon, The invisible dance of CRISPR-Cas9. *Physics Today* **72**, 30–36 (2019).
- F. Jiang, J. A. Doudna, CRISPR–Cas9 structures and mechanisms. *Annual Review of Biophysics* **46**, 505–529 (2017). This article includes informative computer animations of Cas9's structural changes, available on YouTube: https://www.youtube.com/watch?v=XAtZEIyzd7g, https://www.youtube.com/watch?v=Ya_Xoom7YAY.
- Nature Video, CRISPR: Gene editing and beyond, https://www.youtube.com/watch?v=4YKFw2KZA5o (2017).

p. 251 "Three papers published in 2005 . . . announced that the DNA sequences of spacers matched nonbacterial and nonarchaeal DNA sequences"
- F.J.M. Mojica, C. Díez-Villaseñor, J. García-Martínez, E. Soria, Intervening sequences of regularly spaced prokaryotic repeats derive from foreign genetic elements. *J. Mol. Evol.* **60**, 174–182 (2005).
- C. Pourcel, G. Salvignol, G. Vergnaud, CRISPR elements in *Yersinia pestis* acquire new repeats by preferential uptake of bacteriophage DNA, and provide additional tools for evolutionary studies. *Microbiology* **151**, 653–663 (2005).
- Bolotin, B. Quinquis, A. Sorokin, S. D. Ehrlich, Clustered regularly interspaced short palindrome repeats (CRISPRs) have spacers of extrachromosomal origin. *Microbiology* **151**, 2551–2561 (2005).

p. 252 "Reporting the results in a 2012 paper"
- M. Jinek, K. Chylinski, I. Fonfara, et al., A programmable dual-RNA–guided DNA endonuclease in adaptive bacterial immunity. *Science* **337**, 816–821 (2012).

p. 252 "awarded at a celebrity-laden Silicon Valley gala"
- C. Zimmer, CRISPR natural history in bacteria. *Quanta Magazine*, February 6, 2015. https://www.quantamagazine.org/crispr-natural-history-in-bacteria-20150206/.

p. 252 "another research group, led by Virginijus Šikšnys at Vilnius University, Lithuania, submitted a paper on April 6 also describing RNA-guided DNA cleavage"
- G. Gasiunas, R. Barrangou, P. Horvath, V. Šikšnys, Cas9–crRNA ribonucleoprotein complex mediates specific DNA cleavage for adaptive immunity in bacteria. *PNAS* **109**, E2579–E2586 (2012).

p. 253 "the forgotten man of CRISPR"
- S. Begley, Who gets credit for CRISPR? Prestigious award singles out three. STAT (website), May 31, 2018. https://www.statnews.com/2018/05/31/crispr-scientists-kavli-prize-nanoscience/.
- R. Bichell, Science rewards eureka moments, except when it doesn't. NPR.org, November 2, 2016. https://www.npr.org/sections/health-shots/2016/11/02/500331130/science-rewards-eureka-moments-except-when-it-doesnt.

p. 253 "he shared with Doudna and Charpentier the million-dollar Kavli Prize for Nanoscience"
- G. Guglielmi, Million-dollar Kavli Prize recognizes scientist scooped on CRISPR. *Nature* **558**, 17–18 (2018).

p. 253 "the labs of Feng Zhang at the Broad Institute of MIT and Harvard University and George Church at Harvard Medical School demonstrated CRISPR/Cas9 in mouse- and human-derived cells"
- L. Cong, F. A. Ran, D. Cox, et al., Multiplex genome engineering using CRISPR/Cas systems. *Science* **339**, 819–823 (2013).
- P. Mali, L. Yang, K. M. Esvelt, et al., RNA-guided human genome engineering via Cas9. *Science* **339**, 823–826 (2013).

pp. 254–255 On prime editing:
- V. Anzalone, P. B. Randolph, J. R. Davis, et al., Search-and-replace genome editing without double-strand breaks or donor DNA. *Nature* **576**, 149–157 (2019).
- H. Ledford, Super-precise new CRISPR tool could tackle a plethora of genetic diseases. *Nature* **574**, 464–465 (2019).

p. 256 On CRISPR and tomatoes:
- A. Zsögön, T. Čermák, E. R. Naves, et al., De novo domestication of wild tomato using genome editing. *Nature Biotechnology* **36**, 1211–1216 (2018).

p. 256 "researchers at the University of Pennsylvania piped blood cells and circulating immune cells"
- P. Tebas, D. Stein, W. W. Tang, et al., Gene editing of CCR5 in autologous CD4 T cells of persons infected with HIV. *New England Journal of Medicine* **370**, 901–910 (2014).

p. 257 "Another pioneering application in 2015 also involved immune cells . . . placed into a one-year-old girl who suffered from leukemia and was unresponsive to all other treatments"
- S. Reardon, Leukaemia success heralds wave of gene-editing therapies. *Nature News* **527**, 146 (2015).

p. 257 "In early 2020, a person with a rare genetic mutation that causes blindness became the first human to receive direct delivery of CRISPR/Cas9"
- H. Ledford, CRISPR treatment inserted directly into the body for first time. *Nature* **579**, 185–185 (2020).
- M. L. Maeder, M. Stefanidakis, C. J. Wilson, et al., Development of a gene-editing approach to restore vision loss in Leber congenital amaurosis type 10. *Nature Medicine* **25**, 229–233 (2019).

p. 258 "Anti-CRISPRs were discovered around 2012"
- J. Bondy-Denomy, A. Pawluk, K. L. Maxwell, A. R. Davidson, Bacteriophage genes that inactivate the CRISPR/Cas bacterial immune system. *Nature* **493**, 429–432 (2013).
- E. Dolgin, The kill-switch for CRISPR that could make gene-editing safer. *Nature* **577**, 308–310 (2020).

p. 258 "There are a multitude of different anti-CRISPRs"
- N. D. Marino, R. Pinilla-Redondo, B. Csörgő, J. Bondy-Denomy, Anti-CRISPR protein applications: Natural brakes for CRISPR-Cas technologies. *Nature Methods* **17**, 471–479 (2020).
- L. Liu, M. Yin, M. Wang, Y. Wang, Phage AcrIIA2 DNA mimicry: Structural basis of the CRISPR and anti-CRISPR arms race. *Molecular Cell* **73**, 611–620.e3 (2019).
- J. Shin, F. Jiang, J.-J. Liu, et al., Disabling Cas9 by an anti-CRISPR DNA mimic. *Science Advances* **3**, e1701620 (2017).

p. 259 "delivering an anti-CRISPR after Cas9 effectively shuts down further gene editing and hence reduces unwanted genetic alterations"
- M. Nakamura, P. Srinivasan, M. Chavez, et al., Anti-CRISPR-mediated control of gene editing and synthetic circuits in eukaryotic cells. *Nature Communications* **10**, 194 (2019).

CHAPTER 16. DESIGNING THE FUTURE

p. 260 "the horse and the sea-horse, the dog and the dog-fish, the snake and the eel"
- T. H. White, *The Book of Beasts: Being a Translation from a Latin Bestiary of the Twelfth Century* (UW-Madison Libraries Parallel Press, Madison, WI, 2002).

p. 260 "whole ecosystems are also subject to similar rules." See, for example:
- S. B. Carroll, *The Serengeti Rules* (Princeton University Press, Princeton, NJ, 2016).

p. 262 "Editing embryos was successfully demonstrated in mice and zebrafish in 2013"
- H. Wang, H. Yang, C. S. Shivalila, et al., One-step generation of mice carrying mutations in multiple genes by CRISPR/Cas-mediated genome engineering. *Cell* **153**, 910–918 (2013).
- W. Y. Hwang, Y. Fu, D. Reyon, et al., Efficient genome editing in zebrafish using a CRISPR-Cas system. *Nature Biotechnology* **31**, 227–229 (2013).

p. 262 "and in monkeys in 2014"
- Y. Niu, B. Shen, Y. Cui, et al., Generation of gene-modified cynomolgus monkey via Cas9/RNA-mediated gene targeting in one-cell embryos. *Cell* **156**, 836–843 (2014).

p. 262 "Chinese researcher He Jiankui stunned the world with the announcement that his team had used CRISPR/Cas9 to produce the first ever gene-edited babies."
- D. Cyranoski, The CRISPR-baby scandal: What's next for human gene-editing. *Nature* **566**, 440–442 (2019).
- Outrage intensifies over claims of gene-edited babies. NPR.org, December 7, 2018. https://www.npr.org/sections/health-shots/2018/12/07/673878474/outrage-intensifies-over-claims-of-gene-edited-babies.
- D. Normile, Chinese scientist who produced genetically altered babies sentenced to 3 years in jail. *Science*, December 31, 2019. https://www.sciencemag.org/news/2019/12/chinese-scientist-who-produced-genetically-altered-babies-sentenced-3-years-jail.

p. 262 "extremely abominable in nature"
- S. Jiang, H. Regan, J. Berlinger, China suspends scientists who claim to have produced gene-edited babies. CNN, November 29, 2018. https://www.cnn.com/2018/11/29/health/china-gene-editing-he-jiankui-intl/index.html.

p. 265 On baby carrots:
- R. A. Ferdman, Admit it, you didn't know this about baby carrots. *The Independent*, January 13, 2016. http://www.independent.co.uk/life-style/food-and-drink/news/admit-it-you-didn-t-know-this-about-baby-carrots-a6810651.html.

p. 265 On the Haber-Bosch process:
- J. W. Erisman, M. A. Sutton, J. Galloway, et al., How a century of ammonia synthesis changed the world. *Nature Geoscience* **1**, 636–639 (2008).

- S. K. Ritter, The Haber-Bosch reaction: An early chemical impact on sustainability. *Chemical & Engineering News* **86**, no. 33 (August 18, 2008). https://cen.acs.org/articles/86/i33/Haber-Bosch-Reaction-Early-Chemical.html.

p. 265 "About 40% of the earth's land is used for agriculture"
- J. Owen, Farming claims almost half earth's land, new maps show. *National Geographic*, December 8, 2005. https://www.nationalgeographic.com/news/2005/12/agriculture-food-crops-land/.

p. 266 On cane toads:
- *National Geographic*, Cane toad. https://www.nationalgeographic.com/animals/amphibians/c/cane-toad/ (September 10, 2010).
- T. Butler, Cane toads increasingly a problem in Australia. https://news.mongabay.com/2005/04/cane-toads-increasingly-a-problem-in-australia/ (April 17, 2005).

pp. 266–268 On gene drives, including applications to mosquitoes:
- National Academies of Sciences, Engineering, and Medicine, *Gene Drives on the Horizon: Advancing Science, Navigating Uncertainty, and Aligning Research with Public Values* (National Academies Press, Washington, DC, 2016). https://doi.org/10.17226/23405.
- J. Champer, A. Buchman, O. S. Akbari, Cheating evolution: Engineering gene drives to manipulate the fate of wild populations. *Nature Reviews Genetics* **17**, 146–159 (2016).
- N. Wedell, T.A.R. Price, A. K. Lindholm, Gene drive: Progress and prospects. *Proceedings of the Royal Society B: Biological Sciences* **286**, 20192709 (2019).
- M. Scudellari, Self-destructing mosquitoes and sterilized rodents: The promise of gene drives. *Nature* **571**, 160 (2019).
- K. Kyrou, A. M. Hammond, R. Galizi, N. Kranjc, A. Burt, A. K. Beaghton, T. Nolan, A. Crisanti, A CRISPR–Cas9 gene drive targeting doublesex causes complete population suppression in caged *Anopheles gambiae* mosquitoes. *Nature Biotechnology* **36**, 1062–1066 (2018).

p. 268 "Each year over 400,000 people die from malaria and more 200 million are infected"
- World Health Organization, World malaria report, 2018. https://www.who.int/malaria/publications/world-malaria-report-2018/report/en/.

p. 268 "Over a billion male mosquitoes with a genetic variation lethal to their offspring have been set free in Brazil, Malaysia, and the Cayman Islands over the past decade"
- S. Milius, Genetically modified mosquitoes have been OK'd for a first U.S. test flight. *Science News*, August 22, 2020. https://www.sciencenews.org/article/genetically-modified-mosquitoes-florida-test-release.

- L. Winter, 750 Million GM mosquitoes will be released in the Florida Keys. *Scientist Magazine*, August 21, 2020. https://www.the-scientist.com/news-opinion/750-million-gm-mosquitoes-will-be-released-in-the-florida-keys-67855.
- R. Lacroix, A. R. McKemey, N. Raduan, et al., Open field release of genetically engineered sterile male *Aedes aegypti* in Malaysia. *PLOS ONE* **7**, e42771 (2012).

p. 268 "Oxitec . . . notes that their deployment has been successful, causing, for example, an 80% decline in the mosquito population at its Grand Cayman test site"
- N. Gilbert, GM mosquitoes wipe out dengue fever in trial. *Nature*, November 11, 2010. http://blogs.nature.com/news/2010/11/gm_mosquitoes_wipe_out_dengue.html.

p. 268 "a 90% decline in Jacobina, Brazil"
- K. Servick, Study on DNA spread by genetically modified mosquitoes prompts backlash. *Science*, September 17, 2019. https://www.sciencemag.org/news/2019/09/study-dna-spread-genetically-modified-mosquitoes-prompts-backlash.

p. 268 "In 2021, the insects were released in the Florida Keys"
- E. Waltz, First genetically modified mosquitoes released in the United States. *Nature* **591**, 175–176 (2021).
- D. Coffey, First genetically modified mosquitoes released in U.S. are hatching now. *Scientific American*, May 14, 2021. https://www.scientificamerican.com/article/first-genetically-modified-mosquitoes-released-in-u-s-are-hatching-now/.

p. 269 "Oxitec's . . . lethal variation is not in a gene that encodes some biochemically specific activity, but in a transcription factor"
- K. Servick, Brazil will release billions of lab-grown mosquitoes to combat infectious disease. Will it work? *Science*, October 13, 2016. https://www.sciencemag.org/news/2016/10/brazil-will-release-billions-lab-grown-mosquitoes-combat-infectious-disease-will-it.

p. 270 "such as DNA-encoded mechanisms to block or even reverse gene drives in a population"
- M. R. Vella, C. E. Gunning, A. L. Lloyd, F. Gould, Evaluating strategies for reversing CRISPR-Cas9 gene drives. *Scientific Reports* **7**, 11038 (2017).

p. 270 "amateur enthusiasts . . . explore biotechnological crafts from their garages"
- H. Ledford, Garage biotech: Life hackers. *Nature* **467**, 650–652 (2010).

p. 270 On the Open Insulin Project:
- https://openinsulin.org/.

Index

16s ribosomal RNA, 135–37

adhesion between cells, 7–8, 78–79, 123, 126–28
Agrobacterium tumefaciens, 245–46
alpha helix. *See* helices
amphiphilic materials, 77, 79–80
antelopes, 11, 168, 170, 273–75
antibiotics, 147–48, 239, 242
anti-CRISPR, 257–59
ants, 173–75, 179
archaea, 76, 100, 134, 248–49, 252
area, scaling of, 162–65, 174, 176
Aristotle, 1

baby carrots, 264
bacteria, 24, 47–49, 62–64, 72, 76, 81, 100–101, 134–54, 158–61, 164, 234–46, 248–49, 258, 264
basal metabolic rate, 184
Bergmann's rule, 176
beta sheet. *See* sheets
Biobank project, 225, 230
biophysics, briefly described, 3–4, 9
Biosphere 2, 144
blindness, 246–47, 257
bones, 110, 125, 156, 169–72, 191
bovids, 168–72
BRCA genes, 229–230, 237
breast cancer, 229–31, 237
Brownian motion, 91–102, 114–15, 175, 204, 249, 250, 254, 272

Calvin and Hobbes, 240
cancer, 73, 107, 111, 200, 221, 229–31, 237, 257
cane toads, 266
carbon dioxide, 173–74, 265
cargo transport in cells, 34, 99–100

CCR5 gene, 256, 262–63
central dogma of biology, 217
channel proteins, 33–34, 126
chaperones, 36–37
Charpentier, Emmanuelle, 251–53
chickens, 227–28
cholera, 48–49, 144–47
chromosomes, 53, 234, 264, 267
cilia, 161
circadian clock, 70, 72
cities, 193–94
clocks, biological, 70–72, 120–23
Clostridium difficile, 139
cockroaches, 166–67
coefficient of variation, 226–27
computers, 41–42, 73, 98, 248–49, 255
CRISPR/Cas9, 239, 244, 247–59, 261–63, 266
cystic fibrosis, 221, 234

Darwin, Charles, 2, 11
deoxyribonucleic acid. *See* DNA
designer babies, 235–36
diabetes, 138, 141, 229–30, 240, 243, 270
diffusion, 95, 97, 100, 110, 112–13, 115, 118, 125, 175. *See also* Brownian motion
diversity of ecosystems, 135, 150–52
DNA, 15–28; attached to beads, 208–11, 218, 223; double helical structure of, 19–20, 241; electrical charge of, 55, 203, 213–14, 251, 258–59; melting of, 21–24; packaging of, 52–59; reptation of, 204; sequencing of, 135–37, 199–219, 248; stiffness of, 52–54
DNA polymerase, 24–25, 27, 202, 207–9, 211–13
DNA sequencing, 135–37, 199–219, 248

Doudna, Jennifer, 251–53, 259
Driesch, Hans, 106
Drosophila melanogaster. See fruit flies

E. coli. See Escherichia coli
ecosystem engineering, 266–70
ecosystems, 135, 143–44, 150–53, 192, 260
egg cells, 228–29, 231–33, 245
electrical charge of amino acids and proteins, 35–37, 251, 258
electrical charge of DNA, 55, 203, 213–14, 251, 258–59
elephants, 171–72, 176, 184, 192
embryos and embryonic development, 7, 73, 105–23, 129, 233, 245, 261–63. *See also* embryo selection; gene editing in embryos
embryo selection, 231, 234–38; ethical issues related to, 236–38
energy: overall rate of consumption of, 183–222; thermal, 91–94
epigenetics, 74
Escherichia coli, 62, 64, 66, 68, 72, 101, 239–40, 248
ethics, 131, 236–38, 261–63, 269
eukaryotes, 48, 76, 99, 100, 134, 146, 244–45
evolution, 2, 3, 38, 171, 176, 178, 260
extracellular matrix, 129, 132–33
eyes, 6–7, 117, 131

fecal microbiota transplantation, 139
feedback, 69–70
fire ants, 179
first-generation DNA sequencing methods. *See* Sanger sequencing
fish, 2–3, 158. *See also* zebrafish
flagella, 101, 148, 158, 161
flatworms, 194
flies. *See* fruit flies
fluid mechanics, 158–61
fluorescent proteins, 32, 66, 114, 245. *See also* green fluorescent protein
fractals, 187–90
fruit flies, 6–7, 48, 107–8, 113–16, 118–19

Galileo, 172
gels, 204–5
gene drives, 266–69
gene editing, 253–59, 261–64, 267–68
gene editing in embryos, 261–64
Genentech, 243–44
gene regulation, 47, 56, 60–74, 222, 255
genes: definition of, 44–47; editing of (*see* gene editing); mapping onto traits 220–31; regulation of (*see* gene regulation)
genetically modified crops, 246–47, 265
genetic engineering, 199, 239–59, 261–64, 267–70
genetic switches, 68–70
genetic variation, 223–227
genome, human, 17–18, 48–50, 65, 76, 206–7, 213–17, 223–28, 270
genomes, 17–18, 20, 47–50, 107, 201, 205–7, 211–12, 215–18, 220, 225, 228–31, 233–35, 244–49, 256, 266–70
genome selection. *See* embryo selection
germ-free animals, 140–42
GFP. *See* green fluorescent protein
glucocorticoid receptor, 34–35, 47
glucose, 64
golden rice, 246–47
green fluorescent protein, 32, 37, 72, 208
Green Revolution, 227
gut microbiota, 134–54

heart rate, 193
height, 221–27, 235, 272
helices, in protein structures, 30–32, 35, 108
histones, 54–57, 73, 203, 218
HIV (human immunodeficiency virus), 79, 256, 263
homunculus, 105
human genome, 206–7, 213–17, 223–28, 270
human genome project, 206–7, 216–17
human population, 227, 265
humerus, 169–71
hydrophilic materials, 35–37, 76, 79, 80, 81, 83, 181

hydrophobic materials, 35–37, 76, 79, 80, 83, 126, 177, 179, 181

Illumina sequencing, 211–12
immune system, 35, 78–79, 102, 149, 256–57
immune system of bacteria, 248–49, 251
immunological synapse, 78–79
Infant Respiratory Distress Syndrome (IRDS), 180–82
insulin, 240–41, 243–44, 270
in vitro fertilization, 231–33, 264
ion channels, 33–34
isometric scaling, 164–68

Kennedy, Patrick, 79, 181
kinesin, 33–34, 100
Kleiber, Max, 185
Kleiber's law, 185–94
Kuhn length, 52, 55, 57
kuru, 39–41

lac repressor, 62–69, 99, 201
lactose, 62–66, 72, 99
lateral inhibition, 117–18
lifespan, 192–93
light: as used in DNA sequencing methods, 203, 208, 210, 212; limits on resolution of, 200
limb bud, 110
Linnaeus, Carl, 1
lipid bilayers, 77, 78, 80, 81, 83, 85–88, 126
lipids, 76–88, 90, 96, 126, 181–82
liquid surfaces, 176–79
logarithmic graphs, 164–71, 185
lung on a chip, 132–33
lungs, 2, 132–33, 167–68, 173–75, 180–82, 221

mad cow disease, 40
malaria, 268–69
mechanobiology, 126
membranes, 33–34, 75–88, 116–18, 126
metabolic interactions among bacteria, 151–52
metabolic rate, 151–52, 168, 184–194

metabolism. *See* metabolic rate
methylation of DNA, 73–74, 218
mice, 48, 49, 67–68, 119, 128, 131, 140–42, 184, 192–93, 218, 245, 253, 255, 262
microorganisms, 9, 134–54, 158–61
Mojica, Francisco, 248–49
Monod, Jacques, 64, 68
Mullis, Kary, 24–25, 27
morphogens, 111–16, 120–22
mosquitoes, 267–69
mucus, 161, 180, 221
mycobacteria, 81
myoglobin, 30–32

nanopore sequencing, 214–15
neurons, 96–98
neurotransmitters, 97–98
nitrogen, 18, 264–65
Notch and Delta, 118
nucleotides, 18–21, 24–27, 44–48, 55–57, 73, 200–212, 216–18, 223, 252

orders of magnitude, 157
organoids, 129–33
organs on a chip, 128–29, 132–33, 150
oscillators, biological, 70–72, 120–23
oxygen, atmospheric, 175
oxygen consumption, 167–68, 173–75, 184–94

PCR. *See* polymerase chain reaction
phase separation, 85–88
phase transitions, 22–24, 86–87
plasmids, 241–42, 244
polar bears, 175
polymerase chain reaction, 21–28, 202, 241
power laws, 183, 193
predictable randomness, 8, 26, 52–53, 79, 90–102, 115, 152–53, 175, 202, 204–5, 224–27, 260
prime editing, 254–55
prions, 40–41
probability, 112, 224–27, 230, 235, 254, 258. *See also* randomness
prokaryotes, 100

promoter sequences, 61
protein folding, 35–43, 95–96
proteins, 29–43
pulmonary surfactant, 181
pyrosequencing, 207–10

R^2. *See* coefficient of variation
randomness, 2, 8–10, 24, 26, 51–53, 79, 90–102, 110, 115, 152–53, 175, 202–5, 224–27, 228, 231, 235, 239, 245, 257, 260, 266–67
random walks, 51–54, 94, 110–12, 175, 224, 254
reciprocal motion, 158–61
regulatory circuits, 7, 57, 60, 105, 110, 113, 116, 120, 127, 135, 148, 260, 269
repression of genes, 62–72, 112, 127, 255, 272
restriction enzymes, 241–42
reverse transcriptase, 218, 254
Reynolds number, 158–61
ribonucleic acid. *See* RNA
ribosomes, 45
rice, 48–49, 246–47
risk of disease, 221, 229–31, 233–35
RNA, 45–47, 61, 71, 73, 114, 135–36, 217, 249–52, 255–58
RNA polymerase, 45, 47, 50, 61–64, 71, 73, 217
RNA sequencing, 217–19

Sanger sequencing, 202–5
scaling, 8–9, 52, 156, 162–72, 178–79, 183–94, 260
scallop theorem, 161
second-generation DNA sequencing, 207–12
self-assembly, 5–6, 9, 20, 29, 35, 43, 55, 60, 76, 88, 119, 135, 142, 150, 181–82, 214, 233, 260
semiconductor sequencing, 211
sheets, in protein structures, 30
Šikšnys, Virginijus, 252–53
single nucleotide polymorphisms, 223–29, 235
skin, 127–28
SNPs. *See* single nucleotide polymorphisms
soap, 77, 80, 181

soap bubbles, 5–6, 80
somites, 119–23
sonic hedgehog, 107–11
sperm cells, 46, 53, 105, 228–29, 231–33, 266
stem cells, 124–26, 128, 130–31, 133
surface area, 5–6, 163–65, 167, 173–77, 180–81, 186–87
surface tension, 176–82, 209
swim bladder, 2
swimming, 101, 144, 146, 148–49, 158–61, 172
synapse, neural, 96–98

temperature, 4, 22–26, 85–87, 91–92, 94–96, 176, 203
test tube babies, 231–32
third-generation DNA sequencing, 212–15
transcription, 45–47, 61, 64–66, 70–71, 74, 127, 217
transcriptional regulation. *See* gene regulation
transcription factors, 46, 65–66, 69, 100, 111–16, 118, 218, 269
transistors, 211, 216
translation, 45–47, 66, 71, 217
translational regulation, 66
trehalose, 81
tuberculosis, 48–49, 81–85

Venter, Craig, 135, 206
Vibrio cholerae, 144–47, 153
viruses, 8, 57–59, 70, 79, 156, 205, 214, 217–18, 241, 245, 248–50, 256
viscosity, 158–60, 221
vitamin A, 246–47
volume, scaling of, 163–65, 173–74, 176, 178

watermelons, 229
water striders, 176
wheat, 227
wildebeest, 185, 273–74

X-rays, 20, 30, 41, 200–201

zebrafish, 119, 140–41, 143–49, 153, 218, 262

INVENTING NADAR

Sign, Storage, Transmission
A series edited by Jonathan Sterne and Lisa Gitelman

INVENTING
NADAR

A History of Photographic Firsts

EMILY DOUCET

DUKE UNIVERSITY PRESS
Durham and London
2026

© 2026 Duke University Press
All rights reserved
Printed in the United States of America on acid-free paper ∞
Project Editor: Lisa Lawley
Designed by A. Mattson Gallagher
Typeset in Garamond Premier Pro and Futura
by Westchester Publishing Services

Library of Congress Cataloging-in-Publication Data
Names: Doucet, Emily G., [date] author
Title: Inventing Nadar : a history of photographic firsts / Emily Doucet.
Other titles: Sign, storage, transmission
Description: Durham : Duke University Press, 2026. | Series: Sign, storage, transmission | Includes bibliographical references and index.
Identifiers: LCCN 2025038504 (print)
LCCN 2025038505 (ebook)
ISBN 9781478038634 paperback
ISBN 9781478033721 hardcover
ISBN 9781478062196 ebook
Subjects: LCSH: Nadar, Félix, 1820–1910 | Photography—France—History—19th century | Aerial photography | Photography, Artistic | Photographers—France
Classification: LCC TR140.N24 b45 2017 (print) | LCC TR140.N24 (ebook) | DDC 778.3/5—dc23/eng/20251117
LC record available at https://lccn.loc.gov/2025038504
LC ebook record available at https://lccn.loc.gov/2025038505

Cover art: Félix Nadar (negative), Paul Nadar (enlargement), *Premier résultat de photographie aérostatique // Applications : Cadastre, Stratégie, etc // Cliché obtenu à l'altitude de 520 m par Nadar 1858* (First result of aerostatic photography, applications: Mapping, strategy, etc. Photograph taken at an altitude of 520 meters by Nadar 1858), 1889. Collection of the Département des Estampes et de la photographie, Bibliothèque nationale de France, Paris.

THIS WORK WAS MADE POSSIBLE WITH THE GENEROUS SUPPORT OF THE PUBLICATION GRANT FUND OF THE LEONARD A. LAUDER RESEARCH CENTER FOR MODERN ART, THE METROPOLITAN MUSEUM OF ART.

CONTENTS

List of Illustrations *vii*

Introduction: "Who Do You Think Is the Greatest
Photographer in the World?" *1*

1. Collecting the Ideas in the Air:
The First Aerial Photograph *16*

2. Patent Priorities:
The First Photograph by Electric Light *44*

3. Sound Reproductions:
The First Photographic Interview *67*

4. Illuminating Infrastructures:
The First Photographs Underwater and Underground *91*

5. When I Was a Photographer, or The
History of Photography in Photographic Firsts *121*

Epilogue: The History of Photography,
as Told to Me by Nadar *139*

Acknowledgments *153*
Notes *157*
Bibliography *197*
Index *217*

ILLUSTRATIONS

As will be discussed throughout this book, the way that many of these images have been referred to has shifted over time. While most photographs were not titled as such when they were made, many contain handwritten annotations, while others have descriptive titles given to them by the author or by collecting institutions. Here and in the captions throughout, all titles, including descriptive titles, have been italicized.

I.1	Félix Nadar, *Atelier de Nadar au 35, boulevard des Capucines à Paris* (Nadar's studio, 35 boulevard des Capucines, Paris), ca. 1861–72	6
I.2	Félix Nadar, *Panthéon Nadar*, 1854	7
1.1	Félix Nadar (negative), Paul Nadar (enlargement), *Premier résultat de photographie aérostatique // Applications : Cadastre, Stratégie, etc // Cliché obtenu à l'altitude de 520 m par Nadar 1858* (First result of aerostatic photography, applications: Mapping, strategy, etc. Photograph taken at an altitude of 520 meters by Nadar, 1858), 1889	17

1.2	Félix Tournachon (Nadar), *Brevet d'invention de quinze ans déposé le 23 octobre 1858 par Félix Tournachon, dit Nadar, pour un Système de photographie aérostatique (n° 1BB38509)* (Patent for a system of aerostatic photography, patent no. 38509), 1858	22
1.3	Félix Nadar, *Maquette d'hélicoptère à vapeur de Ponton d'Amécourt* (Steam-driven helicopter model by Ponton D'Amécourt), 1863	28
1.4	Gustave Doré (creator of the image reproduced), *Vient de paraître: Le 1er n° de L'Aéronaute, moniteur de la Société générale d'aérostation et d'automation aérienne* (Just published: The first issue of *L'Aéronaute*, journal of the General Society for Aerostation and Aerial Automation), 1863	28
1.5	Bertall (Charles Albert d'Arnoux), "Les Nadaréostats," 1863	29
1.6	Félix Nadar, *Ballonneau gonflé servant à Nadar de modèle pour le tableau "Le trainage du Géant"* (Inflated small balloon used by Nadar as a model for the painting *The Dragging of the Géant*), ca. 1869–79	30
1.7	Honoré Daumier, *Nadar élevant la photographie à la hauteur de l'Art* (Nadar elevating photography to the height of art), 1862	38
1.8	Félix Nadar, *Vues aériennes du quartier de l'Étoile à Paris* (Aerial views of the Quartier de l'Étoile Paris), 1868	40
1.9	Félix Nadar, *1ères [Premières] épreuves en ballon: Trois vues aériennes de Paris* (First photographs from a balloon: Three aerial views of Paris), 1868	42
1.10	Félix Nadar, *1ères [Premières] épreuves en ballon: Trois vues aériennes de Paris* (First photographs from a balloon: Three aerial views of Paris), 1868, verso	42
2.1	Félix Nadar, *Le premier essai de photographie à la lumière électrique au Cercle de la Presse scientifique, avril 1859* (The first attempts at photography by electric light at the Cercle de la presse scientifique, April 1859), 1859	48

2.2	Félix Nadar, *La main de M. D*** banquier (étude chirographique), cliché obtenu à la lumière diurne, épreuve tirée en une heure à la lumière électrique* (The hand of M. D*** banker [chirographic study], exposure taken by daylight, print made in one hour by electric light), 1861	51
2.3	Félix Nadar, *Dr. Trousseau*, ca. 1859–60	53
2.4	Félix Nadar, *Portrait d'Henry Delaage, h.e [homme] de lettres // obtenu à la lumière électrique, avec renvois reflets et intermédiaires en glace dépolie // bd des Capucines 35, en 1859, ou 60 ou 61 par moi* (Portrait of Henry Delaage, man of letters // obtained by electric light, with reflectors and frosted glass intermediaries // boulevard des Capucines 35, in 1859, or '60 or '61 by me), ca. 1859–61	53
2.5	Félix Nadar, *Mario Uchard par la lumière électrique* (Mario Uchard by electric light), n.d.	54
2.6	Félix Nadar, *Dr. Trousseau*, 1859	60
2.7	Pierre Petit, *Autoportrait de Pierre Petit dans son atelier, posant avec son matériel de lumière électrique* (Self-portrait of Pierre Petit in his studio, posing with his electric lighting equipment), ca. 1880s	61
2.8	Henry Van der Weyde, *Mrs. Langtry*, 1880	63
2.9	Alphonse Liébert, *Verso of cabinet card*, n.d.	63
3.1	Paul Nadar (photograph) and Félix Nadar (text), "L'art de vivre cent ans" (The art of living a hundred years), 1886	68
3.2	Paul Nadar (photographs) and Félix Nadar (text), "L'art de vivre cent ans" (The art of living a hundred years), 1886	68
3.3	Paul Nadar (photographs) and Félix Nadar (text), "L'art de vivre cent ans" (The art of living a hundred years), 1886	68

3.4	Atelier Nadar, *Cérémonie du centenaire de Chevreul* (Centenary of Chevreul), 1886	79
3.5	Adolphe Willette, "Un siècle!..." (A century!...), 1886	81
3.6	Paul Nadar, *Meeting of Chinese ambassador [Xu Jingcheng] and Michel-Eugène Chevreul at Atelier Nadar*, 1886	82
4.1	Félix Nadar, *Atelier de Nadar au 21, rue de Noailles à Marseille* (Nadar's studio, 21 rue de Noailles, Marseille), 1899	92
4.2	Imprimerie de F. Raibaud, *Plan du port du Marseille indiquant les travaux projetés* (Map of Marseille showing the planned works), 1859	95
4.3	Félix Nadar, *Travaux sous la mer à Marseille (Cap Pinède)* (Underwater work at Marseille [Cape Pinède]), ca. 1899/1900	97
4.4	"Un chantier au fond de la mer" (A construction site at the bottom of the sea), 1900	98
4.5	Studio Nadar Boissonas et Detaille, *Caisson ordinaire: Execution des déblais* (Ordinary caisson: Execution of the excavations), 1898–1904	100
4.6	"Section of one of the caissons used in the work of enlarging the port of Marseilles," 1897	100
4.7	Unknown maker, print accompanying an article titled "À huit mètres sous l'eau" (At eight meters under water), 1897	101
4.8	Adolphe Terris, *Embarcadère des blocs artificiels (Port de Marseille)* (Pier for artificial blocks [Port de Marseille]), ca. 1875–78.	102
4.9	Félix Nadar, *Égouts de Paris: Chambre du Pont Notre-Dame, n°1* (Paris sewers: Notre Dame Bridge room, no. 1), ca. 1864–65	108

4.10 Félix Nadar, *Égouts de Paris: Éclusée n°1* (Paris sewer: Lock no. 1), ca. 1864–65 108

4.11 Félix Nadar, *Égouts de Paris: Rue du Château d'Eau, n°2* (Paris sewers: Rue du Château d'Eau, no. 2), ca. 1864–65 109

4.12 Félix Nadar, *Catacombes de Paris: Façade n°11* (Catacombs of Paris: Facade no. 11), 1861 112

4.13 Félix Nadar, *Catacombes de Paris: Façade n°12* (Catacombs of Paris: Facade no. 12), 1861 115

4.14 Félix Nadar, *Catacombes de Paris: Mannequin n°10* (Catacombs of Paris: Mannequin no. 10), 1861 116

4.15 Henry Duff Linton, "L'ossuaire (D'après la photographie faite à la lumière électrique par M. Nadar)" (Ossuary [after a photograph made with electric light by Mr. Nadar]), 1865 117

4.16 Félix Nadar, *Catacombes de Paris: Crypte n°9* (Catacombs of Paris: Crypt no. 9), 1861 118

5.1 Louis-Auguste Bisson, *Honoré de Balzac*, daguerreotype, 1842 127

5.2 Dujardin (héliogravure), Chardon-Wittman (printing), Photomechanical reproduction of a Louis Bisson daguerreotype of Honoré de Balzac, 1891 127

INTRODUCTION

"Who Do You Think Is the Greatest
Photographer in the World?"

Many superlatives have been used to describe the French aeronaut, biographer, caricaturist, collector, journalist, inventor, novelist, and photographer known as Nadar (born Gaspard-Félix Tournachon, 1820–1910). In addition to the medium of photography, Nadar is repeatedly associated with discovery, invention, and extraordinary events; he is described as a legend, a master, an artist, and a celebrity.[1] But perhaps most prominently, he is described as "first." This recurring claim repeats again and again across Nadar's images and texts, contemporary responses to his feats, and the historical scholarship devoted to analyzing his work. If he is to be believed, Nadar produced

> the *first* aerial photograph,
> the *first* portraits by electric light,
> the *first* underground photographs of the Paris catacombs and sewers,
> the *first* photographic interview (in collaboration with his son Paul),
>   and, in a final act,
> a memoir of photography's *first* century.

Taking advantage of the unusually broad range of Nadar's claims to priority, this book follows Nadar's firsts, tracking how they entered the historical record via photography and other media. To do so, I play with the multiple meanings of invention—to create something new or, alternatively, something based in fiction—to trace the invention of man and medium.

Across this book, I recount the material specificity and narrative complexity of each of these photographic firsts to demonstrate the extent to which no one image, text, or event can singularly constitute a first, though all can be deployed to narrate one. In doing so, I argue that histories of photography have been—and continue to be—shaped by the history of the photographic first, a highly mediated process that canonizes the invention of novel applications of photography as discrete techniques with single authors/inventors.[2] As we will see, photographic firsts are both fact and fiction, simultaneously understood to be the "facts of the matter" and a "narrative of those facts."[3] So while photographic firsts are the products of complex processes of mediation, to their makers, collectors, compilers, and, yes, also to historians, they speak of a fantasy of pure historicity. This was first, unreservedly. But by demonstrating the operation of a novel photographic technology or photographic application, they also model a specific technicity—this technique was new. Photographic firsts bring these two formulations together: this was new and first *then*; here is visual proof.

It should come as no surprise to the reader that there is another, much less straightforward, version of these stories. Rather than a single iconic photograph resulting from technical innovation, photographic firsts are better understood as media constellations that render an image (or set of images) interpretable as first.[4] They are at once images, objects, events, accounts, evidential results of experiments, and documents. They are constituted via texts, museum collections, patent applications, newspapers, and lawsuits, among other cultural forms and spaces. They can be evidence of an experiment, instigators of talk among colleagues, documents of a performance, gifts to professional organizations, arbitrators of patent disputes and lawsuits, and archival documents, among other uses to be described across this book. They are made by many media and often have many authors. They are historically transient, their meaning rearticulated or reaffirmed over time. Photographic firsts speak volumes.

Nadar's firsts were instrumental in the invention of his public persona and historical legacy. However, they are not the photographic genre that he is most associated with. Nadar was also a great portraitist, known for capturing the likenesses of some of the most memorable characters of nineteenth-century Paris. The various iterations of his commercial ateliers were fashionable destinations and the resulting portraits prized commodities; these same portraits are now fixtures in surveys of the history of photography and illustrative material in French cultural histories. Not incidentally, Nadar has held an extraordinarily privileged place in the history of photography. This is, in part, because of Nadar's voluminous production (including an archive), but also because of where he worked and with whom he associated himself. During most of his career, Nadar worked in Paris—the city Walter Benjamin called the "capital of the nineteenth century"—and was friends with many of the artists, poets, and writers now firmly ensconced in the canon of European Modernism.[5] His networks included politicians, actresses, musicians, industrialists, bankers, and journalists, among others. That he took portraits of so many of these individuals has only solidified these relationships to history post facto. But while his portrait production may have been voluminous, the media generated by his exploits in other domains produced an equally prolific—and enduring—archive. In this book, I put Nadar's portraits to the side (with a few strategic exceptions) to demonstrate how his archive offers one starting point for a media history of photographic history.

This archive has prompted a lot of response. Beginning in the 1930s, just twenty years after Nadar's death, well-known and much-cited figures like the German-born French photographer and writer Gisèle Freund and the philosopher and writer Walter Benjamin were registering his influence with frequent reference to his writings.[6] In the 1950s, just after the Second World War, the Nadar studio archive entered the collection of the Bibliothèque nationale de France (BNF) in Paris, making a vast new corpus accessible to curators and historians. The 1960s saw the advent of monographic publications and exhibitions on Nadar in France.[7] The 1960s and 1970s also saw an uptick in Nadar's reputation in North America, including through the efforts of the prominent collectors André Jammes and Samuel Wagstaff. Nadar's photographs were part of the massive and lucrative market for photographs that developed in Europe and North America in the second half of the twentieth century. His presence in exhibitions multiplied, and he was featured in numerous articles and magazines—including the first issues of *October* and *Artforum* to focus exclusively on the medium of photography. Biographies of Nadar abound, in multiple languages.[8] Nadar is, you could say, everywhere you look (photo-historically speaking).

In each period, Nadar's work has been instructively mutable—he was a sociologically minded eyewitness to the industrialization of European society in the 1930s, a master portrait artist of photography's *age d'or* in the

1950s and 1960s, a media theorist for postmodern critics writing in the 1970s, and a speculative commodity in the market for photographs in the latter half of the twentieth century. His market value was a cutting example for Marxist photography historians concerned with dismantling the bourgeois history of photography in the 1980s, and his enduring fame was a relic of the birth of celebrity culture for curators in the 1990s.[9] I could go on.

It is through these acts of collection, exhibition, and writing that Nadar became central to an emergent canon of photographic history in the twentieth century—a process to be elaborated on in the epilogue of this book. But to be known as a "master" of photography's "first century," Nadar had to become Nadar, a figure crafted via the forms of media that he created, collected, and circulated during his lifetime. These media representations were produced and disseminated in concert with cultural and institutional processes of the period—the patent office, civic and technology museums, the scientific press, photographic societies, and international exhibitions, among other forms—to articulate the originality of his many and varied projects. For Nadar, "originality" was most often figured in the form of the photographic first. This proto-avant-gardism would prove extraordinarily intriguing to audiences in the nineteenth and twentieth centuries. To denaturalize Nadar's centrality to the history of photography, I turn away from his portraits and toward these stories. In doing so, *Inventing Nadar* shows that Nadar's legacy and the history of photography—to whatever extent one can consider it to be singular—have been shaped by many of the same processes of mediation.

Much ink has been spilt on Nadar and his pictures, both by the photographer himself and his subsequent historical interlocutors—this author included. In returning to Nadar, *Inventing Nadar* inevitably invents Nadar, again. So why reinforce this historiographical trajectory by returning to this figure? An important question, for Nadar inarguably participates in what Lorraine Daston, speaking of the history of science more broadly, has called "a form of European self-portraiture," in which his contributions to photography are conceptualized as an expansion of territory and imbricated in a history devoted to defining one man's (and, in turn, Europe's) exceptionalism and centrality to the history of the medium.[10] And yet Nadar's firsts are still regularly cited as facts. This book could be written in the corrective, demonstrating how Nadar was, in almost every instance, "not first." However, to do so would risk replicating the kind of historical narratives that created these stories in the first place: one first in place of another, a game of besting that recapitulates Nadar's original impulse.[11] Instead, this book posits that there is value in understanding the history of such charismatic facts, if only to better recognize them when they appear elsewhere.

The metaphor of portraiture, as a rhetorical stand-in for historical representation and its omissions, is particularly useful in a study of a photographer who so prodigiously produced portraits (of himself and others) in many different media.[12] As will be unfolded in each chapter of this book, each first is both a portrait and a landscape: of self, of medium, and of the cultural surround. Nadar's self-invention via his media production parallels the processes that made his "photographic firsts" historically legible, but his firsts are not his alone. They were cocreated with the cultural processes and institutions within which he was embedded, narratives later shored up by the historical narratives crafted by other writers and scholars. So, to repeat myself (as my subject was so very fond of doing), in returning to Nadar, *Inventing Nadar* invents Nadar, again.

FIRST, NADAR

There is a telling symmetry between the production of Nadar's photographic firsts and the self-fashioning of his public identity. Images, documents, and lawsuits are all part of the media production by which Nadar invented himself, narrated his innovations, and, in doing so created the material through which later historians could assign him priority. Born Gaspard-Félix Tournachon, "Nadar" began using the pseudonym early in his career as a writer, journalist, and caricaturist. He quickly established the name as synonymous with his public persona. This moniker was visually identifiable via the flourish of the *n* in his signature, often rendered in a dashing, brilliant red—perhaps most famously as a gaslit sign on the facade of his studio (figure I.1).[13] Much of the detail we know about the production of this "brand" is, as would become a trend, due to my subject's strategic propensity for using priority as fuel for media spectacle, thereby producing profusive documentation. We begin with a lawsuit and end with an image.

In 1856, Nadar sued his younger brother Adrien over the latter's use of the name "Nadar" in his photography studio.[14] The case was brought before the members of the Tribunal de commerce de la Seine, the judicial body charged with ruling on disputes of a commercial nature. The suit ended with Nadar being legally recognized as the sole proprietor of the name "Nadar," effectively trademarking the pseudonym in perpetuity.

1.1 Félix Nadar, *Atelier de Nadar au 35, boulevard des Capucines à Paris* (Nadar's studio, 35 boulevard des Capucines, Paris), ca. 1861–72. Collection of the Département des Estampes et de la photographie, Bibliothèque nationale de France, Paris.

1.2 Félix Nadar, *Panthéon Nadar*, 1854. Collection of the Département des Estampes et de la photographie, Bibliothèque nationale de France, Paris.

Describing his motivation for the lawsuit, Nadar postulated that, in his estimation, great artists (like him, we must assume) are usually in it for fame rather than money, but, if you *were* to consider it from the point of view of material interest, "it is obvious that the artist's name is of greater value the more and better known that name is. In the arts, as in industry and commerce, reputation is money."[15] Therefore, when Adrien took Nadar's name (even with the appended *jeune* to designate the older brother's seniority), he "commit[ed] a usurpation no less reprehensible and prejudicial than the manufacturer who counterfeits the trademark of a neighboring manufacturer."[16] In order to make his case to the tribunal, Nadar authored a report summarizing the constellation of media that affirmed his priority in the use of the pseudonym and the personal celebrity associated with his use of the name. This document included a list of various instances where he had used the name and included references to images like the *Panthéon Nadar* (figure I.2). As we will see, making multimedia arguments shoring up his legacy was something of a specialty of Nadar's.

Nadar's case may have been made to a set of judges in a commercial court, but the documents he called upon to do so had a much wider address, even anticipating future historical interest in his work. Self-published in 1856, the same year as the initial lawsuit, the *Panthéon Nadar* is one of the many documents that the judges used to rule upon the use of the pseudonym. The monumental lithograph features a mass of portraits of historic figures, crowded together in a long snaking line. Nadar presents the image to a reader far into the future, his own long-legged avatar seated leaning against a rock at center right. We have not yet arrived as the future audience Nadar imagined for this print, for, at the time of writing, we remain in 2025 rather than the year 3607. He speculated that, in the middle of the third millennium, someone would be searching for a rare (and highly valuable) version of this lithograph. Each figure is numbered and labeled except for Nadar himself, who, while numbered, is labeled in the legend with three asterisks instead of a name. Nadar places himself as both a participant in this parade of celebrity and as its author, signaled in the repetition of his name three times across the lithograph.[17]

Called upon as evidence in the lawsuit against his brother, the image does more than affirm Nadar's license to the name. In the middle of the nineteenth century, the lithograph acted as a preconceived "pantheon" of important individuals who would be remembered by history. Placing himself among a pantheon of figures of significance, but also as the author of their likenesses and witness to their gathering, Nadar is both an image maker and a historian—roles he will reprise again and again. The case was precipitated by Nadar's desire to begin a profitable photographic studio, a project hampered (he thought) by his brother's use of the name in his own studio. Likewise, the documents and images brought together to make his case prefigure the media-archival role that the photographic image would come to hold in Nadar's archive, as well as the role that commercial value had in each of Nadar's exhortations to priority.

Further underscoring his belief in the commercial and rhetorical value of the Nadar name, Nadar would later formally authorize his son Paul to use the pseudonym so that he might continue to operate a commercial photography studio under the name. While Paul took over ownership of the studio as of 1895 (and practically much earlier), a 1903 letter from Nadar *père* to Paul outlines how Nadar considered this a formal business handover. Nadar writes: "I, the undersigned, give my son Paul Tournachon authorization to take as his name . . . my pseudonym Nadar which remains mine as well."[18] The recent 2018 exhibition *Les Nadar: Une légende photographique* at the BNF underscored the significance of the intergenerational perseverance of the studio in crafting the legacy of the brand "Nadar." Paul not only kept the studio legacy alive but also, arguably, was much more adept at running the business side of the studio, becoming Eastman Kodak's Paris representative for a time (to be discussed at greater length in chapter 3). The perseverance of the photography studio as a business under Paul and later his daughter Marthe (bolstered by the legacy of Nadar *père*) is a large part of why the Nadar studio archive was so well preserved prior to its entrance into museum and library collections. Nadar's firsts bolstered, and in turn were fortified by, the family business.

CONCEPTUALIZING (AND SELLING) THE NEW

Like Nadar's firsts, photography's inaugural first (usually known as the "invention" of photography) also comprised a set of events, images, texts, and exchanges deployed to cement a series of stories about priority. As many have observed, the invention of photography was not a single, isolated event but rather multiple, unfolding episodes that have been described, narrated, presented, displayed, preserved, and reanimated over time by many different actors at many points in history for many different purposes.[19] Nor was "photography" ever a single process, technique, or method. The literature on photography's origins is therefore necessarily historiographical. Rather than attempt to determine when and where a specific technology or technical process was invented, scholars have analyzed how and why photography's invention was inscribed in other media and narrated by a variety of individuals and institutions. Writing in the afterword to his edited anthology of texts published about photography in the storied year of 1839, Steffen Siegel notes that "it is more from convention than necessity that the attempts to find an origin of photography are directed especially towards these durable photographic products that we refer to as 'images.'"[20] Images—absent their textual and social-historical surrounds—are something of a liability in the history of photography, as the search for a single, original photograph may obfuscate historical complexity. Histories of the invention of photography are thus necessarily also histories of photography and *other* media.[21]

So too with photographic firsts. These historical amalgams are primarily legible through the recurring claims to priority by their authors and champions, arguments made, like those surrounding photography's invention, via a multiplicity of media. It is only with ink, pencil, and glue, through performance, memoirs, collection, circulation, and annotation, that photographic firsts come into view. In its discursive and intermedial formation, the first thus shares ground with "the new"—a key analytic for studies of media in the nineteenth century. While the term *new media* was, from the 1990s, used primarily to describe the development of interactive digital media, scholars have since used the concept to examine how so-called *old* media were once themselves new. This scholarship examines the mutual discursive formation of media and media history.[22] Historical writing about new media is refractive, inevitably drawing on sources that mediate media.[23] "The history of emergent media, in other words, is partly a history of history," as Lisa Gitelman has put it, "of what (and who) gets preserved—written down, printed up, recorded, filmed, taped, or scanned—and why."[24] Media history is always already a history of mediation.

But newness (or firstness) is never a historically neutral descriptor for technology or events. As Wendy Hui Kyong Chun puts it, the new is "a historical category linked to the rise of modernity."[25] Written in the 1890s at the apex of the nineteenth century, Nadar's description of the "appearance of the daguerreotype" exemplifies the ideological power of this historical concept:

> Exploding unexpectedly, totally unexpectedly, surpassing all possible expectations, diverting everything that we thought we knew and even what could be hypothesized, the new discovery indeed appeared as, and still is, the most extraordinary in the constellation of inventions that have already made our still unfinished century—in the absence of other virtues—the greatest of the scientific centuries.[26]

In Nadar's words, the "new discovery" of photography is "the most extraordinary" invention in "the greatest of the scientific centuries." As Chun reminds us is true of the "new" in "new media," here the superlative verbiage used to describe photography's newness "more often than not works to erase X's previous existence," creating a narrative of rupture and discontinuity.[27] In Nadar's words, photography is both a discovery and an invention, something newly found and something newly created. Yet, as Nadar knew all too well, inventions are rarely sudden, usually being the result of painfully iterative processes of failure and experimentation. These processes of self-historicization are not specific to Nadar. Photography emerged alongside the cultural techniques and institutions dedicated to the constitution of science, industry, the fine arts, and history itself as productive forces undergirding France's self-image as a great, industrializing nation.

Nadar's priority-obsessed narratives are, I argue, deeply connected to the commercial and legal nature of many of his inventive claims, but also, and no less importantly, to the greater hagiography proclaiming France as an industrial capitalist power and French science as a motor for that system of exploitation.

The commodity status of new media technologies and the entrepreneurial nature of invention in the nineteenth century are thus a key context for these technologies' mediation. Speculation and promotion are, of course, processes that occur in relation to capital. The history of photography during Nadar's period is no exception, as several key texts on the history of French photography have noted.[28] A challenge to the history of art and the history of photography, Abigail Solomon-Godeau noted in 1994, is "that the object of study exists also as a commodity within a market system."[29] Indeed, drawing on this literature among other work, Steve Edwards has recently gone so far as to ask (again), "Why pictures?," proposing that the study of photography history instead be carried out as a kind of business history.[30] As Edwards writes, "It is intriguing to consider why the history of photography took root in the academic discipline of art history, with its focus on images. Initially, most accounts of photography were written as 'invention stories' and 'recipes' with claims to priority taking central place; few of these works had anything to say about photographs as pictures."[31]

Edwards asks why it is that photography hasn't been central for "historians of the trades, for studies of the commercial dealings of the middle class or for accounts of consumption and 'lifestyle.'"[32] It is within this framework that I suggest we place the photographic first. Nadar's priority claims, and the conceptual history of the photographic first that I am tracing more generally, are embroiled within the history of photography as a business and of invention as a speculative method for the accrual of capital—financial or otherwise. Understood within the history of nineteenth-century photographic capitalism, we might understand Nadar's claims and aspirations—to being a photographic "artist," to being first, and to being remembered—as part and parcel of his, and subsequently his son's, entrepreneurialism. Nadar was, by most accounts, not a very good businessman and even spent time in debtor's prison. Ernestine Tournachon, his wife, kept the studio business up and running and assistants taking and printing portraits while Nadar was off undertaking the various exploits that will be described across this book. And yet the commercial studio (in its various iterations) is the most consistent thread across

Nadar's life. Being bad at business, it would appear, doesn't preclude one from devoting one's life to trying to succeed. So it seems worth asking: Is Nadar's legacy to artistic Modernism an aesthetic or a commercial one?[33]

INVENTING NADAR, AGAIN

Firsts also point to what they negate. *Inventing Nadar* undermines any neutrality that might have previously been ascribed to the Nadar archive (a cultural form often taken to be analogous with Nadar the man). By arguing for the invention of Nadar as a process of archive making, I investigate how archival production and, in turn, archivally focused scholarship reproduce these forms.[34] His photographic firsts must be understood in concert with the cultural techniques that reproduced his priority (and, correspondingly, his legacy)—institutions like the museum, the archive, the newspaper, and patent law. That these same institutions have served to enshrine the history of French photography as central to the history of photography writ large warrants further reflection. It is through these processes that Nadar's legacy became part of France's artistic and cultural capital or, as it is often diminutively referred to, its "heritage."

In each chapter of *Inventing Nadar*, I describe a constellation of media that has come to stand as a "photographic first," suggesting that each episode is characteristic of a particular set of material and social entanglements that have characterized the development and historicization of photographic technology. The first three chapters focus on a series of entwined cultural institutions that coproduced Nadar's firsts: the museum, patent law, and the newspaper. The fourth chapter examines how these three often worked in concert, creating the archives of photo-historical inquiry that form the basis for my study. Paying attention to these sites of mediation and circulation cuts against the grain of Nadar's desire for historical fixity, revealing the processes by which he created his legacy. The last chapter examines Nadar's own observations on historical writing. My epilogue concludes with an examination of what was done with Nadar's legacy after his death in 1910.

Chapter 1 explores the history of Nadar's "first aerial photograph." I examine the constitution of the "idea of aerial photography" across a series of texts, including speculative fiction, patent descriptions, and memoirs. Moving from Nadar's experiments in aerial photography to his photographs of model helicopters and his monumental balloon project the *Géant*, I conclude with an analysis of how the years in which the aerial photograph came into focus as a "first" coincide with the years in which Nadar was dedicated to compiling and cataloging his collection on the history of human flight and arranging for its sale to the municipal museum and library of Paris.

Chapter 2 recounts the embroilment of Nadar's "first" photographs by electric light in a dispute between several commercial studio photographers in the 1880s. As with the story of the first aerial photograph, the multivalent "first" photographs by electric light were stabilized not simply by their execution but also, and even more assuredly, via their relevancy to institutional processes decades later (museum collections, in the case of the aerial photograph). In this chapter, I outline how Nadar's first photographs by electric light were put into relief by a patent dispute between the Paris-based photographers Alphonse Liébert and Pierre Petit regarding their respective legal rights to processes designed to facilitate the production of studio photographic portraits by electric light in the 1880s, decades after Nadar's first images by electric light. In tracing the story of Nadar's "first" photographs by electric light, I show how patents for photographic techniques became a lucrative marketing mechanism for studio photographers, due to both the technical innovations that they afforded working photographers and the prestige attached to the ownership of intellectual property.

Chapter 3 analyzes a collaborative project between Nadar and his son Paul: a photographic interview with the centenarian scientist Michel Eugène Chevreul. As in the previous chapters, this work exemplifies the intersection of historicism and media. The interview was simultaneously a celebration of Chevreul's life in science and an occasion for the introduction of the new "instantaneous" Eastman roll film in France by Paul, Eastman Kodak's Paris representative. The photographic interview indexes conversations about media fidelity and historical objectivity across different media, including recorded sound, photography, and photomechanical reproduction. Marshaling new-media discourse for the ideological program of the Third Republic, the photo-text offers a meta-epistemological reflection about how photography, as a new media format, made history rather than represented it.

Chapter 4 begins with Nadar's images of underwater work on Marseille's port to better read his larger body of images of infrastructure illuminated by electric light. In Marseille, Nadar documented an infrastructure project that employed construction techniques developed by French engineers working to expand the port of Algiers, material processes wedded to the development of infrastructure critical to continued French exploitation of the colony. The images were described in newspapers as the first underwater photographs, though Nadar himself did not make this claim. Decades earlier, however, Nadar had photographed the renovations of the Parisian sewers and catacombs—projects central to the infrastructural renovation of the capital as a site of circulation and the multimedia project of representing the city as an icon of industrial modernity. These Nadar proudly claimed as the first photographs of the newly renovated Parisian underground. Rather than examine these images as single iconic prints, I argue that—compiled in photographic albums by the Corps des ponts et chaussées (France's national civil engineering body) or commissioned by municipal engineers, reproduced in newspapers, and collected by museums—Nadar's photographs of infrastructure became infrastructural themselves in that they created the material substrate for future histories, whether written by state engineers, by Nadar himself, or by photography historians like myself.

In the book's first four chapters, I describe the institutions, individuals, and infrastructures that shaped Nadar's archive and existing historical accounts of it. In my final chapter, I analyze the photographer's own summative, albeit eccentric, version of events, offering a close reading of Nadar's book *Quand j'étais photographe* (*When I Was a Photographer*), first published as a series of texts in his son Paul's journal *Paris-Photographe* in the early 1890s and later reprised as a book in 1900. While several of the individual texts are discussed across this book, particularly those which lay claim to various firsts, this final chapter lays out Nadar's ideas about history and the place of media within it. In so doing, I argue that Nadar's book is best read as a meditation on the nature of the relationship between new media and historical experience. I track the diverse references in several of the texts to demonstrate how, in Nadar's estimation, photography was enmeshed in a larger history of science and technology, the development of which was akin to that of history itself.

Inventing Nadar unpacks the assumption that Nadar is, unreservedly, one of the "masters" of photography's early years by following Nadar around the history of photography and emphasizing the media techniques by which his position in this historiography was assured. The photographic first was central to this process of canonization. The epilogue follows these firsts into the century after Nadar's death, examining how these photographs entered the spaces of the Modernist art museum, the North American historiography of photography, the market for photographs, and

North American museum collections. Constituted by a surfeit of media, photographs dubbed "first" are often necessarily hyperlegible to historians. Rather than deploying the archival materials and historical methods described above as evidence for who and what was first, displacing one figure in favor of another, in this book I make the process of making a photographic first itself the subject of historical inquiry. In doing so, I've been continually made aware of how photographic firsts return us to some of the basic problems of history writing itself: the (non)authority of historical documents, the assertion of facts, and the wrinkles in linear historical narratives. In tracing the invention of Nadar and the photographic first, I do not assume Nadar's role as a prime protagonist in the history of photography but instead examine how it is that others might have come to do so—first and foremost, Nadar himself.

1

COLLECTING THE IDEAS IN THE AIR

The First Aerial Photograph

On a label pasted over the blank space of the sky in an aerial image of the place de l'Étoile in western Paris, you can read several phrases: "First result of aerostatic photography," "Applications: Mapping, strategy, etc.," "Photograph taken at an altitude of 520 meters," and "By Nadar 1858" (figure 1.1). Surrounded by thin, black-inked cursive, the name NADAR is rendered in red capital letters. Below this assignation, several streets and landmarks are labeled in red and black ink, a textual trespass on the surface of the photograph mediating the then-still-difficult legibility of the aerial image. But what of that label in the sky? It all at once claims a title ("the first"), a use (tools of the military and the state), an atmospheric location (a great height), an author (Nadar, of course!), and, finally, a date (1858, why not?). Taken as a whole, the image narrates a story that has become commonplace in the history of photography, including in the wealth of scholarship on the aerial (or vertical) image that has emerged in recent years: Félix Nadar took the first aerial photograph from the basket of a balloon in 1858.[1] Or so the story goes. The evidence, like the media, is mixed.

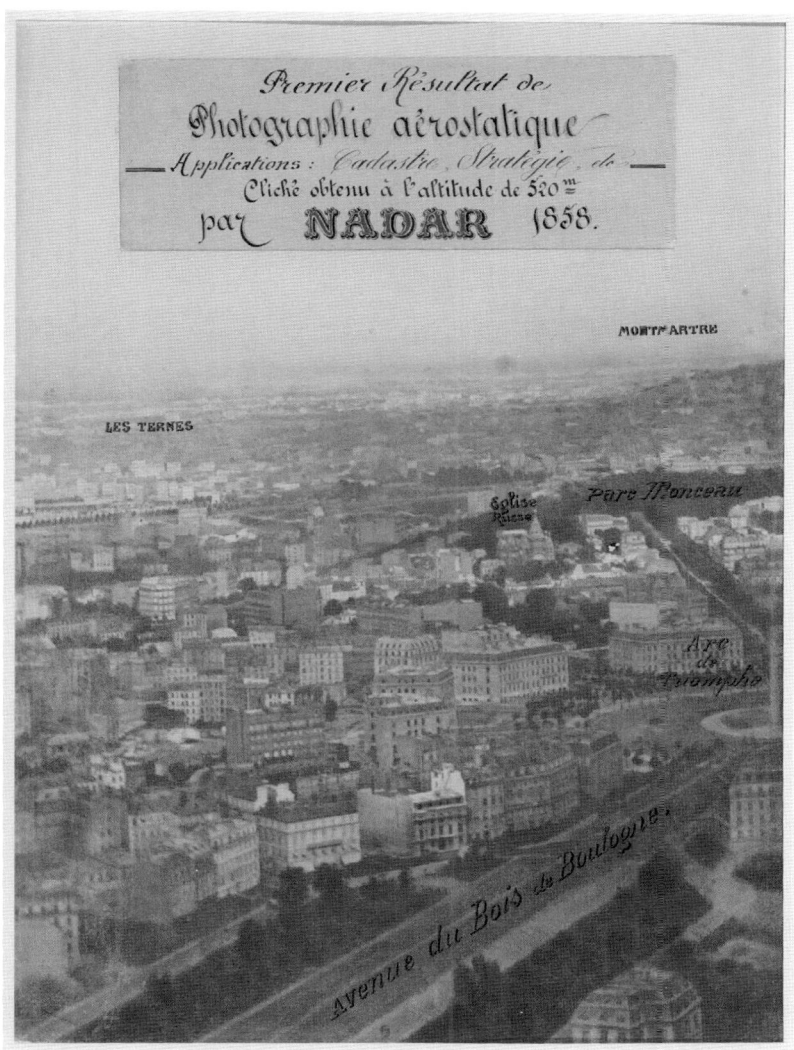

1.1 Félix Nadar (negative), Paul Nadar (enlargement), *Premier résultat de photographie aérostatique // Applications : Cadastre, Stratégie, etc // Cliché obtenu à l'altitude de 520 m par Nadar 1858* (First result of aerostatic photography, applications: Mapping, strategy, etc. Photograph taken at an altitude of 520 meters by Nadar 1858), 1889. Collection of the Département des Estampes et de la photographie, Bibliothèque nationale de France, Paris.

This photograph was in fact taken in 1868, a decade later than most accounts (Nadar's included) date his first experiments with aerial photography. Visual evidence of the 1858 trials—the primordial aerial photograph—has been described as missing, if in fact it ever existed. Historians have questioned the existence of this original image, but its status as a hypothetical photograph has remained uninterrogated.[2] While this single image-object evades apprehension, what has survived is a range of other media that were put to work documenting the invention of aerial photography. Rather than attempt to correct the record or locate firstness in another object, this chapter untangles how the story of the first aerial photograph was produced. This narrative might at first appear to find its primary author in Nadar, but, as the historian of science Simon Schaffer has argued is true of scientific discovery more broadly, when we examine invention narratives, they begin "to look less individual and specific, and more like a lengthy process of negotiation within a set of complex social networks."[3] The first aerial photograph is no different, an inaugural image made recognizable, I argue, through a process of mediation animated by patent applications, personal memoirs, scientific journalism, museum collections, and universal exhibitions. As we will see, this network involved not just people and institutions but also media, including technical drawings accompanying patents, lithographs, caricatures, and, yes, photographs. More than just a deconstruction of a tall tale then, this chapter, and indeed *Inventing Nadar* as a whole, investigates the mediation of invention as a symptom—etymologically, "a departure from normal function or form"—of the cultural experience of novel technological forms.[4] Desirous of an immediate (or unmediated) account of how such innovations have occurred (and would occur in the future), the figures in my study produced a flurry of representations. The first aerial photograph may not be reducible to a single image, but that is not for lack of trying.

Nadar's obsession with all things aerial is well known.[5] What has been less discussed, however, is the relationship between his various roles as a photographer, journalist, historian, collector, promoter, and inventor and the use of visual media in the fields of aerostation and aviation. This chapter tracks Nadar's involvement in this aerial scene, beginning with how the idea of aerial photography was described in the 1850s and 1860s. I read Nadar's own account of his aerial experiments, wrapped in his memoir, published in 1864, about the construction and launch of his balloon the *Géant*, alongside two other documents: a speculative account of the invention of aerial photography written by Antoine Andraud and published in 1855 and the patent text and technical drawings filed alongside Nadar's 1858 patent for photography from the basket of a balloon.[6] These three documents have all mediated how Nadar's contributions to the field of aerial photography have been historicized, and yet, as memoir, speculative fiction, and legal document, they are not uncomplicated sources. Reading them alongside each other reveals the narrative complexity of retrospective accounts of the story of the first aerial photograph, or indeed any photographic first. After describing these three textual representations of the idea of aerial photography, I examine Nadar's use of visual media to represent two other aerial technologies—the helicopter and the balloon—recounting a series of events and images produced in the 1860s, a decade in which Nadar was deeply invested in popular scientific debates around the relative merits of aerostation (machines lighter than air) versus aviation (machines heavier than air). Each of these examples reveals how technical debates played out at the level of the image and signals the impact of this mediation for the retrospective understanding of the history of these technologies—aerial photography included. Finally, I examine the constitution and sale of Nadar's personal collection of materials related to the history of human flight. Sold to a museum devoted to the history of Paris, the musée Carnavalet, in 1898, the collection offers a model for understanding the historicization of the first aerial photograph—exemplifying how Nadar conceived of a history of the aerial and the place of visual media within that history.

Taken as a whole, Nadar's multimedia aerial archive provides crucial context for understanding the historical narrative surrounding his patenting of and experimentation with a system of aerial photography.[7] But the mediation of invention—a symptom, you recall—is of greater significance to cultural history. Throughout this chapter, I linger on moments when Nadar reflects on what such an archive is for. Perhaps contradicting his urge toward self-canonization, the processes of mediation implied by the existence of an archive reveal invention as a refracted and iterative social process, one which he was duty-bound to preserve for future generations. But in canonizing the "first" aerial photograph, Nadar was undoubtedly attempting to historicize his own actions just as much as the creation of a novel application of photography. In the process, the actions of a historical subject and the materiality of a technological form became entangled. This photographic first reveals one way in which an invention (or a technology) becomes an object with an author, a model of media storytelling as history-making that continues to shape public experience of novel technologies.[8]

THE IDEA OF AERIAL PHOTOGRAPHY

Nadar's earliest written account of his experiments with aerial photography is rudely interrupted by a friend's recollection of a curious text which, reportedly and alarmingly, included a description of "aerial surveying by daguerreotype."[9] His friend argues that Nadar was surely not the *first*, for he had just read a book with a reference to the idea of aerial photography days earlier. Shaken, Nadar seeks out said book, even reproducing the front page in the text. His initial alarm subsides, for the text appeared to be speculative fiction, penned in response to a perceived lack in the array of inventions on offer in the Palais de l'industrie during the 1855 Exposition universelle in Paris. The text in question, Antoine Andraud's *Exposition universelle de 1855: Une dernière annexe au Palais de l'industrie*, mimics the format of an exhibition guide, walking the reader through a series of speculative inventions, including new systems of sidewalk paving, covered sidewalks, motorized sidewalks, instantaneous vegetation, and plant meats, as well as, most notably, a system of surveying by daguerreotype, all authored by a certain "M. K."[10] Voiced by Andraud, this fictive inventor recounts his youthful participation in Napoleon I's campaign to remap France: a project to revise the first topographic map of the Kingdom of France commissioned by the Cassini family and created using geodetic triangulation. M. K. (or Andraud) suggests daguerreotypy as a technical shortcut for this lengthy process of triangulation, which took decades and required "an army of engineers, surveyors, chain-measurers, draughtsmen, and calculators."[11] "All this equipment of painful work," which yielded results that were "very incorrect," could be "replaced by poetic walks through the clouds" where one could "look at the earth with the eye of science" as the "earth sends us back its image."[12] As in so many other domains, photography was presumed to be a way to economize an otherwise arduous technical process.

Nadar's 1864 descriptions of aerial photography's potential utility for land surveying and military applications in his book *Mémoires du Géant: À terre & en l'air* (referred to hereafter as *Mémoires du Géant*) are almost word for word those credited to the anonymous inventor of aerial photography in Andraud's text—a tendency toward plagiarism not uncommon for our author.[13] Andraud and Nadar both position aerial photography as a tool of the state, used for military observation and for cartographic surveying. Nadar appropriates (or perhaps doubles) the anonymous surveyor's line of thought to position his own ideas as the germinating seed from which aerial photography would develop, though his nod to Andraud complicates his story. Wondering whether Andraud's publication preceded his patent (it did), Nadar concludes: "There are at certain times endemic synchronisms in human thought, in the moments when our imagination finds itself ordered to respond to our needs. It is in this respect that it was necessary to formulate the saying: *the idea was in the air*."[14] In what had become a common turn of phrase—the idea of an idea in the air was in the air, so to speak—Nadar positions the likeness between his own ideas and that of Andraud's anonymous inventor as indicative of a collective dream of aerial photography.[15] Its invention was inevitable; the ideas in the air had only to become images.

One way that an idea could be made material was through the patent process. Represented by the Parisian patent agent Émile Barrault, Nadar was granted a patent of fifteen years for a system of aerial photography on October 23, 1858. Nadar's patent was accompanied by a descriptive text (as was required by patent law) and a technical drawing. The first article of the standard patent text declares that the patent had been granted "without prior examination, at their own risk and peril, and without guarantee, either of the reality, novelty or merit of the invention, or of the fidelity or the accuracy of the description."[16] This canned disclaimer offers important insight into exactly what kind of technical reality a patent actually represented. In the written description accompanying the patent, Nadar describes a system of photography for making topographic and hydrographic maps and supporting strategic military operations, seeking to turn the basket of a balloon into a photography studio. The camera was to be positioned perpendicular to the earth, affixed either to the side of the balloon basket or to the bottom of the basket (figure 1.2). As in the written description, the technical drawing included as part of Nadar's patent application offers several different options for camera placements, as well as a vague reference to how a string could be used to open and close the camera shutter and a proposed method for transforming the basket of the balloon into a darkroom by controlling the entry of light. No detailed description is provided as to how the camera operator would navigate sensitizing and exposing a collodion plate during the actual flight of a balloon, a highly improbable technical feat. Nadar's patent, like his missing first aerial photograph, remains hazy and imprecise, a document dedicated to cementing his authorship (and ownership) of the "idea in the air" rather than offering evidence of its realization—a prime example of the speculative, mediating role of patents in the history of technology.

The evolution of modern patent law in France in the late eighteenth century redefined the legal definition of invention and in turn radically

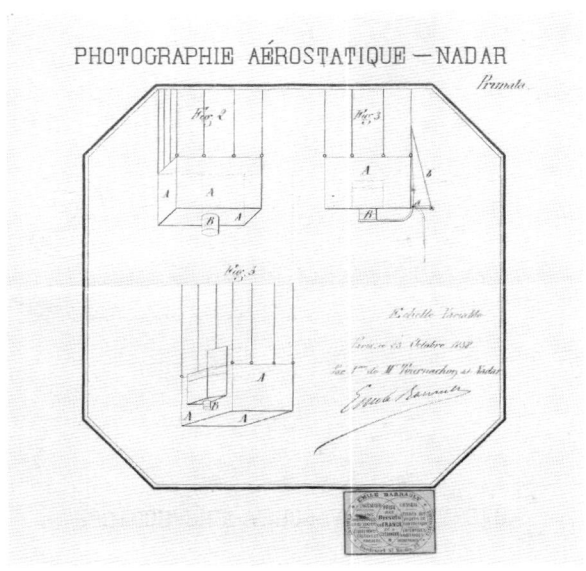

1.2 Félix Tournachon (Nadar), *Brevet d'invention de quinze ans déposé le 23 octobre 1858 par Félix Tournachon, dit Nadar, pour un Système de photographie aérostatique (n° 1BB38509)* (Patent for a system of aerostatic photography, patent no. 38509), 1858. Collection of the archives of the Institut national de la propriété industrielle, Paris.

transformed the representational processes by which inventors were required to articulate their ideas. The first patent act in France in 1791 reoriented the granting of patents from a system of privileges granted by monarchical rulers to a system of intellectual property based on principles of natural law. Patents were now understood as a contract between the inventor and the state. Supporters of the law believed this process would contribute to the public good, allowing for the knowledge of the inventor to pass into the public domain via publication.[17] The short-term monopoly offered by the patent was designed to encourage inventors to publish their ideas. Whereas prior to 1791, inventors were required to demonstrate the operation of their inventions to convince authorities of their utility, after 1791, patent descriptions required lengthy descriptive texts, and in many cases drawings, to describe the *idea* behind their invention—representations that allowed immaterial knowledge to eventually pass into the public domain. These "figures of invention," as Alain Pottage and Brad

Sherman have described the media of patents, are the result of the move from demonstration to description in the patent application process.[18] Indeed, they are central to the philosophy of technology undergirding the modern patent process itself, for, as Pottage and Sherman argue, "it only became possible to visualize and communicate the idea independently of its embodiment when patent lawyers and administrators had access to the specific material media in which inventions could be represented; namely, scale models, formalized texts, technical drawings, and archives."[19] In the process of working out how to represent a technical idea (or retrospectively describe its origin), Nadar and other would-be inventors were enacting a historically contingent understanding of invention as an abstract idea awaiting realization.[20] The media of invention described here are therefore not documents of an inventive act traceable to a precise moment but instead representational efforts to realize an idea. Attempts to define the origin of said ideas accompany moments of profound technological change.[21]

Nadar understood this all too well. "My patents are taken. Now it's just a matter of seeing if I was right," he writes in a glib acknowledgment of this separation between idea and embodiment.[22] And yet his first attempts at taking photographs from the basket of a balloon were unsuccessful, resulting in no permanent image. But in his estimation, the camera had already seen the view from the air, and he just had to unlock the image from its chemical captivity. After numerous attempts during his trials on the outskirts of Paris in Petit-Bicêtre (now Petit-Clamart), Nadar realized that the problem was due to the mechanism of the balloon itself. The hydrogen sulfide that flowed out of the base of the balloon to regulate the height at which it flew was clouding the photographic plate as it was exposed. According to his 1864 account, he decided to make one more attempt, this time leaving the balloon inflated overnight. The next morning, he returned to his balloon to find an unfortunate fog. He forged ahead despite the bad weather. However, the volume of gas filling the balloon had diminished in the night. Nadar quickly emptied the balloon of all unnecessary items, including, reportedly, all his own clothing. After rising to eighty meters, by his own estimation, Nadar exposed the photographic plate. This time, an image! He writes: "The image reveals itself, very faded and very pale, but clear and certain.—It will only be a simple positive on glass, very faint, all stained, but no matter! I emerged triumphant from my improvised laboratory . . . I was right: Aerostatic photography was possible."[23] The idea of aerial photography—a hypothesis about which one could be right—was made material at last, albeit here only in printed text.

These three representations of the idea of aerial photography—Andraud's fiction, Nadar's patent, and Nadar's memoir—all figure aerial photography as an idea awaiting embodiment. But how could this be done? Each text offers variations on a theme. As if anticipating how his text would pollinate Nadar's later writings, Andraud considered the public utility of speculation. He posited that the exhibition imagined in the text "offered refuge to ideas," embodying "future industries in embryo."[24] Filed three years after the publication of Andraud's text, Nadar's patent offered one way to germinate such an embryo. While both patent holder and patent document admitted the idea was speculative, the patent has nonetheless forevermore worked to secure Nadar's claim to the first aerial photograph, despite being a representation of a proposed application of photography rather than a photograph. Finally, Nadar's account of the 1858 trials in *Mémoires du Géant*" retrospectively sculpted his realization of the idea of aerial photography as the inevitable conclusion to the hypothesis set out in the patent. Nadar was right—and first. This formulation of media as the embodiment of technical ideas (or hypotheses) would be central to Nadar's aerial preoccupations in the 1860s. The media representations under examination here are the result of that obsession.

MACHINES HEAVIER (OR LIGHTER) THAN AIR

In 1858, aerial photography was synonymous with photography *en ballon*. Since the late eighteenth century, ballooning had been the subject of much contemporaneous historization. The inaugural public flight of balloons designed by the Montgolfier brothers in France in 1783 produced a profusion of media representations, as did the flurry of launches and experiments that accelerated over the next century. As would later be the case with aerial photography, military affairs and statecraft were imagined as the first applications for aerostation, including the 1794 founding of the French military's Compagnie d'aérostiers, used primarily for aerial reconnaissance during the French Revolutionary Wars. Likewise, the fantasy of a sky without borders prompted by balloon flight was quickly enrolled as part of the French colonial project—for example, as a tool of aerial surveillance on plantations in the French colony of Saint-Domingue, now Haiti.[25] But the view from above remained hazy, and balloon flights were dangerous and unpredictable. As a spectacular first, the Montgolfiers' manned balloon experiments in France demanded narrativization. This first provoked international competition (particularly between England

and France), the canonization of individual inventors as figures of great national and historical importance, and a wealth of writing about the history of human flight—all processes that Nadar was deeply invested in as he began to craft historical narratives about his own firsts.

Nadar's interest in the possibility of human flight was reportedly prompted by witnessing numerous balloon ascensions in his youth, but by the 1860s he was an outspoken critic of the possibility of developing dirigible balloons. In his opinion, a much more promising direction was research into machines heavier than air (what would come to be known as aviation) as opposed to the already achieved yet directionless flight of machines lighter than air (known as aerostation). On his subject, he writes,

> More convinced every day, I am astonished by the blindness and indifference of men before this immense question, the greatest human question in the entire series of centuries,—when it did not even await an inventor like Papin or an explorer like Columbus, when the answer to the problem lay simply in the reasoned application of known phenomena.[26]

The references to Denis Papin, a French physicist and inventor whose research on steam was credited in the development of the modern steam engine, and Christopher Columbus, the Genoese navigator credited with "discovering" the Americas, indicate a curious temporality for the project of the flight of machines heavier than air—positioned, like aerial photography, as an inevitability awaiting realization. As with Papin, the research foundation had been laid; as with Columbus, Europeans laid claim to a so-called discovery. Flight by machines heavier than air would unlock the future of modern aeronautics as steam power had done for navigation and industry but also, in distinctly colonial language, open the skies to conquest by European nations. But for Nadar, the expansion and "reasoned application" of "known phenomena" was, to a large extent, a problem of representation. The documentation and circulation of technical facts and ideas relating to the flight of machines heavier than air would be central to the inevitable invention of a system for dirigible human flight. The texts, images, and events concerned with human flight that Nadar produced across the 1860s all indicate concerted attention on how media might expand the range of "known phenomena." Directly coinciding with the period between 1858 (the "first" aerial photograph) and 1868 (the first aerial photograph redux), these processes of mediation and the theory of technology that they embody provide an essential context for understanding the history of the first aerial photograph.

In a fateful encounter for the fulfillment of Nadar's aerial interests, Nadar met the journalist and writer Gabriel de La Landelle in 1863.[27] De La Landelle had been collaborating with the inventor, numismatist, and archaeologist Gustave de Ponton d'Amécourt on a series of mechanical models designed to lift themselves into the air. Ponton d'Amécourt had developed his models from the known mechanics of the *spiralifère* (a helicopter-like children's toy operated with a pullstring). After numerous experiments with different materials and motors, Ponton D'Amécourt designed a steam-powered model making use of aluminum, the recently discovered and incredibly light metallic substance.[28] The inventor sought to give his project a name, eventually settling on the word *hélicoptère*, a derivative of the word *hélice* or propeller.[29] Thus the helicopter, a machine whose history is darkened by the ensuing violence its use would enable, was "born."

That the story of another "first" is necessary to unfurl the story of Nadar's aerial preoccupations is not incidental, for, though Nadar was not the inventor of the helicopter, he was a fervent promoter of the idea undergirding its mechanics: the flight of machines heavier than air. As if paralleling the production of the first aerial photograph as a historical artifact—through promotion and narration, via representation and collection—Nadar quickly sought to fold Ponton d'Amécourt's helicopter under the sign of his own name.[30] Inspired by the small successes recounted by de La Landelle on behalf of Ponton d'Amécourt, Nadar charged forward in support of the helicopter project. Outlining his intentions, he writes, "I wanted to kill aerostation by replacing it with purely mechanical apparatuses:—I resolved to ask aerostation itself for the means to create the new agents that would kill it."[31] In addition to photographing the helicopter models in his studio, he founded a society to support helicopter research: L'association libre pour la navigation aérienne au moyens d'appareils plus lourds que l'air (Association for Aerial Navigation by Machines Heavier Than Air). The association held its first meeting on July 30, 1863, and according to Nadar's account, "Five or six hundred people came; the main scientific bodies, the railway administration, the press, high society, finance, and even the Institute [Institut de France]!"[32] Retrospectively describing the meeting in 1882, de La Landelle writes, "On the 30th of July 1863, he [Nadar] summoned a large, elite audience to his large photography studio, before whom our pretty little models, powered by clockwork springs, demonstrated the reality of the principle of aviation."[33] A form of public witnessing, this demonstration launched the "idea" of helicopters in more ways than one.

Nadar's photographs of the helicopter models (see, for example, figure 1.3) were likely taken while they were in the studio for demonstration at the first meeting of the association, a visual extension of their public presentation.[34] Stunning and playful in their austerity, the photographs offer a hypothesis to the viewer: Machines like these could and would fly. Suspended with string, the models are photographed as if already in flight. Somewhat paradoxically, the strings remain visible, and the models appear against the plain backdrop recognizable from Nadar's well-known portraits. Nadar's photographs present the models for both public speculation and scientific understanding. Yet the mediatic success of Nadar's photographs of the models are better seen as hedged upon their capacity for remediation. Consider the 1863 poster promoting the journal of the Association for Aerial Navigation (figure 1.4).[35] Drawn by the artist Gustave Doré, the image would become the frontispiece for *L'aéronaute*. Imagining a future for the heavier-than-air flying machines, Doré depicts the model helicopters that Nadar photographed in flight amidst a landscape of fields and clouds. Contrasted with a balloon in the upper-left-hand corner and a steam train traveling across the horizon, the models are inserted within a larger narrative of mechanical locomotion. The engraved helicopters appear more dirigible than the balloon floating whimsically off in the distance, more agile than the train anchored to the terrestrial sphere by the track.

This mobility is significant for several reasons. Describing the widespread public interest in photography in its early years, Nadar would later characterize the refractive quality of modern media this way: "We can say, in modern parlance, that a thing is 'launched' when caricature becomes aware of it and targets it."[36] Judged by these criteria, Nadar's photographic campaign was indeed successful. His photographs of the helicopter models quickly became points of reference for the illustrated press, further working to firmly associate Nadar's name with helicopter research. The helicopters were a popular subject for caricaturists, seen in several engravings in the *Journal amusant* to which Nadar was also a frequent contributor (figure 1.5). Renamed as "Nadaréostats," Ponton d'Amécourt's helicopters become a character in Nadar's extended aerial universe. Most frequently visualizing futures of public transportation with flying carriages crowding the city, these caricatures are both mocking and imaginative, riffing on Nadar's images with a mixture of disbelief and hope.

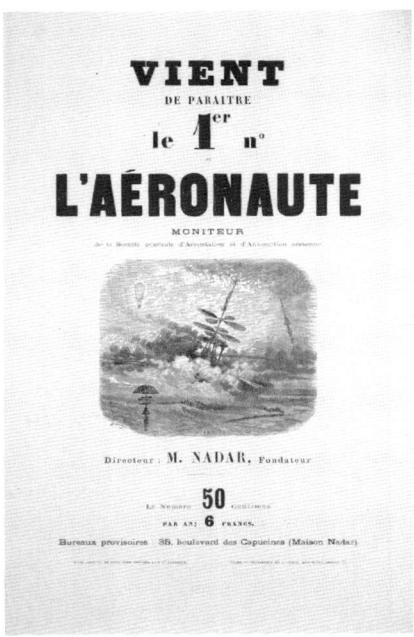

1.3 (*above left*) Félix Nadar, *Maquette d'hélicoptère à vapeur de Ponton d'Amécourt* (Steam-driven helicopter model by Ponton D'Amécourt), 1863. Collection of the Département des Estampes et de la photographie, Bibliothèque nationale de France, Paris.

1.4 (*above right*) Gustave Doré (creator of the image reproduced), *Vient de paraître: Le 1er n° de l'Aéronaute, moniteur de la Société générale d'aérostation et d'automation aérienne* (Just published: The first issue of *L'Aéronaute*, journal of the General Society for Aerostation and Aerial Automation), 1863. Collection of the Département des Estampes et de la photographie, Bibliothèque nationale de France, Paris.

Photographic mediation afforded the helicopter models a mobility that was not only material but imaginative.[37] Disarticulated from their weighty presentness in material form, the models' transformation via photographic technics catalyzed a further chain of mediations. Unbound from the restraints of earthly physics, the photographs were "launched," as Nadar put it, into a more aerial plane of representation via caricature. The helicopters were also firmly pitched, in images at least, as Nadar's "ideas."[38] This process mirrors the articulation of Nadar's aerial photo-

1.5 Bertall (Charles Albert d'Arnoux), "Les Nadaréostats," *Journal amusant*, no. 403 (September 19, 1863): 1. Collection of the Département des Estampes et de la photographie, Bibliothèque nationale de France, Paris.

graphic experiments *as first*, a process of mediation that, like that promoting the helicopter, could not be produced by a single inventive act. Instead, the technical and, it follows, historical significance of these artifacts was promoted via a constellation of media, texts, and events.

As with the authorial annotation on the surface of the first aerial photograph, technical debates frequently played out in the mise-en-scène of Nadar's photographs. Consider, for example, his photograph of a partially inflated balloon inside the studio (figure 1.6). Staged as a reference

COLLECTING THE IDEAS IN THE AIR   29

1.6 Félix Nadar, *Ballonneau gonflé servant à Nadar de modèle pour le tableau "Le trainage du Géant"* (Inflated small ballon used by Nadar as a model for the painting *The Dragging of the Géant*), ca. 1869–79. Collection of the Département des Estampes et de la photographie, Bibliothèque nationale de France, Paris.

image for a painting by Nadar's brother Adrien Tournachon, a balloon is pictured in repose, a broom holding its partial deflation just so. In contrast, Nadar's photographs of the helicopter models designed by Ponton d'Amécourt show them tied up with string as if in flight. Defying gravity, pinning one theory down while propping the other up, Nadar's mediatic refutation of the possibility of navigable balloons would characterize the media campaign surrounding one of his largest projects: the construction and multiple launches of a giant balloon designed to raise funds for research into flight by machines heavier than air. Or in the words Nadar used to describe the triumph of subsequent photographic processes over the daguerreotype: "This would kill that."[39]

The most salient means for this assassination were the techniques of large-scale media spectacle centered around the construction and disastrous flight of a colossal balloon, a vessel referred to as *Le Géant*.[40] The spectacular ascensions and wayward travels of the balloon were planned to raise funds for constructing flying machines heavier than air. By pushing balloon technology to its scalar extreme, Nadar's balloon became a spectacular example of the inherent problems of aerostatic technology—above all, its imprecisions in steering and vulnerability to atmospheric disruptions. By investing in a massive ballooning experiment and documenting its failure, Nadar ended up falsifying the theory of navigable ballooning—a theory which, as Nadar continually emphasized, had already been multiply refuted by the litany of balloon crashes and fatalities prior to 1863.[41]

Built by three hundred women and men under the supervision of aeronaut brothers Louis and Jules Godard and using over 20,000 meters of silk, Nadar's balloon was to be no exception in this grisly prehistory of disasters.[42] Not unlike the wayward trajectories of its predecessors, the balloon's first two journeys, both in October 1863, ended in dramatic denouement. Nadar envisioned reaching Russia in the aerostatic vessel, but such distances were beyond the scope of the balloon and its occupants.[43] Launched on October 4, 1863, the first ascension lasted only a few hours. The balloon crashed violently just outside of Paris, near the town of Meaux; basket and passengers were dragged for a mile before coming to a complete stop.[44] Although the second flight, launched on October 18, managed to stay aloft for much longer, it too ended in a dramatic and gruesome crash near Hanover. Nadar and his wife were among those seriously injured. The spectacular failure of these balloon voyages provided Nadar ample material for his ongoing campaign to "kill" ballooning and render aerostation historically obsolete.[45]

Despite grave injury to his co-aeronauts, Nadar must have been perversely pleased by the *Géant*'s spectacular Hanover crash in mid-October 1863, for this calamity yielded potent evidence against aerostation. His own voyage had confirmed his hypothesis: Balloons were disasters waiting to happen. In this, Nadar participated in a longer tradition of picturing accidents as a means of critiquing technology.[46] And yet the case of the

Géant is particularly instructive insofar as representations of its accidents were crafted explicitly to advance a rival theory of human flight. Not simply the subject of representations and a site for the physical and discursive production of media, Nadar's balloon also operated in anticipation of—and found its raison d'être in—the future archiving of its own failure.

MAKING HISTORY

Whatever is made of Nadar's priority in aerial photography and his spectacular use of media to "kill" ballooning, his production of aerial media cannot be understood outside of a final, lesser known supplement: the collecting of forms of media related to aerostation and aviation.[47] As he noted in *Mémoires du Géant*, Nadar knew of and took archival inspiration from the collector, inventor, and aeronaut Jules-François Dupuis-Delcourt.[48] A confrère of the Montgolfier brothers, Dupuis-Delcourt presented demonstrations of a "*hélice*" (or propeller) in the Orangerie of Luxembourg. Much like Nadar later would, he founded the Société aérostatique et météorologique de France, published the *Manuel de l'aérostier*, compiled a comprehensive collection of materials relating to the history and development of aerial flight, and left at his death in 1864 an unfinished manuscript titled "Traité complet, historique et pratique des aérostats."[49] Described by Nadar as "the most curious, the most instructive, the only Aerostatic museum in the whole world," Dupuis-Delcourt's collection was itself acquired in part by Gaston Tissandier, another major collector of aeronautica active in Nadar's circles.[50]

Initiated around 1864—simultaneous to the years in which his involvement in flight research intensified and just following the period in which his earliest experiments with aerial photography occurred—Nadar's own collection was substantial. In 1897, by then living in Marseille, Nadar wrote to the municipal council of Paris to offer the city his collection of "documentary pieces of all kinds (books, prints, autographs, portraits, various objects, etc. . . .) relating to the history and technology of aerial navigation before and since the Montgolfier era."[51] The council offered Nadar 8,000 francs for his collection of over five thousand items.[52] At the time of its acquisition, a journalist described the purchase as

> a very curious collection, probably the only of its kind in the world . . . This collection consists of: a library offering pretty much everything that has been printed since Gutenberg on the subject of the flight of

birds, on aerostation, on aerial navigation; piles of cardboard boxes where the countless prints, manuscripts, memoirs, autograph letters, drawings, naïve or accomplished sketches, staid or feverish sketches of researchers are methodically grouped from all over the world; newspapers and magazines in bundles: that is to say, the universal chronicle of aeronautics—and there, incidentally, the documented history of the siege of Paris—the earthenware, the jewellery, the medals . . . the mechanisms, the coated taffetas, the fragments of aerostats or aircraft that have burst or burned up, or flapped their wings and sunk . . . ; in short, all pious relics of this duel through the ages between human audacity and the sphinx.[53]

Nadar held his "relics" in high esteem. Correspondence dating from as early as 1881 with Jules Cousin (the first librarian of the Bibliothèque historique de la ville de Paris [BHVP]) indicates that Nadar had long considered the paired musée Carnavalet and the BHVP a sufficiently prestigious home for his collection.[54] A newspaper article from 1897 notes that Nadar felt his collection might have been better housed at the Conservatoire national des arts et métiers (CNAM)—a potentially apt home for a collection so devoted to the history of technology.[55] While his final decision was likely financial, Nadar's consideration of the various possible homes for his collection highlights the numerous histories—urban, national, scientific, and art-historical—into which aviation could be inscribed.

In an extensive catalog organized for the sale, Nadar presented his collection as a series of "Documents relatifs à l'histoire de la navigation aérienne."[56] The collection featured a large array of images signaling the scope of the aerial imagination in the eighteenth and nineteenth centuries. Prints documented the first balloon ascensions in England and France, as well as representations of speculative aerial vehicles.[57] As noted in the quotation above, the collection also contained items documenting the use of balloons and pigeons during the 1870–71 siege of Paris, including an example of pigeon-transported microphotography. Beyond the graphic art and decorative objects, Nadar also collected letters and documents relating to aerial innovation, such as patents and prototype sketches. The collection also incorporated Nadar's personal library, photographs, and documents about the history of human flight, as well as materials related to Nadar's "Manifeste de l'automotion aérienne."[58]

But this archive, crucially, was not simply Nadar's creation. The multitude of media forms included in this collection index the aerial fervor of the nineteenth century. Alongside images, objects, texts, and documents by a variety of authors, many of the letters, plans, and publications included in the collection were likely acquired because of Nadar's 1863 publication of his "Manifesto of Aerial Locomotion." Minutes from the Société's meetings included in the sporadic issues of *L'aéronaute* repeatedly refer to the overwhelming influx of letters, publications, drawings, and plans pouring in in response to the manifesto.[59] Today, the collection is a fascinating resource on the early history of ballooning and aviation, but its structure and scope also offer an opportunity to reflect on the role of visual media in the history of technology—and, in turn, the ways the methods of this book remain entangled with the actions of its subject. Understood here as a compelling complement to Nadar's aerial-mediatic campaign against ballooning, the collection also exemplifies the kind of mediation that has shaped the story of the first aerial photograph and that of the other firsts described throughout this book. But as I am arguing is true of the photographic first more broadly, mediation is not simply, or not only, a component or process in the construction of a particular narrative. It is also an index of change, a way of making sense of, or attempting to explicate, how something has occurred—in this case, human flight. The unruliness of the collected material—no single narrative, many dead ends—attests to the extraordinary expansiveness of the cultural category of invention and its historical import.

For Nadar and his fellow aeronautic enthusiasts, creating and gathering an archive of media on *navigation aérienne* was thus not just about the past. It provided the conditions of possibility for interconnecting past, present, and future. Consider the following three commentaries, all written in the foundational years of the aerial-archive project by collaborators of Nadar. All posit a strikingly nonlinear model of transhistorical technological collaboration. The first is by physicist, mathematician, and astronomer Jacques Babinet. In 1864, as part of his introduction to *Mémoires du Géant*, Babinet endorses Nadar's contribution to the field of aviation science with these observations:

> It is now an idea fallen into the public domain, namely that with a known mechanism, the propeller, and a sufficiently powerful engine, steam, it is possible for man to rise, sustain himself, and progress, and even, up to a certain point, to move in the opposite direction by means of a current of air, that is, by a moderate wind. . . . What then, in the question of human flight, is the specialty of Mr. Nadar, who

repudiates any claim of priority for the mechanical idea? It is this: It is quite simply to put into practice what he conceived, with everyone, I mean with all those who think... there is more merit in realizing a useful idea than in inventing it. Since here the idea already belongs to the public, I don't see how we can argue against the merit of Mr. Nadar's aerial flight, if he succeeds in putting it into practice. He has the propeller and steam, but he also has faith, which is an even more powerful engine.⁶⁰

A year later, in the preface to Nadar's 1865 *Le droit au vol*, novelist George Sand sets out a rhetoric of innovation that affords distinct but interconnected roles for "thinkers" and "makers." She writes:

> Honour to those men whose initiative sets free the first hypothesis, the sovereign induction, from the chaos of dreams, from the thousand gropings of the imagination struggling with the unknown! When those great and generous minds have succeeded in well posing the question to be solved, they have already made a grand step: they have opened the way.
>
> Afterwards come the men of application, not less useful, not less admirable, who, by clever and patient experiments, proceed from the hypothesis to the discovery. From that time, Genius becomes a material force; and the idea which was only a promise becomes the real benefit with which the human race enriches itself.⁶¹

And in a report that same year, de La Landelle argued that

> collecting, classifying, and propagating *the tradition* is one of the principle duties of our **Society for Encouragement**, because nothing is more ENCOURAGING for researchers than the certainty that their research is no longer chimerical, since the invention has already been made, it is only a matter of finding what has been lost. Let us be archivists of progress. Let us resurrect the past, awaken the present, and enliven the future. Let us use our influence to call out to all researchers who still have the weakness to isolate themselves. Let us demonstrate to them that *collaboration is power* and that *there is nothing new under the sun*, not even the art of mechanically rising in the air. The light of history will this time be persuasive, and those who remain sullen and obstinate will finally understand that *progress* excludes only the selfish: "misfortune to those who wish to remain alone!"⁶²

How are we to understand these statements by these self-designated "archivists of progress" and their implications for Nadar's archive and the story of the first aerial photograph? Claims that "the invention having already been made, it is only a matter of finding what has been lost"? That "there is nothing new under the sun"? That Nadar conceived mechanical flight with "the whole world"? That "genius becomes a material force" through the transtemporal "collaboration" of "imagination" with "application"? Recall Nadar's invocation of the idea of aerial photography as one that was "in the air" and about which he just had to "[see] if he was right."

All point to long-standing models of creativity in the arts and sciences that distinguish between an idea and its execution and, later, inform its canonization as an object or event of cultural import. Early modern theories of artistic practice separated the generation of such *invenzioni* from their final material form.[63] A master artist comes up with the idea, which is the most essential moment in the process of creation. Turning that idea into matter is less important, which is why masters can delegate much of their painting to assistants and why copies of a composition can be just as valued as the original.[64] The theory of scientific discovery emergent in the early nineteenth century likewise distinguished between discovery and justification or between the "eureka moment" and the later establishment of said discovery as fact—a model similarly devoted to a form of creative genius. This "assent to the matter of fact" is, as Schaffer has observed, a "complex enterprise" that "generates objects which are then labelled as discoveries."[65] Later, "the story of that process is rewritten" and, in a final interpretative act, "the lengthy enterprise is telescoped into an individual moment with an individual author."[66] Like a photographic first (and indeed all analog photographs), invention is something that needs to be fixed.

Such expansive conceptions of invention were certainly alive for Nadar and his compatriots. The definition of *inventer* in both the 1835 and 1878 editions of the *Dictionnaire de l'Academie Française* begins: "To find something new, ingenious, by the force of one's spirit, of one's imagination."[67] Perhaps this is what de La Landelle means when he writes that "the invention has already been made" (that is, the *idea* of mechanical flight had existed for centuries) and why all three authors present narratives that divide the process of invention between at least two roles: those who come up with new ideas and those who realize those ideas.[68] To quote Babinet again: "There is more merit in realizing a useful idea than in inventing it." And to repeat Nadar on the idea of aerial photography: "My patents are taken. Now it's just a matter of seeing if I was right."

This crowdsourced, transhistorical, and "faith-based" understanding of invention of course had its risks: Unrealized ideas depend on the work of future generations for their realization—and, no less importantly, on citizens of the future to recognize the contributions of earlier thinkers. In the minds of this group of collaborators, scientific research was thus imagined in explicitly historical, archival, and museal terms. De La Landelle's call in 1865 for present-day researchers to bring back the dead ("Let us resurrect the past") recalls an object mentioned in his 1863 *Aviation, ou navigation aérienne*. Describing Henry Bright's design for a propeller-powered parachute (patented in 1859), de La Landelle laments that this "trinket" was stuck "under glass."[69] Trapped there "perfectly asleep under its dome," as de La Landelle put it, "its existence was ignored and without influence on the march of invention."[70] Faced with the prospect of this neglected invention, de La Landelle feared in turn that the helicopter project on which he was collaborating with Ponton d'Amécourt would face a similar fate, forgotten both to history and science, "joining under glass, the parachute-helicopter of the Patent Museum." But such museal oblivion was averted: "Happily, I repeat to you, Nadar arrived!"[71] De La Landelle cheerfully lists the immense press interest in the helicopter models generated by Nadar's publicity efforts, thereby assuring their place in collective remembrance through the production of representations in various media.[72] This final burst of optimism points—again—to the role of media in that "complex enterprise" of scientific justification observed by Schaffer. The story of the movement between idea and fact is nothing less than an account of the nature of scientific and, indeed, historical change: a mediatic storytelling process understood by Nadar and his compatriots as a duty to future generations—though not one without benefits for the construction of a personal legacy.

AERIAL PHOTOGRAPHY, LAUNCHED

Let us return to the media surrounding the "first" aerial photograph, for that is what Nadar dedicated himself to over the remaining years of the nineteenth century. "Affirmed *first*, despite its imperfection, the practical possibility of aerostatic photography: this was above all what I had aimed at," writes our erstwhile aeronaut in 1893.[73] This first contained a caveat in the form of a footnote: "An honorable scientific journal—*Les Inventions Nouvelles*"—had deigned to claim that "the first aerostatic negative was obtained in 1881—by Mr. Paul Desmarets."[74] In the same footnote, Nadar lays out the material of his argument: "The incontestable notoriety of our

1.7 Honoré Daumier, *Nadar élevant la photographie à la hauteur de l'Art* (Nadar elevating photography to the height of art), 1862. Collection of the Département des Estampes et de la photographie, Bibliothèque nationale de France, Paris.

print which had figured in several exhibitions *much before* 1881 and the date of our patents responded in advance to this unexpected assertion, without any need to refer to the year of Charivari where everyone can find Daumier's lithograph reproduced above."[75] The "incontestable notoriety" of Nadar's experiment had been assured through its mediation: prints displayed in exhibitions, patents filed, and a lithograph featuring Nadar as an aerial photographer. Despite the profusion of aerial media produced in this period—including several other successful experiments with aerial photography—Nadar remained convinced that his priority would remain evident.[76] After all, he had proof.

As the final entry in Nadar's litany of "proof," Daumier's lithograph (figure 1.7) shows Nadar *en ballon* "raising photography to the height of art." We see Nadar with camera in hand in the basket of a balloon emblazoned with the label *"photographie Nadar,"* the signature rendered with Nadar's characteristic flourish on the *N*. The inclusion of Nadar's trademark reinforces Nadar's claim, all the while poking fun at his outsized persona, the balloon studio in the clouds nodding to his inflationary ego, rising above the profusion of Parisian photographic studios beneath him. In Daumier's joke, aesthetic hierarchies find a parallel in the aerial view, both (uncertain) pathways to knowledge or understanding. Daumier's image is no documentary proof, but recall that, for Nadar, something is "launched" when caricature becomes aware of it. Aerial photography was ushered into the cultural sphere via Daumier's caricature, with authorship firmly attributed to Nadar.

But what of the images that were supposedly exhibited *"much before* 1881"? Nadar described the first aerial photograph as being made in 1858 in Petit-Bicêtre (now Petit-Clamart), but the images that remain clearly depict the Arc de triomphe at the center of the place de l'Étoile in Paris. As I noted earlier, the set of images now known as the "first" aerial photograph were taken in 1868 in the bois de Boulogne from a tethered balloon belonging to the aeronaut Henri Giffard. The exhibition, publication, and collecting history of the 1868 image from the bois du Boulogne further demonstrate the collapse of the 1868 trials at the bois de Boulogne with the earlier 1858 date. For example, an image exhibited in the "Maps and Geography and Cosmography Apparatuses" section of the 1878 Exposition universelle in Paris was described as "a planimetric survey, obtained by photography, from the basket of a balloon; reproducing the Arc de triomphe d' Étoile and surroundings ... this first attempt of aerostatic photography could be applied to mapping, military strategy, etc. Nadar's latest experiments have allowed him to guarantee an absolute sharpness of focus."[77] The designation of the exhibited image as the "first attempt" and the description of the Arc de triomphe and its surroundings would indicate that the image exhibited was the familiar enlargement of the 1868 negative masquerading as the 1858 "first attempt." Again, for the Exposition universelle in Paris of 1889, the "first aerial photograph" is presented in the form of an enlargement from the 1868 series—now presented via the annotated form that would come to stand in as *the first*. The image was enlarged by Nadar's son, Paul, with one frame selected from a set of eight images taken on a single plate aboard Henri Giffard's tethered balloon (figure 1.8).[78] The image was enlarged and again annotated as if it were evidence of "the first" aerial photograph, this time even including the 1858 date. The first aerial

COLLECTING THE IDEAS IN THE AIR 39

1.8 Félix Nadar, *Vues aériennes du quartier de l'Étoile à Paris* (Aerial views of the Quartier de l'Étoile Paris), 1868. Collection of the Département des Estampes et de la photographie, Bibliothèque nationale de France, Paris.

photograph was evidently malleable: Exhibition display provided cause for remaking—literally—priority arguments via the image, a technique like that deployed in the battle between machines heavier and lighter than air.

Yet another reference to the 1868 print was made by fellow photographer, aeronaut, and collector Gaston Tissandier in his 1886 book *La photographie en ballon*. Tissandier describes Nadar's earlier experiments, citing heavily from *Mémoires du Géant*, but emphasizes that the 1868 experiments resulted in the first *clear* image.[79] This reference would suggest that no legible negative or print from 1858 was known when Tissandier wrote his historical account of the development of balloon photography in 1886. Seven years later, the first publication of Nadar's essay "The First

Attempt at Aerostatic Photography" in the spring of 1893 in his son Paul's journal *Paris-Photographe* (and later included in his book *Quand j'étais photographe*) includes a reproduction of the annotated enlargement made by Paul. The first aerial photograph does not appear to have been lost but *made*.

The image's uncertain origin also made its mark as it entered museum collections. Nadar's 1893 correspondence with Aimé Laussedat, director of CNAM, indicates that the dating of Nadar's aerial photographs was indeed a subject of concern. In a letter from June of that year discussing the acquisition of a print of the "first aerial photograph," Laussedat attempts to clarify whether the image was taken in 1855 or 1865. Laussedat notes the inconsistency of the various accounts, asking whether it was not in fact the image of Petit-Bicêtre that dated to 1855.[80] A subsequent letter from July insists again on the importance of determining a consistent date for the image, this time offering 1858 and 1868 as possible dates.[81] Laussedat's concern was warranted, for the museum had in fact acquired an enlargement by Nadar's son Paul obtained from a second series of images taken by Nadar in 1868 from a tethered balloon in the bois de Boulogne—the image that was most consistently becoming known as the first aerial photograph.

A hypothesis turned fact, this image resists singularity—and not only because of its mechanical reproducibility. Consider one final example: three small photographic prints affixed to a backward L–shaped card and hand-labeled in pencil with the words "*1ère épreuves en ballon*" (figure 1.9). On the verso of this same card, a series of small sketches in gray and red pencil show the field of vision from a balloon—scoping the intended effect of the photographs stuck to the other side (figure 1.10). In much the same way a patent drawing might describe the idea to be patented, the drawings on the verso of the photographs connect the view from the air represented in the three small images to the technique of their production. Crucially though, the images are mediated by a third intervention: the hand annotation describing these images as the "*1ère épreuves en ballon*." Understood together, annotation, photograph, and drawing constitute a multimedia object that can stand in for both the idea *and* the technical proof of aerial photography, all the while making a historical argument about its instantiation at a specific moment in time.

These three small photographs attached to a sheet of cardboard with the words "first prints from a balloon" penciled below were able to operationalize a claim to historical primacy precisely because of the multimediality of the composite object, not despite it. Graphically figuring the aerial

 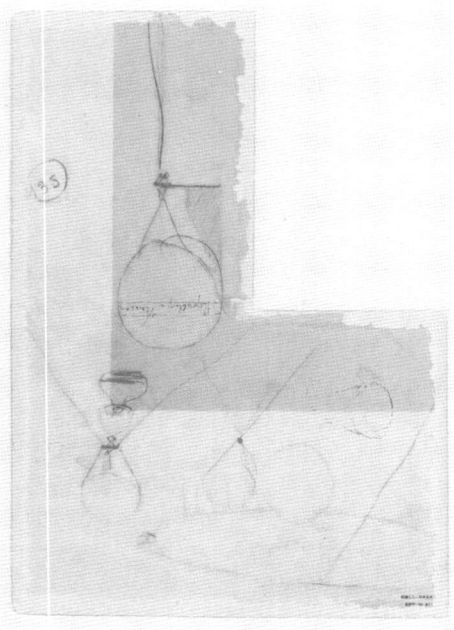

1.9 (*above left*) Félix Nadar, *1ères [Premières] épreuves en ballon: Trois vues aériennes de Paris* (First photographs from a balloon: Three aerial views of Paris), 1868. Collection of the Département des Estampes et de la photographie, Bibliothèque nationale de France, Paris.

1.10 (*above right*) Félix Nadar, *1ères [Premières] épreuves en ballon: Trois vues aériennes de Paris* (First photographs from a balloon: Three aerial views of Paris), 1868, verso. Collection of the Département des Estampes et de la photographie, Bibliothèque nationale de France, Paris.

image as an experimental object, this artifact reinforces my claim that Nadar's "first" could not be contained within a single image or object. But, as Caren Kaplan and others have argued, historical legibility was not the only representational problem that aerial photography faced.[82] In many ways, the combination of text and image that Nadar marshaled to stake his claim to firstness parallels the necessity of text-image composition in aerial photographic interpretation more broadly. The discourse of a "clear view" from the basket of a balloon needed constructing, and if Nadar's account is any indication, the view was not always clear and representational failure was the norm rather than the exception. Even after the technical processes used to take photographic images from the air had vastly improved in the early twentieth century, a complex interpretative apparatus was required to render aerial photographs "useful."[83]

From the 1858 patent to the 1893 essay, the "idea" of aerial photography was made material, through both the act of photographic exposure and this chain of mediations. As should be clear by now, Nadar's (and later his son's) priority claims should not be taken at face value. They do, however, warrant being taken seriously, if only to understand the broader significance of invention as an enduring cultural concept. Nadar's priority claims are part and parcel of a culture that considered invention as a motor for historical change. To have, or to realize, a technical idea was to make history. The genealogy of the first aerial photograph thus does more than just complicate the history of what has previously been understood as a single object. Indeed, inserting multiple, exceedingly mobile, images in place of a single primal image is evidence of more than just Nadar's own grandiosity. The cultural techniques that produced the "first aerial photograph" and "Nadar" as a celebrity figure in the history of photography—exhibitions, scientific journalism, performance, print culture, museums, historical writing, among others—are the same as those that inform the historiography of nineteenth-century French science, as well as culture more broadly. In such environs, historical actors such as Nadar collapsed the act of invention with the making of history. *What will be will have been.* The story of the first aerial photograph captures the cultural infrastructure—and the media—that launched such ideas into the air.

2

PATENT PRIORITIES

The First Photograph by Electric Light

In the years that Nadar was experimenting with aerial photography, he was also trialing methods to decrease photography's reliance on natural light. In 1859, Nadar presented the results of his first experiments with photography by electric light to the Cercle de la presse scientifique—a professional organization of scientific journalists founded two years earlier in 1857. But like the first aerial photograph, the priority of Nadar's first photographs by electric light was fixed not simply by the instance of their execution but even more assuredly by their relevance to institutional processes decades later—museum collections in the case of the first aerial photograph and patent litigation in the case of the first photographs by electric light.[1] As we will see, the story of Nadar's electric firsts was put into sharp relief by a patent dispute years later in the 1880s. This legal battle was mounted by the Parisian photographer Alphonse Liébert against his fellow photographer Pierre Petit, disputing the legal right to commercially exploit processes designed to facilitate the production of studio portraits by electric light. The litigation process occurred decades after Nadar's own first images by electric light, but his images and written testament were drawn into the case as evidence against these claims to priority. Like patents themselves, photographs cannot determine priority alone. The photographic first had to be drawn into a constellation of media, which, together across several decades, were deployed to narrate Nadar's contributions to the origins of the electrified photographic studio.

Like photographic firsts, modern patent law is premised on the idea of novelty. Patentees engage with the patent process to attain a temporary legal monopoly to exploit a process, machine, manufacturing method, or other patentable idea. But to do so, inventions have to be rendered tangible, described through images, models, texts, or demonstrations—processes much like the chain of mediations central to the production of the photographic firsts described in this book. As discussed in chapter 1, the representation of the technical idea behind a given patent application is essential to assessing the "novelty" or "non-obviousness" of a given invention; this is known as "prior art" in the language of patent law.[2] As the owner of several patents, Nadar, I argue, understood his claims to historical primacy—his firsts—in a manner analogous to the modern French patent system, a system in which, through litigation, technical novelty was often determined via the comparison of media (in the form of technical drawings, manuscripts, models, or photographs). The conceptual conversion of a craft process into intellectual property occurred both through the formal procedures and legal strictures of patent law and at the level of media. For those involved, intellectual property was a high-stakes endeavor: Ownership of commercially exploitable photographic processes could be lucrative. In this climate, firsts, too, held promise. The photographic first is therefore as much a product of the economic structure of the photography industry as of Nadar's own force of personality.[3] This chapter explores the consequences of such twinning for photographic history through a media history of one such patent controversy.

We begin with Nadar's early experiments with electric light in the late 1850s. One of the extant documents of these experiments was the result of Nadar's presentation to the Cercle de la presse scientifique, a performance format that was a common precursor to a patent application. A close reading of this composite image opens my analysis of Nadar's firsts as claims to intellectual property. These firsts are then contextualized within the broader commercial problem of the availability of natural light for studio photographers. Electric light—and the patents securing the studio owners' right to use it—was both a compelling draw for would-be customers and an economic proposition for photographers, promising more-consistent and higher-volume production. Finally, we arrive at the main event: an 1880s patent dispute regarding the use of electric light in photographic studios. This case, I argue, illuminates several key phenomena impacting the diffusion of photographic techniques in the nineteenth century and, correspondingly, the way that they have been historicized, including the international circulation of patents and the use of photographic media in patent litigation.[4]

FIRST TRIES

Public presentations cemented Nadar's claims to novelty. But it is the media representations documenting these performances that continue to speak to us of these novel ideas. The media fanfare accompanying the introduction of (apparently) new photographic processes therefore has an archival, and potentially legal, role. Patent claims are often speculative statements (*this will be*). Their litigation is therefore premised on historical argumentation (*this has never been*). This prophylactic production of legal and historical evidence was necessitated, at least in part, by the lack of a legislative requirement for prior examination in French patent law before 1968.[5] As I described in the first chapter, before the first French patent legislation of 1791, inventors were granted royal privileges based on the establishment of the utility and novelty of their invention. However, after 1791, the inventor had to prove neither the utility (now to be determined by the "public") nor the novelty (now to be determined by the courts), reconfiguring the patent process from a system of privileges granted by a ruler to a set of "rights" established between inventors and civil society.[6] These bureaucratic transformations had several important effects. Firstly, a patent guaranteed neither the utility nor the novelty of an invention. Secondly, an invention would now only be legible to the state—and, correspondingly, to the public—via its mediation: in text, image, or model form. Finally, it was up to the public to assess the utility of the invention via its application in industry or commerce and the judiciary to assess its novelty should a claim be contested.[7]

This publicness was a double-edged sword for would-be inventors. Making public the specifics of their invention assured a public record of their priority, but it also opened the door to others to use or replicate their invention. While the patent process remained inaccessible to much of the population, the liberal democratic theory behind the patent process framed patent disclosure as a decentralization of technical knowledge crucial to innovation.[8] This emphasis was very much in line with the expansion of a new type of journalism in the 1850s that aimed to reconfigure the relationship between science and society.[9] Several such journalists, including Louis Figuier, Henri Lecouturier, and Félix Roubaud, banded together in 1857 to create the Cercle de la presse scientifique, an organization that, in their own words, aimed to "strengthen the links between science journalism and [also] contribute to the progress of science."[10] On Saturday April 23, 1859, the Cercle held a meeting to witness three technical demonstrations, including one by Nadar.

An invitation from the Cercle addressed to Nadar's rue Saint-Lazare studio detailed the schedule for the meeting.[11] First, the engineer and inventor Victor Serrin would present his model of a self-regulating carbon arc lamp. Second, the electrician Adolphe Gaiffe would demonstrate the installation and working of a Bunsen battery. The final act was to be performed by none other than Nadar himself, presenting his system for "photography in the salon of the Cercle by electric and artificial light"—a feat facilitated by the two technologies presented by Serrin and Gaiffe. A composite document was produced of the meeting, one which unsurprisingly emerges as central to Nadar's claims to priority in the application of electric light to the production of photographic portraits (figure 2.1).[12] The label "First photographs by electric light by Félix Nadar—1859," written in pencil on white paper, is affixed to a cream sheet of slightly thicker paper, now colored with age and showing signs of liquid damage. On this sheet are four irregularly shaped photographs and several further inscriptions in ink. At the top right is Nadar. Visibly stunned by the glare of artificial light (or impudently emotionless), Nadar's enlarged shadow blends into the darkened foreground, a study in the harsh contrasts that the coming decades of experiments with photography and artificial light would attempt to mediate. At the top left are three male figures that the annotation tells us are representatives of the Cercle de la presse scientifique. The image at the bottom left is a portrait of Nadar's camera operator. The fourth image, placed at the bottom right, shows another member of the Cercle in the act of writing. An annotation reads:

> The first test of photography by electric light—Cercle de la presse scientifique (as far as I can remember: rue de Richelieu).
> In the collections of the newspaper of the Presse Scientifique we can find a trace of this meeting. Nadar.[13]

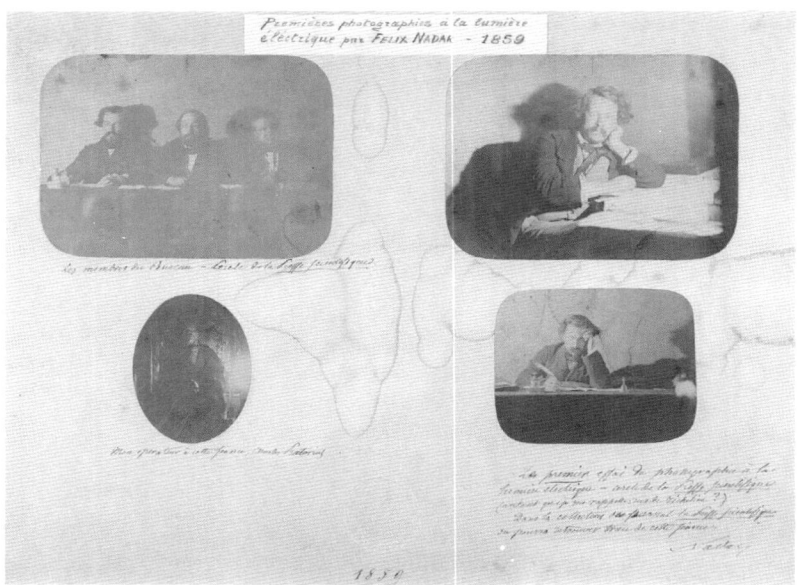

2.1 Félix Nadar, *Le premier essai de photographie à la lumière électrique au Cercle de la Presse scientifique, avril 1859* (The first attempts at photography by electric light at the Cercle de la presse scientifique, April 1859), 1859. Collection of the Société française de photographie, Paris.

Under the sign of Nadar's signature, we are to understand that this sequence of images—brought together with paper, glue, and words inscribed in pencil and ink—constitutes a photographic first. A composite media-object like the first aerial photograph, this photographic first is instructive in that it gathers four images capturing Nadar's performance of invention alongside portraits of the witnesses, the operator, and the archivist or transcriptionist, together with descriptive annotations to declare a first, however uncertain ("as far as I can remember"). The set of images bears the imprint of the concept of invention legislated by the patent process. Intellectual property was delineated by the mediation of an invention and the assignation of its inventor/author.[14]

What (if any) "first" is documented here? The demonstration before the Cercle was not, it would seem, the first time Nadar had experimented with photography by electric light. Describing this episode years later, Nadar writes:

> As mediocre and even detestable as these first negatives were, the news of the attempt had spread in our photographic microcosm, in which everyone kept an eye on his neighbor, and I was soon invited to give a talk for the Circle and journal *La Presse Scientifique*....
>
> Having immediately transported all my bulky equipment to rue Richelieu, I obtained various negatives—among others, the group of the President and his two assessors at their office—, negatives whose positives I made right away with my electric light source.[15]

Indeed, it would hardly seem prudent to present a new experiment to a group of journalists without having cemented the procedure (or at least practiced it) in advance. But there is a further sphere of influence at work; the "mediocre," "detestable" "first negatives" themselves (to which I will turn in a moment) were apparently what generated the invitation to present the system in front of the Cercle. That is, the experimental images themselves reportedly generated the opportunity to cement the first in front of witnesses. This attention created the conditions for the interpretation of Nadar's system for the application of electric light as innovative, assessed as such by the professional estimation of the members of the Cercle. This intrigue also provided the opportunity to create a document of that "first," which has been preserved for historians—like me—to observe. Like a patent, the "first" here points to the moment in which an invention becomes legible via its mediation rather than the moment it becomes technically possible. In other words, Nadar's "first" photographs by electric light became first by way of the performance of their production rather than by simply the fact of having been made.

Emphasizing the process of production was, for Nadar and other enterprising photographers, a method for making an intellectual property claim. In France in the 1850s, photographic images themselves could not yet be copyrighted (rendered as property via artistic authorship). Through the patent system, photographers could, however, be granted temporary monopolies for specific photographic processes or apparatuses that they claimed to have invented or created for the first time. Asserting one's inventions was, like authorship, a property claim.[16] This was certainly true for Nadar. In his retrospective account (authored in the 1880s) discussing the outcome of the experiment described above, he notes that it was not only photographic exposure that could benefit from the application of artificial light but also, and no less importantly, the photographic printing process. This was clearly an important distinction for Nadar, who had filed a patent in 1861 that outlined a process for printing photographs by artificial light, two years after this presentation of the "first" photographs by electric light.[17] The patent, titled "Printing positives by electric or gas light," stated that Nadar's innovation was the application of an artificial light source to a method for printing photographs.[18] Much like a *mémoire descriptif* filed as part of a patent application, Nadar's retrospective account of these firsts compiled a narrative of priority by referencing a series of documents and images, including the images of Nadar's presentation to the Cercle and his patent.

Accustomed to writing with a historical bent and determined to assess a gap in the market, Nadar describes the currently unsatisfactory nature of positives enlarged with electric light in the *mémoire descriptif* filed on his behalf by his patent agent Barrault. Nadar argues that photographers would do well to continue to put their mind to this question. Electric light would allow photographers to reduce their reliance on the unpredictable availability of natural sunlight, offering both practical utility and commercial possibilities. After describing the necessity of such a system, Nadar narrates how, following his experimentation with a variety of methods, he settled on combining electric or gas light with Albert Moitessier's process for printing positive prints. Nadar argued that his method had a number of benefits: It made it possible to create prints with contrasting shades paralleling those produced with natural sunlight, to correct the balance of poorly exposed negatives, to increase the speed of production for photographic prints, to replace the sun as the main source of light, to obviate the need for a special laboratory specifically designed to enhance access to natural light, and, finally, to liberate photographers and the photographic industry from the inconsistencies of the atmosphere. As Nadar summed it up: "This new photographic process allows me to obtain superior prints in the day, at night, and in all rainy or foggy weather in less time than in the most beautiful, diffuse diurnal light, since I obtain prints in one and two seconds, which has practically never happened before now."[19] As we shall see, Nadar was slowly building his case for priority, from the "detestable" first negatives produced sometime prior to the September 1859 meeting to the "first photographs by electric light" created during that meeting and the patent text filed two years later.

As the "first photographs by electric light" would indicate, Nadar's priority, like the specifics of the process itself, was not always legible at the level of the image. Annotation was critical to articulating Nadar's innovation, as was the entry of those firsts into the collections of trade associa-

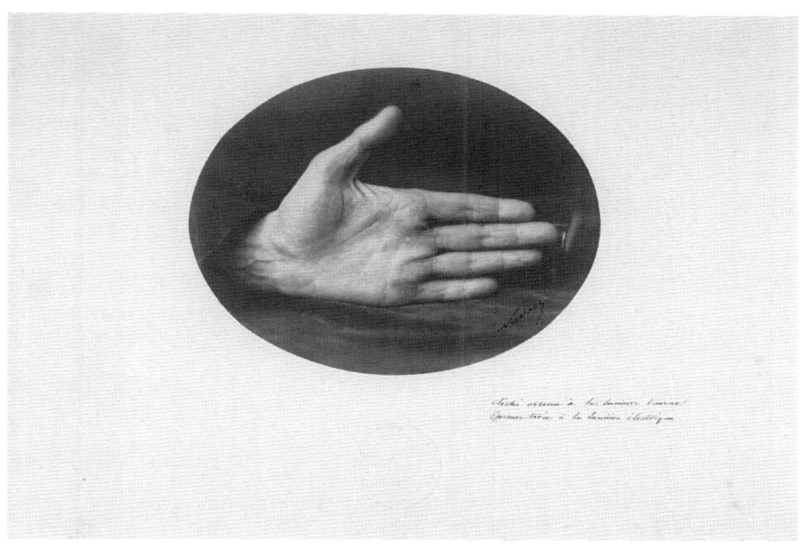

2.2 Félix Nadar, *La main de M. D*** banquier (étude chirographique), cliché obtenu à la lumière diurne, épreuve tirée en une heure à la lumière électrique* (The hand of M. D*** banker [chirographic study], exposure taken by daylight, print made in one hour by electric light), 1861. Collection of the Société française de photographie, Paris.

tions like the Société française de la photographie (SFP). In the collections of the SFP, there is another print taken by daylight but printed by artificial light, also annotated to describe the method of its production (figure 2.2). It is a delicately rendered image of the palm of a hand, captured against a textured, dark textile surface much like the surrounds of many of Nadar's portraits from this same period. Displayed at the SFP in 1861 (the same year as his patent application), the print is a particularly useful demonstration of Nadar's capacity to print sharp, tonally balanced positives with artificial light. The surface of the palm has only slight variations in tone and shade to demarcate the topography of the lines and surfaces—information supposedly linked to the character of the human attached to the hand according to the then-novel study of chiromancy. Described as the hand of a banker, the image is proof twice over.[20] The empty outstretched hand of the banker is presented as evidence of both the technique's ability to clearly render detail and the commercial possibility inherent in the decoupling of photographic production and natural sunlight—a novel photographic process making legible the hand of capital.

The switch from sunlight to electric light was not, however, without its drawbacks. Compare the effect of artificial light on Nadar's portraiture to a portrait of Dr. Armand Trousseau from a similar date, clearly labeled as having been taken in daylight (figure 2.3). The portrait of Henri Delaage by electric light (figure 2.4) is marked by dramatic shadows such as those under the sitter's right elbow, and the features of the subject's face are almost washed out in the sea of white light. By contrast, the image taken in daylight has the subject's features carefully rendered, visible perhaps most prominently in the delicate ombré of the subject's graying hair and the stark white of his sideburns. To navigate these pictorial discrepancies, the extraordinarily bright light of the carbon arc lamp needed to be dampened to re-create the gentle gradation of tones and dramatic chiaroscuro for which Nadar's portraits had become renowned.

Describing his desire to mimic the illumination of daylight more accurately, Nadar writes:

> In order to perfect [the photographs], we needed a second source of softened light, penetrating the shaded parts. I tried bursts of magnesium; but we did not yet have the propitious lamps invented later, and the use of magnesium, let alone of smoke, presented a number of problems.
>
> I tried to attenuate my light by placing frosted glass between the lens and the model, which did not help me much; then, more practically, I arranged reflectors on white twill, and finally a double set of large mirrors reflecting intermittently the luminous light source of the shaded parts. I thus managed to restore my exposure time to my daily average; finally I was able to obtain negatives with equal speed and of totally equivalent value to that of the negatives executed daily in my studio.[21]

Nadar's description of the aesthetic problems posed by the application of artificial light to portrait photography is draped in technical description, further supporting his claim to experimental priority in this domain. A portrait by electric light of Mario Uchard appears to have benefited from these experimental strategies (figure 2.5). While Nadar still struggled to balance

2.3 (*above left*) Félix Nadar, *Dr. Trousseau*, ca. 1859–60. Collection of the Département des Estampes et de la photographie, Bibliothèque nationale de France, Paris.

2.4 (*above right*) Félix Nadar, *Portrait d'Henry Delaage, h.e [homme] de lettres // obtenu à la lumière électrique, avec renvois reflets et intermédiaires en glace dépolie // bd des Capucines 35, en 1859, ou 60 ou 61 par moi* (Portrait of Henry Delaage, man of letters // obtained by electric light, with reflectors and frosted glass intermediaries // boulevard des Capucines 35, in 1859, or '60 or '61 by me), ca. 1859–61. Collection of the Département des Estampes et de la photographie, Bibliothèque nationale de France, Paris.

tones in rendering the faces of his sitters, his compositions had become much more tonally balanced since the 1859 experiments discussed above. The figures no longer disappear under the harsh glare of the light source.

In each of these scenes, Nadar's supposed priority in the application of artificial light to photography was mediated by the patent process. Nadar's "first" images by artificial light, viewed alongside his patent application and an example of his use of the patented process, show the

2.5 Félix Nadar, *Mario Uchard par la lumière électrique* (Mario Uchard by electric light), no date. Collection of the Département des Estampes et de la photographie, Bibliothèque nationale de France, Paris.

parallel strategies involved in maintaining the photographer's historical legacy (he was first) and his entrepreneurial identity (he would be first to exploit). But, as with many of the other applications of photography that Nadar was interested in, the use of artificial light in commercial studio portrait photography was predicated on a complex confluence of technologies.

EXCESSIVE LUMINOSITY

An artificially illuminated photography studio requires a steady source of actinic light, the quality so prized in sunlight. The nonactinic and often flickering light of gas or candle flames had not been able to replicate the qualities of natural daylight. As opposed to the "flash" of light often associated with artificially illuminated photography, the relatively steady light of the carbon arc lamp was Nadar's light source of choice.[22] The development of the carbon arc lamp is enmeshed in a longer history of scientific investigations into the nature of electricity and experiments designed to harness its power for human use. In a lecture delivered to the Royal Society—a performance format that has mediated many discovery stories—in November 1809, the British chemist Sir Humphry Davy demonstrated that, when a current from a large grouping of voltaic piles (electric batteries) was sent through two pieces of charcoal, there arose "a most brilliant flame, of from half an inch to one and a quarter inches in length.'"[23] The phenomenon Davy demonstrated would later be developed as the carbon arc lamp.[24] With these experiments, ongoing throughout the nineteenth century, we see the beginnings of the lighting technology that Nadar would use to produce his photographs by electric light. But with the introduction of the carbon arc lamp, the issue of too little light was rapidly transformed into a problem of too much light, a technical problem that photographers would attempt to address throughout the nineteenth century.

Picking up on the principles of producing light by carbon combustion outlined by earlier experimenters, Nadar's co-presenter to the Cercle, the electrical engineer Victor Serrin, had developed a self-starting and self-regulating carbon arc lamp in 1857. Serrin had patented the lamp in 1859, which was likely the reason for his presentation to the Cercle alongside Nadar. Serrin's innovation was the development of a model of an arc lamp that was easy and safe to light and extinguish, further recommending its application to a wide variety of uses.[25] The main problem preventing the widespread application of arc lamps like Serrin's was the availability of a reliable energy source. Luckily, the third presenter, the journalist and scientific popularizer Adolphe Gaiffe, demonstrated a solution (however temperamental) to the crowd at the Cercle: the electric Bunsen battery. The Bunsen battery had been developed by the German chemist Robert Bunsen in 1841, replacing the more expensive platinum electrodes of previous battery models with a cheaper carbon electrode.[26] Though more powerful than previous batteries, Bunsen's model still required many batteries to achieve the desired amount of power. The combination of the carbon arc lamp and the electric battery provided the technical infrastructure for Nadar's application of electric light to photography.

As we saw in the case of the portraits described above, these technologies were not without their defects. The light was harsh and the equipment unwieldly. "These first negatives came out coarse, with clashing effects, the blacks opaque, cut without details in every face. The pupils either extinguished by excessive luminosity or brutally sticking out, like two nails," writes Nadar.[27] Recall Nadar's description of his experiments with "bursts of magnesium," "placing frosted glass between the lens and model" and "reflectors on white twill," with "a double set of large mirrors reflecting intermittently the luminous light source of the shaded parts."[28] He goes on:

> We did not have the precious portable accumulators, or the intermediary generators of Gaulard, or all the other equipment we have now, and we were forced to accept all the cumbersome inconveniences of the Bunsen battery. We had no choice.
> I therefore had an experienced electrician install, on the solid part of my terrace on the Boulevard de Capucines, fifty medium elements that, as I had hoped, turned out to be sufficient in providing me with the required light.[29]

It was only with this elaborate system in place that Nadar could hope to return to his "daily average" (read: portraits, prints, and revenue). The artificially illuminated portrait studio required not only a consistent light source and the means to power it but also a studio apparatus that could mediate said light in a manner appropriate to produce portraits up to Nadar's production standards and business requirements. This transformation of Nadar's studio practice represented a substantial renegotiation of studio infrastructure. Nadar's addition of fifty cumbersome Bunsen batteries to his studio apparatus should therefore be understood within the larger trajectory of reconfigurations of the photographic studio as an architectural space providing adequate light for photographic exposure.[30] The different studio spaces that Nadar both imagined and built over his career exemplify this search for light.

In 1860, Nadar moved from his previous studio on rue Saint-Lazare to a much more lavish one at 35 boulevard des Capucines.[31] The building that he rented had been constructed in 1855 and had previously housed the studios of the photographers Gustave Le Gray and Adolphe Menut (known as Alophe). Ever the entrepreneur, Nadar undertook a massive reconstruction project financed by selling shares to investors, completely renovating the facade of the building and adding two new stories.[32] The new face of the building was encased in glittering glass on a frame of cast iron, designed to communicate the innovative spirit of the photographer whose studio it housed. The enormous central studio was thus visible from the outside, allowing as much natural light as possible to pour into the studio. Affixed to the facade was the pièce de résistance, an enormous sign in the form of Nadar's iconic signature, glowing red, gaslit at night, and designed by Antoine Lumière, father of Auguste and Louis Lumière.³³ Nadar, however, was not the only photographer pursuing light-chasing studio renovations.

In the notoriously cloudy cities of London and Paris, access to light was at a premium for studio photographers. Glass-walled studios and top-floor skylights gave photographers access to as much sunlight as possible to expose and develop their products. Like Nadar, other photographers were experimenting with different forms of artificial light to render their commercial production more predictable and profitable. After experimenting with several different light sources, the London-based studio owner Henry Van der Weyde developed a system using an early type of battery, a refractive lens originally used in lighthouses, and a carbon arc lamp.³⁴ The results were evidently satisfactory, for Van der Weyde was ready to declare in an 1882 publication: "I continue to photograph without daylight—in fact, I never took, and never will take a photograph by daylight."³⁵ Like Nadar, Van der Weyde was concerned with the aesthetic quality (and the commercial viability) of photographic portraits created using artificial light. Surveying previous experiments with artificial light, Van der Weyde declared that, though some "were very remarkable, they were all, from an artistic point of view, failures" due to the fact that "even the best results suggested a metallic or varnished surface, with glittering high lights, dense shadows, and ghastly reflections."³⁶ He realized that the "question was not [simply] to discover a better artificial light but to turn and twist its diverging rays from a point, so as to concentrate them, and, so to speak, make them embrace instead of strike the sitter."³⁷ After experimenting with magnesium light and limelight, he began to work with a "Grove Battery with a four-foot-wide Fresnel Diotropic Lighthouse Lens and Serrin Lamp with a silvered-copper reflector."³⁸ He quickly determined that the metallic interior was not a good choice due to the uncomfortably high temperature it created for the sitter and instead whitewashed the interior of the reflector. The results were satisfactory, producing in his words an "artistic portrait with tender crisp highlights which did not need retouching, and full of modelling and transparency in the shadows."³⁹ In 1878, Van der Weyde filed a patent specification for "an improvement in illuminating objects to be photographed, and the interior of public and other buildings," which consisted of using a concave reflector screen in order to diffuse the overly harsh direct light from an artificial source.⁴⁰ This patent would later form the basis of a dispute in which Nadar's own "*premiers essais*" would become embroiled.

Photographers in Paris were likewise converting their studios to enable the provision of electric light. Alphonse Liébert opened a photography studio equipped with electric light in Paris in 1879, capitalizing on a trend for high-society portraits taken at night, with wealthy individuals desiring to have their portraits taken in their opera best. The novelty of photography by electric light was a major selling point. Describing Liébert's techniques for tempering the harsh glare of electric light, photographer and science journalist Gaston Tissandier writes:

> The novelty of the system adopted by Mr. A. Liébert is that the electric light does not fall directly on the model. This light is first projected on a reflector which, in turn, sends it back onto the walls of the cylinder which are dazzlingly white, so that the luminous rays, thus dispersed, thus divided, are equally spread across the person whose picture is to be reproduced.
>
> The clarity is superb. The face is softly lit, without hardness, without exaggerated shadows.⁴¹

By the time that Liébert electrified his studio in 1879, energy sources for the provision of electric light had significantly improved compared to Nadar's description of the cumbersome Bunsen battery setup. Writing about a tour of Parisian studios in the *Photographic Times*, a Mr. E. L. Wilson described the entrance to Liébert's studio: "Before ascending the stairway to his studio, I saw at the right hand of the door a powerful little gas engine busily at work, and close by it a Gramme electric machine, which reminded me of the fact that Mons. Liébert practiced portraiture by means of electric light."⁴² As in Nadar's experience, however, and as the description by Tissandier would indicate, provision of artificial light by no means guaranteed its utility for the creation of a well-balanced portrait in the desired style. Techniques involving reflectors and screens to balance the provision of light were necessary.

With light and the energy to sustain it, photographers such as Nadar, Van der Weyde, and Liébert sought to expand the potential for financial profit in each of their respective studios. Sunlight—that prized photographic ingredient—was a fickle collaborator. Though photographers remained reliant on the natural world for the materials of their trade, the electrification of the portrait studio was imagined to be a process of standardization, akin to the industrialization of other processes of production.[43] Light was an essential component in the production of a photographic image, and if its provision could be rendered consistent and reliable, photographers in command of such a system could stand to profit greatly. In its early years, the use of electric light in photography studios (and, more specifically, the precise methods for using electric light to satisfactory aesthetic effect) was thus also a patentable technology. This meant that these procedures could be owned and sold—they represented property relations that were ripe for dispute. Van der Weyde, Liébert, and Nadar were just three of the many photographers interested in exploring and experimenting with the application of electric light to photographic portraiture. Indeed, the cross-pollination of techniques for adapting electric light to commercial portraiture was a cause of some concern for these entrepreneurial photographers seeking to signal their primacy in the race to electrify their studios.

PATENTLY FIRST

Recall the range of dates ascribed to the first aerial photographs as they were collected and exhibited. The date for the first photographs by electric light was similarly elusive, even to their author. Nadar retrospectively annotated several portraits by electric light, including an image of the journalist Henri Delaage (figure 2.4) and a second image of Dr. Trousseau (figure 2.6). The caption underneath Delaage's portrait reads: "Portrait of Henry Delaage, man of letters / obtained by electric light, with reflectors and frosted glass intermediaries / 35 boulevard de Capucines, in 1859, '60 or '61 by me / Nadar." Likewise, the annotation on the portrait of Dr. Trousseau outlines the system of reflectors and closes with the words "certified to be true" above Nadar's signature. The annotation is dated January 16, 1885. But why was Nadar relabeling his photographs in 1885? Much like how the processes of photographs entering museum collections prompted a renarrativization of the "first" aerial photographs in chapter 1, Nadar's first photographs by electric light were brought into historical relief because of their utility in the litigation of a patent dispute in the 1880s. That is, these "electric sketches," as they were called by the president of the Cercle de la presse scientifique, became "first" through the institutions of the patent process. Nadar's photographic primacy was here constituted through intellectual property law.

2.6 Félix Nadar, *Dr. Trousseau*, 1859. Collection of the Département des Estampes et de la photographie, Bibliothèque nationale de France, Paris.

One further figure brings this property dispute to light: the Paris-based photographer Pierre Petit. Like Nadar, Van der Weyde, and Liébert, Petit had also been experimenting with a system for photographic portraiture by electric light. Petit's system for portraiture by electric light is outlined in a self-portrait with his apparatus labeled with descriptions of each device (figure 2.7). Petit had exhibited photographs reproduced with the aid of electric light, a system he called linography, at the 1881 Exposition internationale d'électricité alongside Liébert.[44] At the exhibition, both Liébert and Petit demonstrated systems diffusing electric light with reflectors to facilitate the production of photographic portraits.[45] The face-off between the photographers was not in and of itself unusual; the fair was full of exhibitors clamoring to demonstrate the seemingly endless applications of electricity to modern life. A moment of synthesis for a nascent field, the exhibition (which we

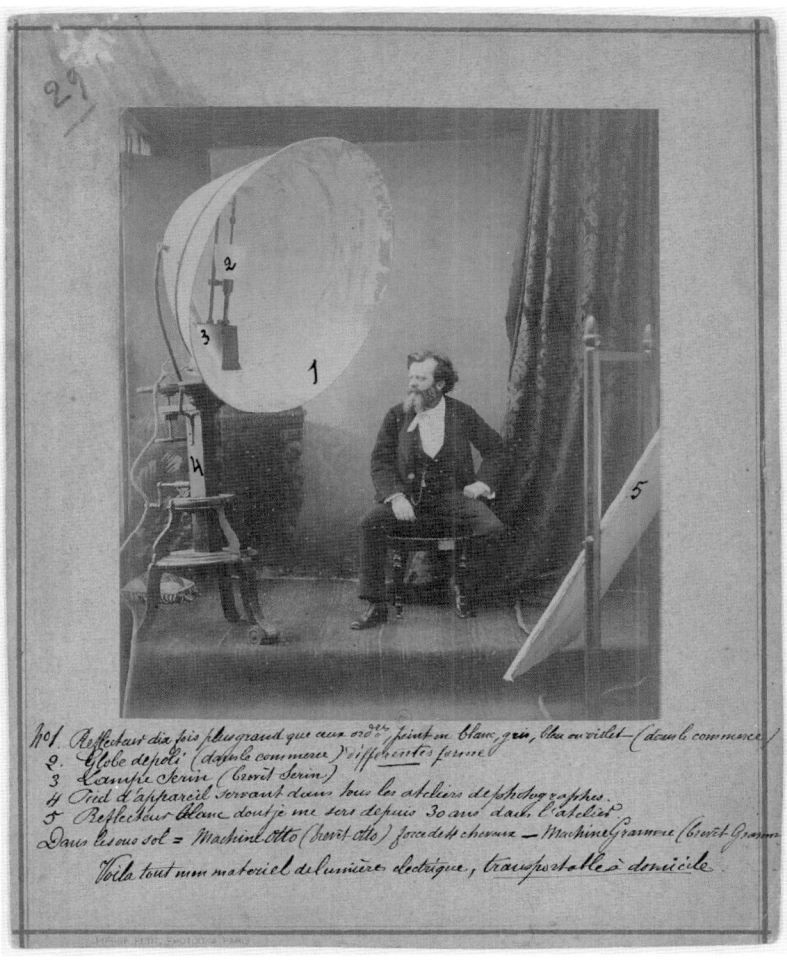

2.7 Pierre Petit, *Autoportrait de Pierre Petit dans son atelier, posant avec son matériel de lumière électrique* (Self-portrait of Pierre Petit in his studio, posing with his electric lighting equipment), ca. 1880s. Collection of the Département des Estampes et de la photographie, Bibliothèque nationale de France, Paris.

will return to several times across this book) also presented an opportunity for electricity enthusiasts of all stripes to consider past, present, and future sources and applications of electricity and included a historical section on early electricity development and views of the entire system of the production of electrical power, from generators to electric light bulbs. The story of electricity was being told (and sold). The facts, however, were not yet fixed. The exhibition launched a dispute involving the three photographers (Liébert, Petit, and Van der Weyde) over who could claim priority in the application of electric light to photography in France—a claim that not even Nadar could make.[46]

If, as the Marxist legal historian Bernard Edelman has put it, photographers have to become authors because "the relations of production will demand it," here, in the absence of those authorial rights, photographers became patent owners to assert their right to capitalize on specific photographic techniques.[47] After the London-based Van der Weyde patented a system for photography by electric light in 1876 in England, he subsequently filed for the French right to the patent in 1877. Liébert purchased Van der Weyde's claim to the French patent in 1879. In a series of proceedings over the next decade, Liébert and Petit fought over the validity of Liébert's patent. The penultimate appeal in 1886 ruled that Liébert's patent was null and void.[48] This controversy reveals the intensity of competition between studio owners and the lucrative place accorded to individual patents in the commercial photography market.

Property relations are marked on the photographic prints themselves. Liébert's studio advertisements specifically claim Van der Weyde's invention as part of the novelty, and therefore the value, of what the studio could offer customers. We can observe the patent claim pass between the two men at the level of their commercial product. A cabinet card from the Van der Weyde Studio in London (figure 2.8) notes that the portrait had been taken by "the van der Weyde light," through which "daylight [was] superseded by electricity." The verso of a cabinet card from Liébert's studio (figure 2.9) notes that the portrait had been taken at night by means of a "new system of artificial light," patented "S.G.D.G" (an acronym standing in for the qualifier "sans garantie du gouvernement") and "invented by Van der Weyde." Nodding to the temporary technological monopoly assured by the patent, Liébert asserts that his was the only studio in France selling portraits made with this system.

This statement was, however, as the accompanying acronym put it, "without guarantee from the government." This provision was an artifact

 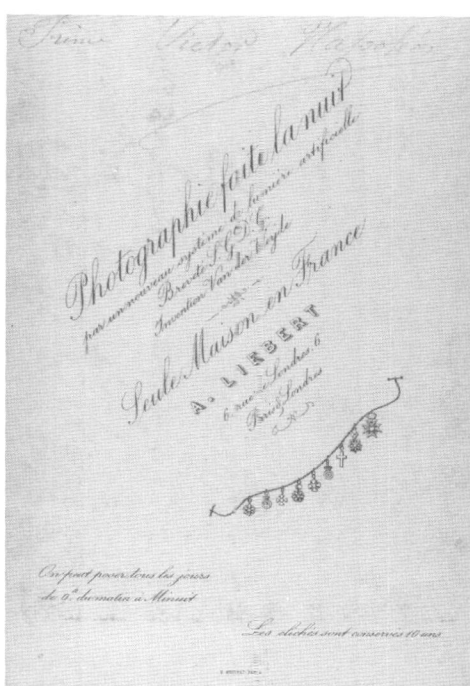

2.8 (*above left*) Henry Van der Weyde, *Mrs. Langtry*, 1880. Collection of the Getty Museum, Los Angeles.

2.9 (*above right*) Alphonse Liébert, *Verso of cabinet card*, undated. Collection of the New York Public Library, New York City.

of the 1844 revision of the 1791 patent laws by the French state. The standard text of each patent contained the qualifying phrase "Without prior examination, at their own risk and peril, and without guarantee, either of the reality, novelty or merit of the invention, or of the fidelity or the accuracy of the description."[49] In other words, an awarded patent meant little more than that someone had filed for a patent. It was only through litigation that the novelty and priority of patented inventions were really examined. Like the patent claims that they contested or sought to uphold, these legal cases often produced a profusion of media representations—a process that here parallels the photographic first.

In 1882, three years after purchasing Van der Weyde's French claim to the patent, Liébert attempted to enforce his patent against the Paris-based studio photographer Pierre Petit. While the first case was decided in favor of Liébert, a series of appeals over the next few years allowed Petit to slowly make his case, often via the very media and techniques under scrutiny in the patent case. The self-portrait of Petit in his studio shows the photographer authoritatively positioning himself as the author-inventor—or one of the author-inventors—of the pictured process (figure 2.7). Petit's annotation of the surface of the image mimics the genre of technical drawings that were required to accompany patent applications, with numbers or letters indicating key components and an associated legend. Breaking down the system into its individual components, this representational mode attempted to articulate the novelty of the process or machine under examination—or, in the case of Petit, to demonstrate the lack of novelty that characterized Liébert's claim to a patent for a system of photographic portraiture by electric light. As with the media surrounding the first aerial photograph, the media gathered in defense of each photographer's claim show the mediatic process by which an invention becomes a technical "artifact" rather than an intangible idea, something that could be owned by an individual, a state, or a corporation.[50]

This process also prompted the recirculation of old media, indexing the flurry of electrical activity in photographic studios in the decades leading up to this case. As Liébert would learn, communities of practice could intervene in and produce evidence against individual claims to technical novelty. Documents solicited by Petit in support of his case against Liébert included a letter from Nadar stating that Liébert's claims to priority were predated by Nadar's own experiments fifteen or sixteen years earlier, thereby rendering invalid Liébert's prohibition on other photographers attempting to use electric light in their studios.[51] In order to make his argument, Nadar describes examples of his own images made with electric light, noting that while he never took out a patent for this process, "the question of priority, *which cannot be doubted*, would seem to me to render the patents null and void."[52] A statement from Émile Lemmonyer of the Cercle also attested to Nadar's priority in the application of electric light to portraiture, with specific reference to the images made during Nadar's presentation to the group. Nadar offered to send a sample of prints of his first experiments with electric light, annotated precisely with how they were photographed and retrospectively dated as proof in 1885.[53] The portrait of Delaage with the uncertain annotation dated to "1859, '60 or '61" was one of these images (figure 2.4). The image formed part of a larger set of media in support of Petit's case. The dispute prompted photographers and journalists around Paris to reinterpret old documents, recount past presentations, and reexamine old letters, acting like historians in search of an invention story. But like many contemporary historians of technology and media, they disputed the notion that there could be a single "inventor" of the idea of taking photographs by electric light. Nadar's firsts could thus be repurposed as legal proof that someone else was distinctly *not* first.

Liébert didn't stand much of a chance. Upholding much of the original patent system designed in 1791, the 1844 amendments of the French patent law were formulated as a response to what legislators imagined to be the demands of a growing industrial society. One amendment was the new provision that no patent could be filed for "all discoveries, inventions, or applications that, in France or abroad, had received sufficient publicity to have been executed."[54] The legislators defined "publicity" as including "all public use of an invention, like the publication of an article, a lecture, or technical tests." This meant that Liébert had to contend with an onslaught of media produced by other photographers, journalists, and legislators to make his case. Liébert's claim to the patent was overturned in 1886, allowing Petit and other photographers to continue to freely commercialize the use of electric light in portrait studios. As this episode indicates, intellectual property law relies on media, performance, and rhetoric to establish claims. Public responses to patent filings elicit their own forms of media production, processes of mediation like those that produced Nadar's firsts.

RELATIVE PRIORITY

Let's make a second lap around the moves (and media) behind the "first photographs" by electric light. Nadar experimented with portraiture by electric light in his studio, facilitated by a carbon arc lamp and a complex installation of Bunsen batteries. The circulation, demonstration, and exhibition of the resulting images provoked interest among journalists and photographers, prompting a letter to be sent from the Cercle de la presse scientifique to Nadar containing an invitation to demonstrate his application of artificial light to photography. During that meeting, photographs were taken, events were witnessed, and something was written down. Afterward, the proceedings were transcribed in the organization's journal. In 1861 or thereabouts, four images from the meeting were attached to a piece of cardboard and annotated with ink. These would now be "the first photographs by electric light," dated to 1859 and given to the SFP. Years later, in the 1880s, images of a similar vintage and technique were reannotated and "certified to be true" to demonstrate that other photographers were *not* first, with photographs, ink, and paper collaborating as legal evidence.

This series of mediations could be a footnote beneath the claim that Nadar made a series of photographs by electric light beginning in the 1850s. However, this episode is instructive for historians of photography and visual media in that it demonstrates the way that this archive has been regularly reconfigured in relation to different institutional processes. Patent litigation, much like the collections and exhibitions described in the previous chapter, instigates the annotation or reanimation of these images in relative linearity. News of an awarded patent can provoke community responses as to the real novelty or even the "non-obviousness" of the patent in question. These processes are bolstered—and occasionally foiled—by Nadar's own reminiscences and archival practices: "taken by electric light" in "1859, '60 or '61 by me / Nadar." These events and responses are, in turn, rich materials for historians working through the history of these media. Nadar speaks to us on the surface of his images. They are, however, instrumental media, deployed here to make legal arguments in support of a commercial right to exploit a specific photographic process and elsewhere to buttress Nadar's own studio name (also legally protected, you remember). As responses to the legal formation of technical ideas—in this case, photographic processes—these documents reify intellectual property. Priority had real commercial value; so, too, did the photographic first.

3

SOUND REPRODUCTIONS

The First Photographic Interview

While the first aerial photograph and the first photographs by electric light came into relief ex post facto, the first photographic interview came prepackaged: already first. The process of asserting priority could take place even more quickly by the 1880s—a speed which paralleled advancements in the exposure speed of photographic film. This accelerated tempo was conferred grand historical implications. "With the Nadar system there is no interpretation," wrote journalist Thomas Grimm in *Le petit journal* in 1886, "one has a document that is absolutely exact."[1] The "exact" document referred to a photographic interview published in *Le journal illustré* on September 5, 1886 (figures 3.1–3.3).[2] The interview was a collaboration between Nadar and his son Paul featuring photographs and a transcript of a conversation between Nadar *père* and the prominent French scientist Michel Eugène Chevreul (1786–1889) on the occasion of Chevreul's 100th birthday.[3] The "Nadar system," as conceived by Paul, consisted of a series of "instantaneous" photographs taken using the new high-speed Eastman roll film alongside a stenographer's transcription of the interview.

3.1 (*above left*) Paul Nadar (photograph) and Félix Nadar (text), "L'art de vivre cent ans" (The art of living a hundred years), *Le journal illustré*, September 5, 1886, front page. Collection of the Département Philosophie, histoire, sciences de l'homme, Bibliothèque nationale de France, Paris.

3.2 (*above right*) Paul Nadar (photographs) and Félix Nadar (text), "L'art de vivre cent ans" (The art of living a hundred years), *Le journal illustré*, September 5, 1886, 284. Collection of the Département Philosophie, histoire, sciences de l'homme, Bibliothèque nationale de France, Paris.

3.3 (*right*) Paul Nadar (photographs) and Félix Nadar (text), "L'art de vivre cent ans" (The art of living a hundred years), *Le journal illustré*, September 5, 1886, 285. Collection of the Département Philosophie, histoire, sciences de l'homme, Bibliothèque nationale de France, Paris.

This pairing of visual and oral documents was envisioned as nothing less than "the journalism of tomorrow."[4]

Like the firsts explored in the last two chapters, the "journalism of tomorrow" rested on a cluster of cultural practices and technologies, including the interview format, recorded speech, instantaneous photography, and photomechanical reproduction. As part of the broader media production of the Chevreul centenary, the photographic interview also participated in a broader process of historical commemoration. A future-oriented, priority-obsessed hagiography colored the enthusiasm surrounding the use of new (and old) media in the production of the interview. The image-event was also conceived by Paul Nadar—newly appointed Eastman Kodak's French sales representative—as a chance to promote the novel high-speed Eastman roll film. So-called instantaneous photography gave the resulting publication a greater purchase on reality, a spurious claim useful for both commemorative and commercial ends. However, as previous scholarship on the interview has determined, Nadar essentially rewrote the interview for publication, cherry-picking particularly evocative moments in his conversation with Chevreul and pairing them with twelve of a total of fifty-eight photographs taken by Paul of the meeting.[5] Compressing into mere months the rhetorical (and archival) sleight of hand that took several decades in the case of the firsts explored in the previous two chapters, the first photographic interview is another example of the media-historical formulation of photographic firsts that I am tracking across this book.

The photographic interview was a historically important event because it could be, and was, photographed. The interview represented a growing acknowledgment that historical events would be legible in the future by way of their mediation. As the editorial framing of the interview makes clear, the project prompted extended conversations about the relationship between media fidelity and historical epistemology and, in the process, purported to transform the experience of media production and consumption. Formerly "readers," newspaper "spectators" were now participating in a form of historical witnessing intimately tied to technical mediation.[6] But what the newspaper "spectators" were seeing and reading was not (or not simply) a document of a unique historical event but a vivid testament to the desire for a faithful reproduction of history itself—whatever such a thing might mean. This was not a neutral process. "Re-production," as Jonathan Sterne has written of the history of sound reproduction, "results in the creation of a distinctive form of originality: the possibility of reproduction transforms the practice of production."[7]

More than a document *of* history, the interview was a hypothesis for how history *could* be made. The interview exemplified how, from now on, the press would "collect and preserve optically and acoustically, in incontestable historical documents, all successive scenes of any action worthy of interest, public or private."[8] Media production was to be understood as a form of historical preservation.[9]

In this, Nadar's choice of conversation partner was not incidental. The interview was published amidst the large national celebrations that accompanied Chevreul's 100th birthday in 1886, festivities that generated a mass of commemorative visual and material culture. The image of the chemist Chevreul, a figure by then practically synonymous with the productive industrial capacities of French science, was a powerful one, and one that the Nadars understood well to associate themselves with. Chevreul's living canonization was part of a broader culture of commemoration, devoted to historicizing and enshrining figures and events deemed central to the history of the French Republic. The scientist's centenary in 1886 anticipated the parallel festivities of the Exposition universelle of 1889 and the 1789 centennial, events which, while controversially connected, explicitly paralleled the economic, industrial, and revolutionary histories of France.[10] Like the interview, these events were designed to celebrate historical achievements while also producing these commemorations as future history—resulting in a flurry of representations in a variety of media. Just like the events they represented, these media forms had new historical value. A concurrent historiographical shift emphasized the importance of original documents in historical analysis alongside the establishment of museums, libraries, and archives to preserve and house such documents.[11] More than just documenting the events of their time, those involved with the Nadar–Chevreul photo-interview saw themselves as responding to a future need for historical documents; they used relatively new journalistic techniques, such as the interview and the instantaneous photograph, to prospectively craft source materials for future historians.

In turn, historians have been kind to them, often displaying the same fascination with historical immediacy that animated those behind the interview. Indeed, this type of quest for unmediated truth became, as James Carey memorably put it in the title of a 1974 article, "the problem of journalism history."[12] Historians of journalism, Carey argued, have, to the detriment of their scholarship, internalized the profession's own gospel of objectivity, resulting in historical accounts of the formation of the profession as a story of progress toward objectivity—a Whig history of the news.¹³ Such is the case of the photographic interview. The format represented a faith that the interview could offer its readers a greater connection to the world beyond the pages of the newspaper, or at least more so than any stuffy repurposing of government publications or, alternatively, obviously opinionated columnists. But subsequent attempts by photographic historians to determine whether the interview was, in fact, a faithful representation of reality have implicitly upheld the idea that there could be such a document. To depart from this mode of analysis, this chapter examines the construction of the photographic interview's priority rather than its objectivity. In doing so, I argue for a parallel between these two concepts: *the truth* and *the first*. Claims to objectivity and priority both rely on strategies of mediation while simultaneously evoking a fantasy of immediacy—whether epistemological or historical.

The journalism of tomorrow was to be the archive of the future. Anticipating how historians would rely on newspapers as source material in the century to come, those involved in the production of the interview held lofty aspirations as to future interest in the interview. They weren't wrong. But crucially, this is because the "event" that the published interview captured was not primarily the conversation between Nadar and Chevreul but the production of the interview itself. In fascinating ways, the photographic interview indexes conversations about media fidelity and historical objectivity across different media, including recorded sound, photography, and photomechanical reproduction. To begin, this chapter traces the individual technologies used in the production of the photographic interview. Departing from the claims to historical objectivity made on behalf of these technologies, I examine the relationship between media production and historicism in the context of the Chevreul centenary. Finally, I examine the metareflection on the nature of evidence contained within the Chevreul–Nadar conversation while considering the claims to instantaneity, fidelity, and priority made on behalf of the photographic interview.

THE JOURNALISM OF TOMORROW

But what did the "journalism of tomorrow" look like? The first page of the newspaper holds a full-page photographic portrait of the centenarian Chevreul (figure 3.1), an eminently wrinkled face that recalls Victor Hugo's line from *Les Misérables* that "the face of the century is made up of the lines of the years."¹⁴ The photograph shows Chevreul in a dark suit against a light-colored backdrop, no furnishings or props in sight. Chevreul's lined face, haloed with a crown of white frizzy hair, holds a look of contemplation, if not disassociation. The article is titled "L'art de vivre cent ans: Trois entretiens avec Monsieur Chevreul" and includes twelve of a total of fifty-eight photographs taken of the conversation alongside an edited version of the interviews and an editorial introduction.[15] Each page of the published interview features sets of four photographs sequenced with accompanying captions, followed by a longer version of the interview text. On the first page of photographs, Chevreul is shown writing in Nadar's studio guestbook (figure 3.2). The final image of this set is captioned with an engraved reproduction of Chevreul's handwritten message in the book, a quotation from the philosopher Nicolas Malebranche: "One should strive for infallibility without claiming to have achieved it"—a would-be warning to the Nadars.[16] Bringing together mechanical reproductions of the photographs of Chevreul, engraved reproductions of his handwriting, and a stenographic transcription of his speech, the interview presents these (copies of) autographic traces as a historical portrait of the elderly man. The two subsequent pages show Nadar and Chevreul in animated conversation, with photographs sequentially arranged to show the broad range of hand gestures and facial expressions enacted throughout their conversation, the animacy of hands and faces speaking to the power of the high-speed Eastman film to capture the potency of their dialogue. The intimate details of their faces are rendered for posterity in all their spirited forms. And yet, for all the enthusiasm that characterizes the framing of the interview, the resulting images are relatively banal, picturing two elderly men seated with blankets and slippers.

The advanced age of the conversation partners aside, the photographic interview was, according to its publisher, distinctly new. In the editorial introduction, the editors lay out their rationale for the novel format:

> Today, the *Journal illustré* offers its readers a new and unexpected work which will go down in the annals of human discovery . . . The French public is rightly in love with the news; . . . they demand that when an interesting event occurs, when it involves a man who merits attention . . . , that this event be accurately reported, that this man be faithfully portrayed.
>
>   To meet this legitimate need for information, which has become a modern necessity, journalists resort to practical means: they themselves interview the people that fate or valor has brought to light.[17]

The photo-interview was, according to the editorial introduction, "new," "unexpected," and a discovery of historical significance. The "new" and the "news" are doubled here; as the editors would have it, the novel demand for different kinds of information required a novel kind of information gathering (the interview) and a new format (the photographic interview). In these terms, the editors describe the interview as an experimental model of information dissemination, one capable of satiating the French public's reported hunger for the news. Meeting this "legitimate need for information," that "modern necessity," the production of the interview became a technological spectacle on par with, if not exceeding, the actual content of the publication. In the stagecraft of the interview, "The reader becomes the spectator."[18] The production of the interview was an event in and of itself.

Interviews were a relatively new addition to the journalist's toolkit, one which, as the editors note in their introduction, was inspired by the use of the format by journalists in the United States.[19] The embrace of the interview format by journalists across the Atlantic Ocean was part of a professionalization of journalistic practice that saw the emergence of the "reporter," an ordained figure mediating events for newsreaders.[20] This represented a significant transformation in the information-gathering practices of journalists.[21] Most "reporting," as Michael Schudson has argued, then amounted to "nothing more than the publication of official documents and public speeches, verbatim."[22] Journalists might speak to sources, but these exchanges were rarely mentioned in print. For journalists reporting on political speeches or legislative sessions, stenography (or shorthand) had long been considered an essential skill, but, in the case of an interview, journalists cautioned that taking notes could distract both the interviewer and the interviewee and risked putting too much attention on minute detail at the expense of the larger significance of the conversation.[23] Interviews, then, were regularly based on the recollections and individual testimonies of reporters rather than on a transcript. These practices were precisely the kind of journalistic mediation that the creators of the photographic interview sought to avoid.

The Nadars sought a different, mechanical, kind of mediation. Reportedly, they would have dearly liked to have been able to record the conversation using a phonograph.[24] Instead, the interview was transcribed by a stenographer, an embodied practice of listening and writing that is also a form of mediation. Stenography was itself the site of experimentation in nineteenth-century France; for example, in the field of music notation.[25]

But, as J. Mackenzie Pierce has argued, "While stenography could facilitate commercial or personal record keeping, the prestige accorded to real-time recording derived largely from its use in law and politics."[26] However, stenography was an established method for recording *speech*, not sound. And while stenography was the technique used to record the conversation, all those associated with the project associated the interview with roughly contemporaneous innovations in sound reproduction. The editorial introduction even goes so far as to suggest that recent inventions such as the phonograph or photophone were themselves the fulfillment of Nadar *père*'s own earlier prophecy, recounting that "this prophecy of Nadar ... we have seen progressively realized, first by the eminent M. Lissajous, who showed sonority to the Academy; then, by the ingenious and modest Mr. Ader, who transmitted to the astonished visitors of our last Universal Exhibition the songs of our operas and the dialogues of our actors, with the hubbub of halls and the applause of distant spectators."[27] In citing the fulfillment of this "prophecy" of Nadar's, the editors reference two quite different technologies: Jules Antoine Lissajous's apparatus for visualizing sound waves with a series of mirrors and vibrating tuning forks and Clement Ader's théâtrophone, a system for transmitting Parisian operas and concerts via telephone.

This "prophecy" first appeared in Nadar's writing in a newspaper article in 1856 with a reference to the possibility of an "acoustic daguerreotype."[28] As of 1856, according to Nadar, "everything was possible"; a method for sound recording was just around the corner, soon to be available for the recording of plays, operas, and parliamentary proceedings, offering "an exact, faithful, and dispassionate account like mathematics."[29] Reprising this prophecy eight years later in 1864 in *Mémoires du Géant*, Nadar described how this idea had been realized in the form of the "phonograph": "Something like a box in which melodies would be fixed and retained, just as the dark room catches and fixes images."[30] Nadar reportedly learned of the invention via the academician and painter Auguste Couder during what was apparently Nadar's one and only drawing lesson. Couder recounted how the Académie des sciences had been thrown into turmoil, for they had "seen sound!!!"[31] But, in 1864, despite his use of the word, Nadar was not describing the yet-to-be-developed phonograph but likely, instead, the aforementioned apparatus for visualizing sound waves developed by Lisssajous. Lissajous had presented his method for the optical study of vibratory movements to the Académie des sciences on April 6, 1857.[32]

Within the textual apparatus of *Mémoires du Géant*, this acoustic digression offers a way for Nadar to attach himself to the history of other technologies, characteristic of his fascination with how technical ideas are connected to their realization (as discussed in chapter 1). What is curious however, is that the technologies under discussion are not systems for reproducing sound in a way that would have supported the presentation of the interview in a newspaper format, at least not in a form legible to most newspaper readers. In the introduction to the interview and in the secondary literature about the interview, the possibility of *reproducing* sound (in the form of the phonograph) is often collapsed or used interchangeably with the process of *transmitting* sound (with reference to the théâtrophone or photophone)—the distinction would come later.[33] While Nadar's direct reference is to Lissajous's ephemeral, light-based visualization of sound waves, he describes it as a phonograph—a different technology that didn't exist in 1864. In fact, the technological wish that Nadar was expressing is somewhat murky. Given the flurry of research into sound and the experiments with different mechanisms for representing or reproducing sound in the decades surrounding the publication of *Mémoires du Géant* (and preceding the publication of the interview), Nadar could also have been describing something like Édouard-Léon Scott de Martinville's phonoautograph, a method for the visual representation of sound that was invented in 1857, a relevant reference as Scott de Martinville had described the phonautograph as an "automatic stenographer."[34] Although they do not reference the phonautograph, the editors do describe the inventor Clément Ader's théâtrophone. The théâtrophône was a system for telephonically broadcasting live musical performances in Paris, first presented by Ader at the previously discussed 1881 Exposition internationale d'électricité in Paris—a performance likely relatively fresh in the minds of the editors.[35] Ader's théâtrophone has also become the subject of a process not unlike what Nadar's iterative reflections on the process of recorded sound offered his readers in 1856, 1864, and 1886. Dubbed by twenty-first century commentators variously as the "19th-century iPod" and "the first music streaming service," the rhetorical practice of lining up these inventions as events in an unfurling chronicle of technological progress, each a predecessor for some future, better thing, is exemplary of how the first (whether photographic or not) operates as a metahistorical device.[36] Like the théâtrophone, the interview was imagined to be a hypothetical predecessor for some future realization of a fantasy of complete archival fidelity.

Considering all the speculation regarding the possibility of recorded sound, Paul Nadar's use of a new kind of photographic film risked being superseded, warranting only one brief, if dramatic, passage in the editorial introduction. For Paul Nadar, write the editors, "the time had come to ask science itself for the 'Proof' we lacked," for science (unlike those involved in the project of prophesying recorded sound) "rejects indeterminacy and hypotheses and relies only on established experimental facts."[37] In this formulation, "science" was the developments in photographic technology that had produced instant photography (or the "proof"). The "established experimental facts" that allowed the project to reject "indeterminacy and hypotheses" were photographs with high-speed exposure times. The interview photographs were taken with a shutter speed of 1/133rd of a second, a rapid exposure time enabled by the Eastman Company's novel stripping film or "American film." Designed to do away with the burdensome collodion glass plates commonly used by photographers, the roll system had been under development by George Eastman and William Walker from the early 1880s.[38] Working from existing roll-film models such as Leon Warnerke's roll-film system developed in the 1870s, Eastman developed a workable roll film (also called a stripping film system) with a photosensitive gelatin layer applied to a base of Rives paper early in 1884.[39] After patenting the process in 1884, Eastman began marketing the product in a number of Western European cities from 1885 onward, including in Paris.[40] Paul Nadar began selling the "American film" in 1885 and, in 1886, became the exclusive Eastman representative for all of France and its colonies, founding the Office général de photographie to house his ventures into the sale of cameras and photographic materials.[41] Published the same year as the formal expansion of Paul's business, the photographic interview should therefore also be understood as part of the broader publicity efforts that Paul undertook to demonstrate the new Kodak products. In a presentation to the Société française de la photographie (SFP) during their January 7, 1887, meeting several months after the publication of the interview, Paul framed the interview project within his ongoing promotion of Eastman products as the company's representative in France, writing that

> the thought had occurred to me for a long time that with the help of the extraordinarily fast processes at our disposal today it should be possible to join photography and stenography. My dream is to see the photographic apparatus recording the different attitudes and changes

in the speaker's physiognomy, just as the phonograph stores his words and thoughts to transmit them to us.[42]

Whereas previously photography had been a metaphor for those experimenting with the reproduction of sound, here recorded sound becomes a measure for photographic instantaneity.[43]

While the novelty of photographic instantaneity was compelling enough to be newsworthy, the publication of the photographs in the newspaper represented yet another innovation. The interview was published in *Le journal illustré*, a publication that would become known for the quality of its photographic reproductions.[44] The development of photomechanical printing processes enabling the publication of verisimilar photographs (as opposed to engravings *after* photographs) in newspapers was itself a new technology. Stanislas Krakow had patented the system in 1884.[45] As Thierry Gervais has noted, despite being greatly concerned with the novelty of the various technical aspects of the project, none of those involved immediately remarked upon the significance of a practical method for including photographs in newspaper format. In the presentation to the SFP months after the publication noted above, Paul Nadar briefly outlined the significance of this kind of innovation for journalism: "The series of portraits of the eminent M. Chevreul will, in future, ensure the immediate and direct use of photography in the illustration of the book and the newspaper."[46] Photomechanical reproduction was the last link in the chain of media technologies used in the production of the photographic interview—each cultivating a sense of immediacy and proximity in their own way. Combining the journalistic interview with stenography, instantaneous photographs, and the reproduction of photographs in the newspaper, the photographic interview became a photographic first, a novel kind of document that, according to its creators, would offer a new kind of material evidence for both historians and reading (and looking) publics. This priority claim was imagined to be self-evident. But, as the discourse swirling around each individual component of the interview would indicate, these claims were anything but. If, in 1864, a phonograph was to be a daguerreotype of sound and, in 1887, an instantaneous photograph was hoped to become a photographic-phonographic recording, then the photographic interview of 1886 drew its novelty claim precisely from this intermediality. It was new—and first—because it marshaled recorded speech, high-speed photography, and photomechanical reproduction *together*. In this way, the interview demonstrates the inextricability of the "new" from the "historical"; the "invention" of the photographic interview is what would make it historically legible as an event of distinction.

L'ART DE VIVRE CENT ANS

But precisely what kind of event was the interview meant to capture? Chevreul had lived long enough to persist as a blur at the center of an "instantaneous" photograph taken in 1886, a century after his birth. So-called instant photography could obscure just as much as it revealed. An image from the Nadar studio collection of the Chevreul centenary ceremony highlights precisely why the interview was staged as it was; high-speed film still had difficulty capturing live events in crisp detail (figure 3.4). The heaving blur of men at the center back of the image becomes one mass, their dark coats blending in with the darkened surrounds and their white buttoned shirts merging with their light skin and hair. Taken at a moment of transition, everyone in movement is indexed as out of step with their seated contemporaries. This image was part of a profusion of visual and material culture produced as part of the national celebrations organized to commemorate Chevreul's centenary—a mass of representations that necessarily forms a crucial iconographic context for the photographic interview. Depictions of the centenarian scientist in different media helped construct a legend of his achievements and personal longevity alongside the legacy of nineteenth-century French science. Chevreul witnessed an incredibly tumultuous and transformative century of French history, born in 1786 just three years before the beginning of the French Revolution. But just as importantly to those constructing his hagiography, Chevreul was a chemist whose work on animal fats had transformed the industrial manufacture of soap and candles and whose research into the chemical nature of color as the director of the Gobelins Manufactory (a tapestry factory) influenced the production of textiles and the color theories that inspired the Impressionist and Neo-Impressionist painters.[47] For the French state under the Third Republic, Chevreul and his research presented a useful ideological icon, deemed representative of the development of French science, industry, and culture across the nineteenth century.

Famously abstemious, a "water-drinker," Chevreul's numerous accomplishments (including his advanced age) were, the official line went, the result of his personal commitment to reason and truth as embodied in experimental science and Republican education. As a monument to man

3.4 Atelier Nadar, *Cérémonie du centenaire de Chevreul* (Centenary of Chevreul), 1886. Collection of the Département des Estampes et de la photographie, Bibliothèque nationale de France, Paris.

and method, the public festivities organized in celebration of Chevreul's centenary were in the style of the French Republic's civic festivals.[48] As Avner Ben-Amos has described, these public spectacles were pedagogical propaganda designed to instruct and educate the citizenry in the Republican ideals of "secularism, patriotism, and republicanism."[49] Chevreul had been elected to the French Académie des sciences in 1826 and the Légion d'honneur in 1844. The centenary in 1886 was thus one further event canonizing the scientist's prominent place in French Republican memory. Chevreul was also a useful symbol of how experimental science and scientific education could reinvigorate French industry and capital, especially considering the French defeat at the hands of the Prussian army over a decade prior and the widespread concern that French science had suffered a decline since the early part of the century.[50]

The spectacle of the festivities, alongside an effusive print culture and several monuments, marshaled stone, print, and parades to lionize

Chevreul. The scientist was the subject of two sculptures, one designed by Léon Fagel and installed in the Jardin des plantes of the Muséum national d'Histoire naturelle in Paris, where he served as a professor, and the other designed by Eugène Guillaume and installed in the gardens of the Musée de Beaux-Arts in Angers, his birthplace. As part of the celebrations on August 30 and 31, 1886, the sculpture by Guillaume was unveiled in the zoology gallery of the Muséum d'Histoire naturelle.[51] The two-day celebration of Chevreul's centenary featured a banquet and nighttime procession.[52] The scientist was also commemorated with a wax sculpture exhibited at the Parisian musée Grevin and featured in the Panorama le "Tout-Paris" at the 1889 Exposition universelle.[53] Upon his death in 1889, Chevreul was given a state funeral, an event also heavily documented in the illustrated press. The photographic interview participated in this Republican propaganda. But, importantly, it did so in both form and subject, laying claim to French technical priority by using new forms of mechanical reproduction to document the presence of a figure deemed representative of the nation's illustrious scientific history.

These assorted monuments to Chevreul often compared the span of his life with that of the nineteenth century, dwelling upon the fact that he would have "witnessed" the revolution of 1789 at the tender age of three. In a special issue of *Le courier français* devoted to Chevreul's centenary, a print by Adolphe Willette makes this most explicit (figure 3.5). Titled "A Century," the print is a tumultuous scene of activity, watched over by the figure of Chevreul in the center right. Chevreul's lower half is indistinguishable from the roots of a large tree that towers over the battle scene at center. Napoleon charges in at the center left, while the pyramids of Egypt, a balloon, and a glittering guillotine fly along an indistinct horizon line in the upper left. The middle and lower planes of the image appear to move chronologically from left to right, with certain momentous episodes of nineteenth-century French history figuring prominently. Just to the left of the crease of the spine, a working-class man mounts a barricade to hold up a tattered flag emblazoned with the year 1848. At center, the figure of Marianne is being shot at, the Republic trampled and fallen under a pile of bricks. Like the photographic interview, Willette's image attempts to spatialize time, visually compressing the events of the nineteenth century into one panoramic summary. The events of a century unfold across the page from left to right like a sinistrodextral sentence. In the Nadars' interview, the sequential arrangement of images attempts to represent the time frame of the conversation, connecting the minute physiognomy of

3.5 Adolphe Willette, "Un siècle! . . ." (A century! . . .), *Le courrier français*, August 31, 1886, 8–9. Collection of the Département des Estampes et de la photographie, Bibliothèque nationale de France, Paris.

Chevreul's face to the contours of a history of French science discussed in the conversation between Nadar and Chevreul.

In the printed interview, as in the Chevreul centenary writ large, the role of the individual (man) as an agent of history is analogized via a canonization of Chevreul's individual accomplishments as representative of the historical developments of the nineteenth century. Compared to these at times classicizing or Romantic representations of the scientist, Paul Nadar's portraits of the two men in conversation in the interior of the Nadar studio appear exceedingly modest, though flashy in their claims to novelty. But the constellation of photographs from which they were drawn reveals a broader range of possibilities for how Chevreul could have been presented in the published version of the interview. As noted above, the twelve images included as part of the published interview were chosen from a larger group of images of the interview in progress, as well as other images, including several of Chevreul's home and his study and portraits of him, his son, and his assistant. The broader image set from which the images for the published interview were drawn also included images that featured a wider cast of characters than those selected for publication,

3.6 Paul Nadar, *Meeting of Chinese ambassador [Xu Jingcheng] and Michel-Eugène Chevreul at Atelier Nadar*, 1886. Collection of the Getty Research Institute, Los Angeles.

including images of Chevreul and his son with Nadar, Nadar and his son with Chevreul, and, in the same setting, Chevreul in conversation with Xu Jingcheng (sometimes differently transliterated as Shu-King-Chen), the Chinese ambassador to France, alongside another unidentified man (figure 3.6).[54] A meeting of statesmen, these images further serve to position Chevreul as a representative of France's national scientific culture.

The images of Chevreul in conversation with Xu Jingcheng would have had particular purchase in 1886, taken as they were after the end of the Sino-French War in 1885.[55] In a period of tense Chinese-French relations, Xu Jingcheng was presented by the French media as a mediator between the two empires and became a popular figure in the French illustrated press. The photographic interview was originally destined for publication in *L'illustration*, which had previously featured Xu Jingcheng on its cover in August 1885, shortly after the Treaty of Tientsin was signed in early June, bringing the war to an end. In addition to participating in the larger documentation of the centenary, the images may therefore have also had another purpose: demonstrating the ongoing negotiation of Chinese-French relations, particularly at a moment of Republican and imperial foment in France. Unlike the images with Nadar *père*, no conversation between these men was recorded, leaving to the imagination what these men would have discussed. Newsworthy, and therefore worth remembering, these interviews or meetings are imagined to be always already historical. But, again, the uncaptioned images of Chevreul and Xu Jingsheng show something of the problem that the Nadars fantasized themselves to be at work solving. Without the stenographed transcript, how are we to interpret the meeting of these two men?

In their strategic absences and presences, these documents offer a meta-epistemological reflection on the audiovisual archive. Even if the sound didn't register in the format they wished, the Nadars ask how history might speak. This is representative of an anxiety about historical legibility that frequently hides at the bleeding edge of media technology. But, perversely, these speculations surrounding greater media fidelity often reveal that there is no historical experience without representation. This is certainly the case when we reflect on the symbolic and imaginative space between the different representations of Chevreul described here. Absent the Romantic allegory, the images of Chevreul presented in the photographic interview speak to a different conception of history and its images. For, as the poet Paul Valery would later put in on the occasion of photography's centenary in 1939, "The mere notion of photography, when we introduce it into our meditation on the genesis of historical knowledge and its true value, suggests this simple question: *Could such and such a fact, as it is narrated, have been photographed?*"[56] In short, Nadar and Chevreul's conversation was of historical significance because it was photographed. Like the canonization of Chevreul's own scientific career, the historical visibility of the photographic interview as *first* is thanks to the media spectacle attending the accomplishment of its production. And yet this mediating spectacle is also representative of a desire for historical immediacy. This tangle—mediation and immediacy—is, I have been arguing, characteristic of the photographic first. For the (photographic) moment, Chevreul was, rather remarkably, still alive. And yet, figured as the mouthpiece of history, it was like he was speaking from beyond the grave—a chasm breached by the assemblage of the photographic interview.[57]

INSTANT HISTORY

In the minds of its makers, the photographic interview was one solution to the epistemic instability of historical evidence, for, in their words, "the study of what is called history is nothing other than the teaching of uncertainty."[58] This, the editors argue, is due to an overarching lack of material evidence. In this formulation, the photographic interview is the "proof" previously absent from historical analysis. The project's authors demonstrate an ideological and epistemic commitment to positivism, a philosophical system in which all knowledge is derived from experience. But much like in scientific experimentation itself, the "facts" and the "experiment" produced in the form of the photographic interview were designed specifically to demonstrate the power of the method, performing the capacity to represent the previously unseen (and therefore unknown) rather than revealing any lasting fealty to historical truth. History was being made because the event could be documented by a photographer and transcribed by a stenographer, nothing more.[59]

The interview was published amidst roughly contemporaneous discussions on the role of photography as documentary evidence and the nature of historical source documents more broadly. The centenary of the revolution in 1889 and the years leading up to it were a politically charged period that witnessed the tremulous maintenance of France's Third Republic (1870–1940). This period saw a growing interest in the history of the revolution, a by then quasi-mythological event whose symbolic power could be used to shore up support for successive Republican governments.[60] Debates around the planned Exposition universelle of 1889 and the commemoration of the centenary of the revolution coincided with the growing professionalization of historical study and an emphasis on historical method and objective analysis of primary sources.[61] As Catherine Clark has put it, university-trained historians sought to replace the "whimsical romantic imaginings and reconstructions of the past that had dominated nineteenth-century historical practices from Salon painting to lectures at the Sorbonne" with "an increasing emphasis on scientific evidence, proof, and rigor in historical research."[62] There was, however, no distinct break between these two approaches to the relation of images and objects to history. Photographs often "straddle the line between imagination and evidence."⁶³ Truth, slippery objective that it is, could take many forms. Looking closely at the production of the interview—as published object, as "proof," and as "first"—points to the faultline at the heart of this reorientation of historical study. Metaphysical interpretation, positivists would argue, obscures the truth offered to us by the data of experience. But it is only through interpretation—the selection of photographs, the translation of shorthand, the abridgement of lengthy transcripts, and, now, my own analysis of closely guarded archives and secondary scholarship—that the interview can become legible as anything close to a document.

This epistemic uncertainty was not shared by those responsible for publishing the photographic interview. Reflecting on the nature of historical documents more broadly, the unpublished manuscript version of the introduction directly highlights the utility of this kind of document for the commemoration of national and, more specifically, Republican history. The editors ask the reader to imagine being able to view photographs of iconic moments of 1789, such as the Tennis Court Oath, an event most famously represented in sketch form by the painter Jacques-Louis David. Unlike the producers of the photographic interview, David necessarily produced his image of the event after the fact, thereby, in the minds of the editors, reducing the capacity of the image to accurately represent the event.⁶⁴ This reference to the possibility of photographing the 1789 revolution was not included in the newspaper version of the interview but is provocative in its allusion to the possibility of visual documents of key moments of the revolutionary period. In acknowledging the lamentable lack of photographic evidence of the events of 1789, the editors position the photo-interview as a remedy to a recognizable problem: the representation of contemporary events as always already historical.

But how to do so? In the photographic interview, instantaneity was rendered synonymous with contemporaneity and the evidentiary. The novel speed of the Eastman roll film offered a solution to the problem of the distance between documents such as paintings and the events they purported to represent. In its ability to freeze motion and render the long, stiff poses of earlier portraits unnecessary, the instantaneous photograph was necessarily of the moment and thus could (supposedly) more accurately represent a given event for future audiences. The objectivity of high-speed photography and, later, cinematography was predicated precisely on its speed—faster than the human eye can see, as the saying goes. As historian of science Jimena Canales has shown, in the nineteenth century, the distinction between so-called scientific or objective and artistic or subjective representations of movement was often predicated on a tenth of a second, which scientists such as Jules Janssen had posited was the limit of human vision.[65] Proponents of efforts to standardize scientific photography held that the ability to capture events faster than this threshold was photography's unique contribution to the scientific investigation of subjects such as animal movement and human physiology. But, as Canales demonstrates, the archives of scientific observation, and particularly those of astronomers, reveal that "their main weapon, photography, had failed to completely demystify this moment."[66] Photography produced consistently inconsistent results. For example, during the widely observed transit of the planet Venus in 1874, the resulting photographs from different astronomers produced in different locales with different equipment diverged so greatly as to be incomparable for scientific purposes.[67] While, in published form, the interview's authors covered their tracks slightly more neatly than the disappointed astronomers of 1874, the resulting publication is no less evocative of photography's instability as a form of historical evidence.

Capturing an instant on film was not a guarantee of increased fidelity to nature. This was, in part, because the concept of instantaneity was a moving target that shifted over the course of the nineteenth century.[68] For investigators of natural phenomena (such as the astronomers described above), instantaneous photography was of interest because of the possibility of representing things beyond human perception, such as the path of a bullet, the hooves of a galloping horse, or the motion of a bird's wings in flight.[69] Much like research into photomicroscopy, instantaneous photography was imagined as a process capable of extending the capacity of the human eye to observe natural phenomena.[70] The instant was a moving target, for the word itself defies definition. An instant, a slice of time that is present and at hand, is a relative category. In the case of the photographic interview, however, instantaneous photography was employed to render a series of physiological expressions that were already entirely visible to the naked eye, if not immediately accessible to the intended observer (the newspaper reader). If instantaneous photography in the context of human and animal physiology sought to render visible a previously invisible set of movements to the scientist's eye, what did Nadar's photographic interview make visible? Quite simply, history itself. The photographic interview mounted the Nadar–Chevreul conversation as a newsworthy event and in doing so attempted (rather successfully, I might add) to enter the interview into the grand annals of history. A staged endeavor, the photographic interview employed the camera as eyewitness, foreshadowing a modern age of photojournalism in which photographic proximity determines authenticity.[71] The interview is thus also an example of what Daniela Bleichmar and Vanessa Schwartz have called "instant history," a dimension of "visual history" that "depicted the present as the future's past."[72] The project both recorded and produced a historical event—a photographic first—using rhetoric enhanced by the relative novelty of instantaneous photography.

Nadar and Chevreul, not incidentally, spent much of their conversation discussing the nature of scientific evidence and its relationship to the history of photography in the nineteenth century. In the section of the interview relating to the history of photography, the pair touch upon a topic that has long been a subject of debate among historians of photography: the invention of photography. Offering a familiar if slightly reframed version of the story, Chevreul argues that history must incontestably attribute the invention of heliography to Nicéphore Niépce.[73] In Chevreul's estimation, Niépce should be credited with the "mother discovery" and fellow Frenchman Louis-Mandé Daguerre and the British Henry Fox Talbot should be understood as the "children" of Niépce's original invention.[74] Ideas are, after all, different from their technical execution, as evidenced by the material distinction between the unique reflective metallic surface of Daguerre's daguerreotype and the soft, reproducible paper surface of Fox Talbot's calotype. The fact that much of France and the world had come to credit Daguerre and Talbot with the invention of photography proper was, according to Chevreul, simply a problem of evidence. The centenarian scientist argued that one of the reasons why the scientist and academician François Arago had determined that Daguerre was worthy of compensation from the French state for his invention in 1839 was because Arago had simply not been able to see Niépce's 1827 heliographs. Considering this material absence, Arago's conclusion in favor of Daguerre, and the resulting purchase of the daguerreotype process by the French state, was based on an incomplete body of historical evidence.[75] This paucity of information had forever altered the historiography of photography. This cautionary tale, Chevreul argued, reinforced his call for the application of the a posteriori method to historical inquiry.

This story of the semi-omission of primary author-inventors from the stories of their inventions also haunted the producers of the photographic interview. As discussed regarding the long history of the production of the first aerial photograph as "first" in chapter 1 and the story of the race toward the electric illumination of the photographic studio recounted in chapter 2, those involved in the production of the photographic interview did not leave their association with the project, nor the public understanding of the project's novelty, to chance. In each of these cases, the photographic first was positioned as a witness to its own process of production, the mediation of which precipitates historicization. Describing this kind of phenomenon, Ulrich Keller has noted photography's complicated status as a historical document, arguing that if photography ultimately came to be understood as the sui generis eye of history, this was due to "its ability to fuse, in ways impracticable for the competing media, the *performance* of historical events with their visual *representation*."[76] The production and dissemination of the interview thus offers a kind of object lesson in the construction of objectivity as a shared ideal (or ideology) between the professions of journalism, science, and history. However, in each of these disciplines—journalism, science, and history, not to mention photography, stenography, and sound reproduction—objectivity is a historically specific phenomenon. For "objectivity," Lorraine Daston and Peter Galison remind us,

> preserves the artifact or variation that would have been erased in the name of truth; it scruples to filter out the noise that undermines certainty. To be objective is to aspire to knowledge that bears no trace of the knower—knowledge unmarked by prejudice or skill, fantasy, or judgement, wishing or striving.[77]

This kind of "epistemic virtue," as Daston and Galison term it, becomes a kind of performance, different in key to the dramatic speculations and spectacles unfurled across the last few chapters but not in substance.[78] In the case of the photographic interview, the resulting document was understood as first because of its (supposedly novel) pretensions to unmediated reality—an epistemic construction that, perversely, reveals the interview as a hypermediated performance.

VIVRE CENT ANS, 100 YEARS LATER

As with the technical missteps and speculative constructions that characterized Nadar *père*'s experiments with aerostatic photography and electric light, the photographic interview was both a future-oriented pronouncement on the possibility of simultaneous sound and image recording and an implicit acknowledgment of the limits of contemporary technology. In the text of the interview, Chevreul unwittingly acknowledged the speculative tension embodied in a photographic first like the first photographic interview. He articulated a conceptual distinction between ideas and their realization in the history of science and technology, proffering that "there have always been people who have claimed invention when they had only imagination, which is far from being the same thing."[79] Embodying just such a leap as criticized by Chevreul, the published interview, the edited manuscript, and the larger series of photographs, annotated prints, and annotated versions of the newspaper interview, here understood as one project, are conceived of as an imaginative compendium of future history, brought together by an as yet unrealized technological union of photography and recorded sound.

As it turns out, the Nadars' vision of the interview as an audiovisual document would be realized (in a fashion) in a 2011 short film directed by the Australian filmmaker Dennis Tupicoff.[80] The film mimics the tremulous, scratched celluloid of early cinema and features two actors (as Nadar and Chevreul) reenacting the photographic interview based on the photographs and the transcripts. Following an introduction narrated by famed French director Agnès Varda that surveys Nadar and Chevreul's biographies, images of the published interview fade into the filmed reenactment. As in the newspaper version, the short film opens with Chevreul seated alone at a table, Nadar's studio guestbook open in front of him. Chevreul signs the book with what he describes as "his first philosophical principle," borrowing the words of Malebranche cited earlier. In the words of its twenty-first-century makers, the film "brought to life" the "first media interview, recorded in words and images for a mass audience," 125 years after its creation. More than a simple reenactment, the 2011 film positions itself as a realization of the hypothesis of the original interview, yet again positioning the Nadars as diviners of technological futures.

Even without Tupicoff's more recent contribution to the genealogy of the photographic interview, the interview could be described as protocinematic. But just as that prefix—*proto*, or first—jumps to the tip of the tongue, I pause lest I, over a century later, allow Nadar (and son) to slip, yet again, into the role of soothsayer. Prophecies, predictions, or speculations must be mediated to be remembered as the germinating seed from which a later realization emerges (whether as technology or event). In the mind of its makers, the photographic interview was thus both a novelty and a prophecy of what could be: a mechanized and synchronized record of image and sound. The technical novelty of the interview and the predetermined historical significance of the conversation that it pictured were coproduced. Indeed, the "first photographic interview" and the Chevreul centenary both offered the opportunity to craft narratives of priority that shored up France's, Chevreul's, and the Nadars' self-images. In the next chapter, I turn to Nadar's images of French infrastructural projects and examine how the strategies deployed to produce the photographic firsts explored across the last three chapters could work together to position Nadar's photographs as first even when he did not explicitly claim them as such. From monuments of men to civic infrastructure projects, the photographic first was a legacy project.

4

ILLUMINATING INFRASTRUCTURES

The First Photographs Underwater and Underground

In 1897, Nadar reprised his signature sign over the awning of a new studio in the southern French port city of Marseille (figure 4.1). The photographer installed himself on the city's rue de Noailles at the age of seventy-seven, the reins of the Paris studio firmly in the hands of his son Paul. This iteration of the Nadar studio was established just a short walk from Marseille's Vieux-Port—a location that would become a site of fascination for Modernist photographers in the twentieth century.[1] As in Paris, the studio primarily produced portraits, but a commission from a local newspaper would allow the photographer to experiment with photography in a new and challenging milieu. In 1899, *Le petit marseillais* asked Nadar, working alongside the Swiss photographer Frédéric Boissonnas and his assistant Fernand Detaille, to document an underwater construction project at the city's rapidly expanding port.[2]

This was not the first time that Nadar had ventured below the surface of the earth. In the 1860s, Nadar completed two separate series of photographs documenting the recent renovations to the Paris sewers and

4.1 Félix Nadar, *Atelier de Nadar au 21, rue de Noailles à Marseille* (Nadar's studio, 21 rue de Noailles, Marseille), 1899. Collection of the Département des Estampes et de la photographie, Bibliothèque nationale de France, Paris.

catacombs—part of a massive program of public works in Paris that resulted in the razing and reconstruction of much of the city across the nineteenth century.[3] In both Marseille and Paris, Nadar used his expertise in working with electric light to illuminate these infrastructural spaces. The port, the sewer, and the catacombs were all state-led infrastructure projects managing the circulation (respectively) of shipping vessels, goods, and materials; human waste; and human remains. The representation of these sites was part of the national narrative of an ever-modernizing France. But for Nadar and his historical legacy, these were also photographic firsts: the first photographs of underwater construction work in a caisson and the first photographs of the Paris sewers and catacombs. These two historical narratives—French modernity and Nadar's photographic innovations—were constructed using similar materials.

As images of infrastructure *and* photographic firsts, the series surveyed in this chapter assert the facts of their existence by revealing how they are constructed or maintained. In doing so, the production of the photographic image becomes coextensive with the physical functioning of infrastructure. For these images to exist, underwater caissons must hold the ocean at bay, the rectilinear stone architecture of the renovated sewers must facilitate space for underground movements, and human bones must support the crumbling underground tunnels of Paris. And in each case, labor—digging trenches, pulling sluice carts, and carting human remains—clears the way for the photographer and his camera. In pursuing these infrastructural images, Nadar also regularly elected to reveal the apparatus behind the production of his own images. All three series were made possible by electric light (the subject of the previous chapter). As if emphasizing these luminous apparatuses, Nadar included the lamps and wires within the frame of many of the images. In its reliance on a network of action, both for its physical existence and its maintenance over time, the photographic first finds its rhetorical twin in images of infrastructure.

Nadar's images of underground Paris have become iconic illustrations in cultural histories of the transformation of nineteenth-century Paris. Nadar's images in Marseille, however, have rarely been exhibited and even more rarely been written about. This elision is likely in part due to the photographer's primary association with Paris, his studio being a foundational part of many of the legends of European photographic Modernism and his images central to surveys of French art and visual culture. Being late images produced just a decade before his death, the images of Marseille don't fit neatly into the narrative surrounding Nadar's centrality to the story of Parisian Modernism, though of course Paris was itself reliant on the economic flow enabled by such ports. By attending to these sets of images together, this chapter argues that there is a parallel between the representation of civic infrastructure projects and the fabrication of Nadar's photographic firsts. In both cases, the category of the new was a value proposition and a media effect. Whether describing a sewer or a dock, or a photograph of a sewer or a dock, these processes of mediating the new depended on the circulation of the image to document (and sometimes to construct) a rupture with what had come before—one of the founding mythologies of European modernity. Tellingly, all three sets of photographs picture the novel technologies and infrastructures that brought them into being. But, just as importantly, the images bear the mark of the institutions and individuals that conscripted them to ideological—and historical—ends: the state, the newspaper, the museum, and, last but not least, Nadar himself.[4] Nadar was a particularly effective engineer of such materials, but like most engineers, he didn't work alone.

UNDERWATER WORK AT MARSEILLE

Nadar's move from Parisian metropole to southern port city echoed the mobility afforded by the massive French civil engineering projects of the nineteenth century, with new rail lines connecting Paris to Marseille.[5] Marseille's connection to the rest of France, and the world, was a subject of much speculation and planning throughout the nineteenth century. Commercial interests debated how Marseille's infrastructure and maritime economy might be developed in response to new developments in transportation technology and the extractive economy fueled by French imperialism. It was imagined that "Marseille will one day be the Liverpool of the Mediterranean."[6] Marseille had benefited from the French Atlantic colonial economy throughout the eighteenth century, but its long-held trade connections with the Levant and North Africa insulated it from the economic downturn experienced by other French ports at the end of the eighteenth century in light of world historical events like the Haitian revolution.[7] France's invasion and occupation of Algeria in 1830 and, later in the century, the opening of the Suez Canal in 1869 would only solidify this economic position.[8]

Marseille had long been a site of shipping and trade, featuring a fortuitous rocky inlet used as a harbor. However, the shift from sailing ships to steam ships enforced different demands on receiving ports, most notably

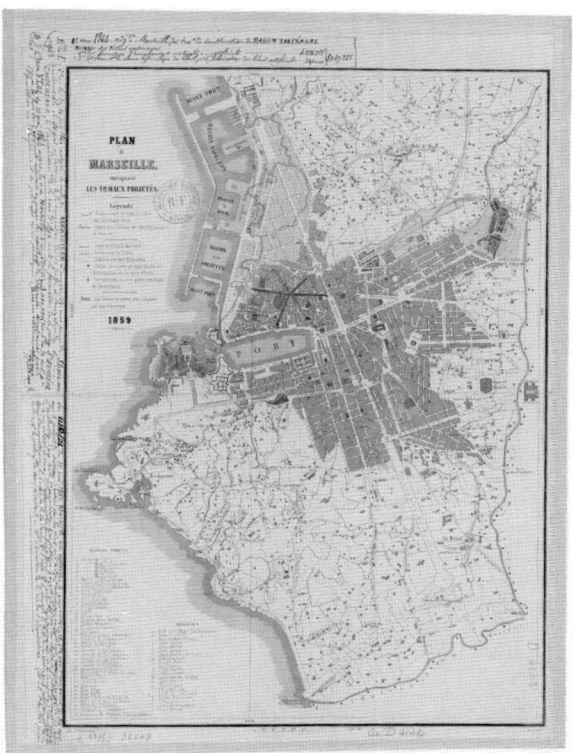

4.2 Imprimerie de F. Raibaud, *Plan du port du Marseille indiquant les travaux projetés* (Map of Marseille showing the planned works), 1859. 31 × 45 cm. Collection of the Département des Cartes et plans, Bibliothèque nationale de France, Paris.

in the size of the ships and the quantity of the cargo to be processed. Infrastructural expansion was deemed necessary to support Marseille's (and France's) economy. Beginning in the 1840s, a series of new docks were commissioned and built along the quai Joliette to the north of the Vieux-Port (visible in the upper left along the coast in figure 4.2).[9] Construction of the Pinède basin—which Nadar, Boissonnas, and Detaille would photograph—began in 1897, the fourth in a series of docks designed to expand the city's port along the coastal exterior.

The images of the construction site circulated in multiple formats. Originally used as source material for engravings published in a newspaper, the images Nadar and his colleagues took of the construction site were also bound into an album documenting the construction project that is now held in the archives of the Corps des ponts et chaussées, the national civil engineering body responsible for this kind of infrastructure development, and featured in cabinet cards printed by Paul Nadar emblazoned with the Nadar signature in gold, and in stand-alone albumen silver prints. Moving between newspapers, bureaucratic archives, and the Nadar studio collection, the photographs of the port under construction simultaneously became part of the story of the expansion of France's global trade and Nadar *père*'s photographic oeuvre. Images of infrastructure supported both France's economic power and the Nadar studio business.

In the handful of images of the construction project authored by Nadar, we see the labor, apparatus, and material techniques required to build infrastructure rather than the spectacle of its completed form. We see a crew of men working (figure 4.3). In the foreground, a man with a raised implement stands above another who is kneeling, while others look on in the background. The crew is building a walled trench with a wooden structure overhead. The exposure is hazy, the motion and clouded air a testament to the labor of these men and the uncertain atmosphere that they work in. The scene is harshly lit, though not, it would appear, by the light hanging at the top center of the image. While this image might appear to have been taken "underground," in fact the photograph depicts the interior of an underwater caisson (also known as a cofferdam), a watertight, compressed air chamber used in work on underwater construction projects, such as dams, ports, or ship repair.

The images were also, reportedly, yet another first. Describing an engraved reproduction of another of the photographs of the site in the newspaper *Le petit marseillais* (figure 4.4), the journalist Léon Larouaire writes:

> Our sketch represents the underwater construction site at work. It is a reproduction of a photographic print from M. Nadar, the master artist, who had already distinguished himself in 1858 by photographing the panorama of Paris from a balloon at an altitude of 350 meters. This was the first time—apart from Jules Verne's novels—that a photographer's lens had been aimed 12 meters into the depths of the sea, and the fact deserves to be emphasized.[10]

Larouaire argues that Nadar's underwater images represented the first time a photograph had been taken at twelve meters underwater. This is one of

4.3 Félix Nadar, *Travaux sous la mer à Marseille (Cap Pinède)* (Underwater work at Marseille [Cape Pinède]), ca. 1899 or 1900. Collection of the Département des Estampes et de la photographie, Bibliothèque nationale de France, Paris.

the only "firsts" in this book that Nadar himself did not claim, though there is no evidence that he disputed it. The fact that a newspaper could articulate these claims to priority independent of Nadar's own efforts is a testament to the mediatic power of the "first." However, as the reader will by now understand, Nadar's reputation for photographic innovation was something of a speculative project.

Like the first aerial photograph, the first portraits by electric light, and the first photographic interview, this first must be qualified, for it is

4.4 "Un chantier au fond de la mer" (A construction site at the bottom of the sea), *Le petit marseillais*, April 8, 1900. Fonds Nadar, YB3–2340, Box 7. Collection of the Département des Estampes et de la photographie, Bibliothèque nationale de France, Paris.

even more uncertain than those explored in the preceding three chapters. There are numerous examples of underwater photography dated earlier than Nadar's images. Two sets of images that are both more visibly aqueous and dated earlier are the underwater photography experiments undertaken by William Thompson in Weymouth Bay, Dorset (1856), and the experiments by Louis Boutan at the Arago Laboratory in Banyuls-sur-Mer, France (beginning in 1893, with a remarkable series dated to 1899). The latter even published a book on the subject.[11] But despite these predecessors, as *Le petit marseillais* would have it, Nadar was not only first but also making good on a tradition of literary prophecy exemplified in the scientific romances of the French novelist Jules Verne.[12] No matter the veracity of such claims, Nadar hung his hat on associating his name—trademarked, you will recall—with spectacular projects.

In the Corps des ponts et chaussées album, Nadar's images of the caisson interiors are placed in sequence with images by Boissonnas and Detaille documenting the various technologies in use and sites under transformation throughout the Pinède project.[13] Together, the images form a processual narrative describing infrastructural development. Unlike other prints of Nadar's caisson images, the prints included in the album are annotated with labels describing the technology represented, semantically pushing human labor to the background. For example, in one image contained within the album (figure 4.5), the writing beneath the image describes the photograph as of an "ordinary caisson," while the smaller text below this title describes which step in the construction process is pictured. The images are of technology in action and infrastructure in development, a temporality implied in the textual relationship between technology (the caisson) and construction technique (excavation). For a Third Republic French civic engineer, we can imagine that the images might act as evidence of the link between modern technology, like the caisson, and the progressive possibility of modern infrastructure facilitating the expansion of the docks and increasing the shipping capacity of the port.

The design of the Pinède docks used a foundation of concrete blocks based on a system originally developed to prolong the sea wall at Algiers.[14] In order to install these concrete blocks, the construction site made use of the aforementioned underwater caissons.[15] Images published in local and international journals (figures 4.6 and 4.7) documented the works on the Pinède docks and outlined the mechanics of the underwater caissons for their readers.[16] The rectangular lower space visible in the published diagrams is the space where Nadar was tasked with photographing the

4.5 Studio Nadar Boissonas et Detaille, *Caisson ordinaire: Execution des déblais* (Ordinary caisson: Execution of the excavations), 1898–1904. Collection of the École nationale des ponts et chaussées, Paris.

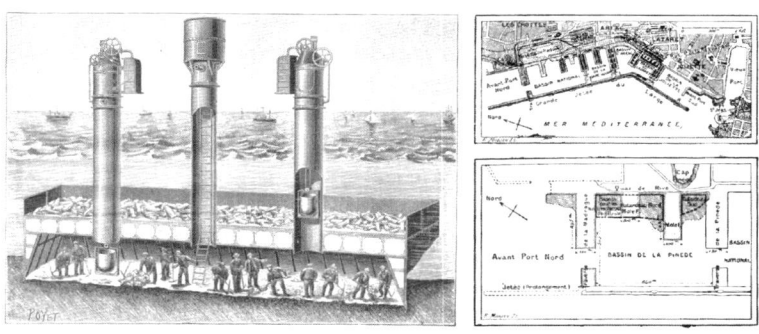

4.6 "Section of one of the caissons used in the work of enlarging the port of Marseilles," *Scientific American Supplement*, no. 1147 (December 25, 1897): 18331.

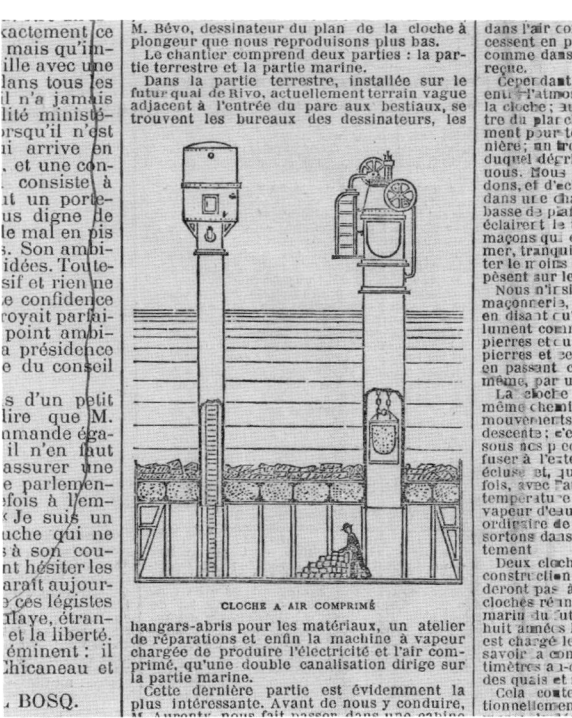

4.7 Unknown maker, print accompanying an article titled "À huit mètres sous l'eau" (At eight meters under water), *Le petit marseillais*, July 12, 1897. Fonds Nadar, YB3-2340, Box 7. Collection of the Département des Estampes et de la photographie, Bibliothèque nationale de France, Paris.

construction workers by artificial light. Press coverage of the construction project highlighted both the significance of the expansion for France's maritime economy and the novelty of the construction techniques used. Nadar's photographs extend this diagrammatic exposition of the relationship between infrastructure and France's economy. Journeying into the interior of the caisson with his camera, Nadar pictures the "technological environment" of the work site.[17]

The images of the caisson interior were put in dialogue with the broader narrative of the construction project. In the compiled album, the photographs of the caisson interior are placed between images of coastal quarries and the other machines and processes used in the construction of the

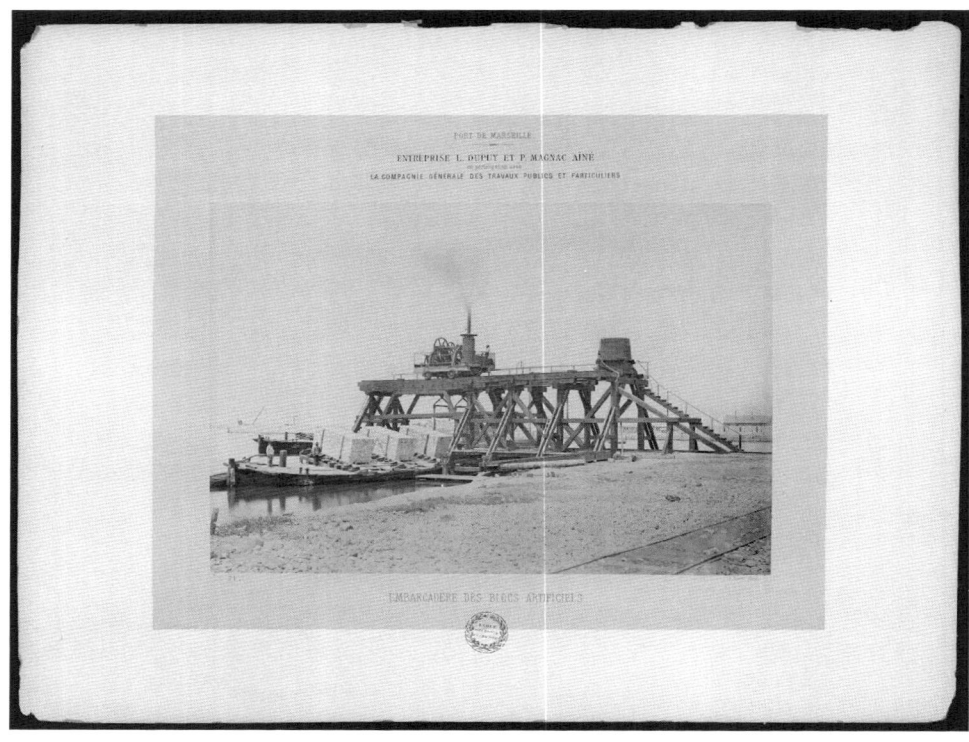

4.8 Adolphe Terris, *Embarcadère des blocs artificiels (Port de Marseille)* (Pier for artificial blocks [Port de Marseille]), ca. 1875–78. Collection of the École nationale des ponts et chaussées, Paris.

docks. Each image is labeled as to its utility within the construction project. This practice of narrating infrastructure projects like ports and railways in the form of the photographic album was widespread.[18] Another album created between 1875 and 1878 with photographs by Adolphe Terris offers a companion example of this genre of infrastructural album, also depicting the ongoing expansion of the Marseille port in the latter part of the nineteenth century (figure 4.8).[19] Like the album of the Pinède project, Terris photographed the technologies that facilitated the modernization of the Marseille port, depicting the construction of the National Docks. The annotation at the top of the image notes the source of the financial capital that underwrote the project, sandwiching the image between a description of its site (the Marseille port), the construction companies undertaking the project (Enterprise L. Dupuy et P. Magnac Aîné), and a label describing the infrastructure and material depicted (the pier and the artificial concrete blocks). Likewise bound into an album for the Corps des ponts et chaussées, Terris's images are marshaled to mediate civic engineering projects in bureaucratic spaces—a fate that parallels that of Nadar's images of the caisson interiors described above.

Both albums use the arrangement of photographs and text to describe the construction of infrastructure projects as the product of civil engineering and the financial expenditure of private or state interests. This method of presentation articulates the novelty of the project by sequentially depicting its coming into being—an iconographic symbol of the physical impact such infrastructure projects had on the environment. In the second half of the nineteenth century, French civil engineers were increasingly fascinated with the medium of photography. This was, at least in part, due to the addition of photographic instruction to the curriculum of the École des ponts et chaussées in 1858.[20] Photography became, as Sean Weiss has described, "a generative medium for the construction of the metropolis," not only documenting construction sites in urban centers and coastal ports but also, crucially, acting as a point of mediation between construction sites, engineers, organs of state bureaucracy, and publics.[21] Photographs could be reproduced and circulated, used to garner public support for new projects and to bolster political buy-in for ongoing modernization initiatives—even facilitating the raising of funds for building projects.[22] As part of this trend, Nadar's contributions to the Pinède album participated in a broader national project seeking to promote the role of engineering and public works in service to the French nation state via the expansion of France's largest Mediterranean port.[23] That one of Nadar's photographic firsts was tied up in this story is significant.

Characteristically unsatisfied with contemporary notoriety, Nadar also integrated the images of the Pinède project into a longer historical trajectory. In doing so, he drew his own exploits into a heroic history of innovation in underwater exploration and civil engineering originating in the ancient world. An unpublished manuscript by Nadar titled "L'emploi de l'air comprimé dans les travaux publics" describes the history of compressed air chambers in the construction of public infrastructure.[24] Nadar begins his text with a reference to such technology in the "time of Aristotle," referring to the polymath Greek philosopher's writings on the problem of divers' eardrums bursting.[25] After briefly summarizing several innovations in underwater exploration, such as the eighteenth-century English engineer John Smeaton's diving bell (which had also been used to facilitate underwater construction in an infrastructure project), Nadar's history jumps to developments in underwater construction in the nineteenth century. For Nadar, this history provides a set of precedents for the technologies at work in the Pinède construction project.[26] Marginalia in the manuscript include several small sketches outlining specific construction techniques associated with the project, indicating Nadar's interest in the engineering requirements for this kind of underwater construction project. Nadar's involvement was thus to be considered innovative not only in his use of electric light to photograph underwater but also in his choice to foreground the specific technological conditions that enabled such a feat to be undertaken, not to mention the challenges of photographing in such an environment.

But what kind of image of infrastructure does Nadar show us? Lit up by the glare of electric light, the scene is not observably aqueous. The disconnect between the marine environment hinted at beyond the frame of the photograph and the relatively dry scene pictured serves to further underscore the act of picturing (and the infrastructure project itself) as a supposed triumph over the natural world. But, more precisely, what we see are workers toiling underwater in the interior of the caisson, digging into the seabed off the coast of Marseille and installing concrete blocks to stabilize the new docks. Wires hang from the wooden rafters, powering the light necessary both for the work to occur and the image to appear. Workers and photographer alike subsist on the air provided by the technological environment of the caisson, not unlike the divers' helmets and suits (or Nemo's submarine) that allowed for other nineteenth-century experiments with underwater photography. In sum, these are images of workers in the mud, laboring in an apparatus designed to transform the natural world to create infrastructure to facilitate the exchange of capital. This is not as tidy a description as the "first" underwater photographs at a depth of twelve meters, but it may be a more accurate one.

While contemporary scholars of infrastructure have often suggested that infrastructure is mostly invisible until it breaks down, in the nineteenth century, infrastructure was a spectacular emblem of the historical period known to its promoters as modernity.[27] Infrastructure had to be revealed to later be concealed. And if modernity was to be legible as a radical break with the past, it had to be pictured.[28] Images of construction projects were a very tidy way to do so. Nadar's images of the coming into being of infrastructure, lit up by electric light and described as "first," become a self-portrait of this project, of construction, of picturing. The "underwater" images at Marseille were thus part of the construction of both France's *and* Nadar's "modernity"—reputations that were incomplete and under construction.

THE PARIS SEWERS

As with his photographs "underwater," Nadar's photographs of Paris sewers parallel the production of the image with the working of infrastructure—this time the putrid inner workings of Paris's expanding sewer network.[29] In an account of his underground journey, Nadar acknowledged the doubly messy reality of picturing working infrastructure and working with electric light underground:

> But I couldn't say how many times our work was interrupted, stopped, for one reason or another. Sometimes the weakened acids were not sufficiently replenished and we had to stand at attention during these unpleasant delays, suspending all operations. Twice I had to change the operator whom we had contracted to supply our light. Should we still talk about our disappointment, our anger, when, after many efforts to overcome a difficult problem, at a moment in which, having taken all precautions and eliminated or circumvented all obstacles, our decisive operation was nearing its end—suddenly, in the last seconds of the exposure, a cloud rising from the canal came to veil our photograph—and how many imprecations, then, against the beautiful woman or the good man above us, who, without suspecting that we were there, chose that exact moment to replenish the water in their bathtub![30]

These were working sewers, already integrated into the daily life and movements of urban citizens and their waste. The "weakened acids" of the Bunsen batteries powering the lights illuminating the scene had only just been replenished when wastewater in the form of a "cloud rising from the canal" moved to obscure the image. These textual descriptions do more than provide context for the making of the photographs. They animate the still images as moments grabbed during the active operation of civic infrastructure, but they also highlight a contradiction: It is often difficult to stage an image of infrastructure while it is working as intended. To create the image, the batteries, lights, and camera had to work together with the architecture of the sewer and the movement of the water, not to mention the photographer, assistants, and others working above ground to ensure a consistent power source for the lights.[31] Figuring sewer and camera, wires and water, Nadar's texts and images register the parallels between the production of the image and the infrastructure undergirding social life in urban Paris.

As with Nadar's aerial photographs, his photographs of the Paris underground are often misdated (including by himself), and his two series of underground photographs (of the sewers and catacombs, respectively) often have been collapsed into a single project. These chronological inconsistencies have been clarified by Sylvie Aubenas, who convincingly dates the photographs of the catacombs to 1862 and those of the sewers to 1865.[32] As Aubenas notes, the elision of the two projects was first suggested by Nadar himself, who, in the collectively authored *Paris-Guide*, published in 1867 for the Exposition universelle in Paris, recounts the two distinct projects as if they were the result of one long, arduous underground journey, not unlike the collapse of multiple balloon journeys into one described in chapter 1.[33] The photographs of the sewers were executed with the support of Eugène Belgrand, the engineer appointed director of sewers and waterworks by Baron Haussmann in 1855. The experiments also came, however, at great personal expense to Nadar, estimated by him at about "75,000 francs—equally divided between each project—although at the eleventh hour M. Belgrand glimpsed, or rather guessed, that I was financially desperate and offered me the cooperation of the city, which was just then in the process of taking measures to auction my furniture to pay my delinquent tax bill."[34] Rescued (if only just) from insolvency, Nadar would continue his project of representing the Parisian underground as part of French history writ large and, of course, as part of his personal record of photographic accomplishments.

Belgrand repeatedly affirmed the significance of the sewer renovations—"this magnificent network"—in the crafting of a modern Paris, situating the nineteenth-century renovations within a longer archaeology of the Parisian underground in a manner similar to Nadar's historicization of the use of compressed air in underwater construction.[35] Until the fourteenth century, the largest waste-disposal system in urban Paris consisted of open sewers, simple gutters running down the middle of the street.[36] The first underground sewer was built underneath the rue de Montmartre in 1370, draining into a tributary of the Seine.[37] After centuries of halting renovations, and spurred on by multiple cholera outbreaks, rapid development of the city's waste-management infrastructure was undertaken in the nineteenth century. The renovations were vast; while the Second Empire renovations of Paris saw the total length of streets double (424 kilometers to 850 kilometers), the underground sewer system grew by a magnitude of five times (143 kilometers to 773 kilometers).[38] One of the most important innovations in the sewer expansions was architectural, replacing hewn stone with millstone and cement mortar, allowing for rounded floors, increasing circulation, and preventing material buildup. The expansion of the network also included the addition of various instruments and vehicles designed specifically for managing the sewer, such as sluice carts or boats developed to help clear sediment from the bottom of the sewer canals, elements that figure prominently in many of Nadar's images.[39]

Lest we forget, the sewer network was only perceptible to Nadar's camera thanks to electric light—a photographic technique that, as explored in the previous chapter, Nadar was only newly used to working with. The apparatus enabling the illumination of civic infrastructure is front and center in many of the images of the sewers. In one photograph (figure 4.9), the light provided by the arc lamp whites out the walls of the sewer gallery, accentuating the dark tunnels beyond and visibly positioning the source of illumination as one of the primary subjects of the image. This expansive view is contrasted with a handful of images that focus on details of specific technologies making up the modern sewers, such as details of pipes along the walls and ceiling of the sewers. Though we don't necessarily have a sense of intended sequencing, this oscillation between the part and the whole recalls the relationship between technology and infrastructure in the album depicting work on the Pinède docks. The smooth linearity of the gallery is accentuated by the tidy functioning of mechanisms designed to ensure proper circulation through the spaces of the sewers. In another image, the relative safety of the modern sewer is emphasized via the visible metal handrails and walkways, innovations unique to Belgrand's renovations (figure 4.10). In the same image, the mass of wires powering the electric light illuminating the image parallels these architectural forms. These images are imbued with cleanliness and order. Indeed, Haussmann was initially reticent to allow for the entrance of human feces into the renovated system, a component of the contents of a sewer that we might take for granted.[40]

The maintenance of the sewer network was, like the construction site described above, dependent on human labor. However, Nadar struggled

4.9 (*above left*) Félix Nadar, *Égouts de Paris: Chambre du Pont Notre-Dame, n°1* (Paris sewers: Notre Dame Bridge room, no. 1), ca. 1864–65. Collection of the Département des Estampes et de la photographie, Bibliothèque nationale de France, Paris.

4.10 (*above right*) Félix Nadar, *Égouts de Paris: Éclusée n°1* (Paris sewer: Lock no. 1), ca. 1864–65. Collection of the Département des Estampes et de la photographie, Bibliothèque nationale de France, Paris.

to include live figures in his images of the sewers due to the dim light and still lengthy exposure times of the 1860s. Labor is instead represented by mannequins dressed up as sewer workers (figure 4.11). We might explain away the difference as one of technology, the intervening years speeding exposure times capable of representing live bodies. But Nadar's deployment of leaden mannequin workers is an uncanny acknowledgment of the hidden labor of municipal infrastructure maintenance. This disembodied image of labor offers a glimmer of a radical politics of the underground. Nadar, like so many commentators, understood the underground to be a space of political possibility, a space where the repressed parts of society might burst forth.[41] Describing parts of the sewer as "coliseums," Nadar later noted that they might "offer quite useful points for the concentra-

4.11 Félix Nadar, *Égouts de Paris: Rue du Château d'Eau, n°2* (Paris sewers: Rue du Château d'Eau, no. 2), ca. 1864–65. Collection of the Département des Estampes et de la photographie, Bibliothèque nationale de France, Paris.

tion of forces in some contingency," lamenting that the leaders of the Paris Commune of 1871 had not made use of this vast underground network to fortify their opposition to state violence.[42] Whether full of water, excrement, or socialists, the sewer was a revolutionary place of circulation.[43]

Photographing the new sewers in all their glory was itself considered an achievement. Like the images of the underwater caissons, different prints of Nadar's images of the sewers bring different elements into focus. Annotations on different extant prints emphasize the different parts of the image or the process pictured, functioning almost like intertitles in early cinema. For example, a series of prints of the images of the catacombs and sewers donated to the Conservatoire national des arts et métiers in 1884 added additional information in the form of annotations.[44] Each print in the collection featured a handwritten note with either "*catacombes*" or "*égoûts*" "*de Paris*," accompanied by the recurring phrase "*premier essai de photographie part la lumière électrique* (1860)." Below this qualifying phrase rests Nadar's eponymous signature. On one of the prints of the sewers, the date 1859 is crossed out in favor of the parenthetical (1860)—a rhetorical uncertainty that spotlights the strategic role of these annotations in the larger media-historical narrative of Nadar's electric firsts.[45]

These images were part of a group of photographs given by Nadar to the Conservatoire national des arts et métiers in 1884, part of the museum's photography collection amassed under the supervision of director Aimé Laussedat beginning in 1881.[46] In a letter to Nadar from 1884, Laussedat notes that he had just returned from Nice, where he had seen Nadar's "first" aerial photograph exhibited alongside several images by electric light. This display, which must have been at the Exposition universelle that took place from January through May that year in Nice, prompted Laussedat's request for Nadar to donate copies of each print to the museum. From the sewer to the museum, the first becomes an ouroboros in form and in fact: A museum director sees an image labeled as first and wants one of his own; the photographer obliges and, in the process, labels other images as firsts of their own. In other words, the first cannot exist at the level of the image alone but instead is created through the exhibition, collection, and circulation of the image *as first*.[47]

The photographic first is unique in that, as a descriptor, it describes not only an object but also a relationship between objects: This, not that, is first. Likewise, as Brian Larkin has put it, the "peculiar ontology" of infrastructures "lies in the fact that they are things and also the relation between things."[48] Like the Pinède docks, the sewers were infrastructural forms designed to facilitate systems of circulation. Nadar's interest in photographing these sites was thus twofold: capturing new infrastructural forms and revealing the photographic techniques and apparatus necessary to picture these spaces. By tracing how these images of infrastructure were defined as "first," we can see the contours of how media have been used to articulate the "newness" of events, images, and infrastructure, both real and imagined. Nadar pictures apparatus and infrastructure, carbon arc lamp and smooth sewer wall, and in doing so he also pictures the conceptual faultline central to many representations of infrastructure: They present objects rather than systems.[49] But follow the battery wire along the edge of the image, the wastewater down the tunnel, or the proxied body of the municipal laborer, and you see a networked set of relations emerge. Both municipally sponsored images of infrastructure and Nadar's photographic firsts thus stage a confrontation between the new and the already existing, a dialectical tension that can be interpreted either as evidence for novelty or proof of the emergence of the new from the skeletons of old.[50]

THE PARIS CATACOMBS

The sewers were the second of two underground Parisian sites that Nadar photographed in the 1860s. First, he had photographed the macabre, cavernous spaces of the city's newly opened catacombs (see, for example, figure 4.12). Nadar's electrically illuminated photographs of the catacombs were a response to the perceived failure of other techniques of representation.[51] The mining and quarry engineer Ernest Lamé-Fleury had suggested that Nadar consider photographing the catacombs, writing to Nadar in 1861:

> An amateur has just published a small study under the title, *The Catacombs*, that I naturally procured to see what was being said about my obscure domain. A dreadful lithograph gives the most inaccurate idea of the ossuary. I would be happy, sir, if you would allow yourself to be tempted by the enterprise of applying your beautiful electric photography to create an exact and picturesque representation of one of the most unusual Parisian curiosities.[52]

The suggestion to photograph the catacombs was prompted by Lamé-Fleury's awareness of Nadar's expertise in the domain of "electric photography," discussed at length in the preceding chapter. As with the description

4.12 Félix Nadar, *Catacombes de Paris: Façade n°11* (Catacombs of Paris: Facade no. 11), 1861. Collection of the Département des Estampes et de la photographie, Bibliothèque nationale de France, Paris.

of Nadar's images of the interior of the underwater caisson as the "first" time a camera had operated twelve meters under water, here Nadar's "first" experiments with electric photography provided the infrastructure for another "first": the "first" photographs of the catacombs. Firsts beget other firsts.

Nadar and Lamé-Fleury were not alone in their fascination with the catacombs. Nadar's photographs participated in a larger project to define the symbolic and literal uses of *Paris souterrain* past and future.[53] In the case of the catacombs, the future was quite literally to be arranged out of the bones of the past. The renovated catacombs were part of the broader modernization of Paris over the course of the nineteenth century.[54] The literary historian Philippe Muray has gone so far as to state that the nineteenth century only truly began when municipal workers began to transfer human remains from urban churchyard cemeteries to the underground quarries-turned-catacombs.[55] Epochal transformation or not, the catacombs were initially designed to address two significant problems facing the Parisian municipal administration in the late eighteenth century: the lack of space available for new burials in church graveyards and concerns over the instability of the city's underground quarries.[56] The quarries had been used since antiquity as a source of materials for the city's expansion but, by the nineteenth century, had begun to cause monumental collapses of infrastructure, such as the sinkhole that appeared on the very aptly named rue d'Enfer in 1774.[57] As the streets now hold the possibility of opening up beneath one's feet and seizing life, mass public burial sites within the city limits began to return the dead to the living in unsettling ways. In 1780, the inhabitants of the streets surrounding the Cemetery of the Holy Innocents saw their houses invaded by human remains that had broken through the basement walls.[58] Public health concerns eventually led the government to close this cemetery, and later all other urban cemeteries.[59] On April 7, 1786, the quarries were consecrated and transformed into the Paris catacombs.[60] While the catacombs were drawn into grisly confusion and disrepair throughout the revolutionary period, they were subsequently renovated and transformed into a public monument that opened in 1809.[61]

Nadar's images of underground Paris are rarely discussed without reference to his essay on the subject, first published in the 1867 *Paris-Guide, par le principaux écrivains et artistes de France*.[62] After describing the entry into the underground ossuary, Nadar recounts the modern history of the renovated catacombs much like I have here. For him, the reorganization of the material substrate of the Parisian past had resulted in an "egalitarian confusion of death."[63] Nadar fears that the repetition of human skeletons arranged into different forms may bore guests, for, as he puts it, "the points of view are not varied," and that they might "have already discovered, like me, that a few steps in these subterranean passages are enough to satisfy your curiosity."[64] Nadar's text positions his work as somewhere between a public service and a professional obligation, the underground completing the immense panoramic and topographical project initiated by his aerial experiments.[65]

As Nadar's images and the guide text announce, the underground is (like the view from above) a necessarily mediated space that, absent firsthand experience, is not visible or accessible without representational or technological intervention. "The subterranean world was opening up an infinite field of operations no less interesting than the telluric surface," writes Nadar. "We were going to penetrate, to reveal the mysteries of the deepest, the most secret caverns."⁶⁶ With this language Nadar joins other writers, scientists, and thinkers in the nineteenth century who described journeys into the underground as part of a process of acquiring knowledge about the past.⁶⁷ For "the subterranean environment," writes Rosalind Williams, "is a technological one—but it is also a mental landscape, a social terrain, and an ideological map."⁶⁸ Excavation, Williams notes, becomes a metaphor for intellectual inquiry. Inquiry, mediation, excavation, and representation all involve, in one way or another, the journey from known to unknown, the transfer of knowledge or a movement through space via a mediating material. The underground, like the underwater space discussed in the first section of this chapter, is therefore a particularly resonant site to observe the processes of mediation that, I argue, underwrite the photographic firsts traced across this book.

For Nadar, this mediation occurred, in part, by way of the illuminated glare of electric light and his camera lens. The photographer does not recount the specifics of his electric light apparatus in photographing the catacombs (as he does with the sewers), but we can infer from glimpses of the lighting setup in the photographs that the arrangement was like the combination of an arc lamp attached to Bunsen batteries with wires used in both his studio portraiture and the later photographs of the sewers. In several of the images, the glaring quality of the electric light and visible parts of the lighting apparatus are accentuated, rendering the series as much a portrait of the power of electric light as the catacombs.⁶⁹ This dialectic between infrastructure and apparatus becomes the subject of several of the images. One particularly dramatic image pictures a face-off between the lighting apparatus, a monument constructed of skulls and femurs, and a stone cross featuring a Latin Bible verse flanked by two obelisk-like columns (figure 4.13).⁷⁰ The verse comes from the Book of Daniel, a text containing an eschatological prophecy from the Old Testament. The electric light illuminates this text while also affirming its correlative power to unveil—biblical prophecy juxtaposed with the revealing glare of electric light. Depicting a set in action, the image documents its own production.

Alongside the production of electricity necessary for photographic exposure, the organization of labor required to render the quarries into catacombs is visible in the juxtaposition of several images of the catacombs. The decorative order of neatly stacked skulls and bones in some images is contrasted with the disorderly pile of human remains in others. This labor is further represented by the inclusion of mannequins in several

4.13 Félix Nadar, *Catacombes de Paris: Façade n°12* (Catacombs of Paris: Facade no. 12), 1861. Collection of the Département des Estampes et de la photographie, Bibliothèque nationale de France, Paris.

of the photographs, as in the images of the sewers, standing in for the municipal workers employed to arrange and organize bones deposited in the catacombs (figure 4.14). In Nadar's words,

> I had judged it advisable to animate some of these scenes by the use of a human figure, less in order to render them picturesque than in order to give a sense of scale, a precaution too often neglected by explorers in this medium, sometimes with disconcerting consequences. For these eighteen minutes of exposure, I found it difficult to obtain absolute, inorganic immobility from a human being. I tried to get around this difficulty by means of mannequins, which I dressed in worker's overalls and arranged on the stage with as little awkwardness as possible; this detail did not complicate our tasks.[71]

4.14 Félix Nadar, *Catacombes de Paris: Mannequin n°10* (Catacombs of Paris: Mannequin no. 10), 1861. Collection of the Département des Estampes et de la photographie, Bibliothèque nationale de France, Paris.

Like the inclusion of the wires and lights, this admission reveals both technique and apparatus. The description of the lengthy exposure time and the problem of representing scale in so unusual a site is intended to fortify the reader's perception of the adventure of photographing underground while also drawing attention to technical choices made by the photographer. The available photographic technologies condition the possibility of representing labor, but the decision to describe the inclusion of labor as a product of a desire to represent scale rather than work is part and parcel of the ideological work of these images. Despite himself, Nadar draws attention to the daily labor of infrastructural maintenance.

The material mobility of photography offered engravers working for newspapers such as *Le monde illustré* a new basis from which to develop

4.15 Henry Duff Linton, "L'ossuaire (D'après la photographie faite à la lumière électrique par M. Nadar)" (Ossuary [after a photograph made with electric light by Mr. Nadar]), *Le monde illustré*, October 21, 1865, 261.

images of the catacombs. Working from Nadar's images of mannequins qua workers arranging human bones in the catacombs, the resulting engravings restore life to the mannequin (see figure 4.15). Other scenes are populated with fashionable visitors to the underground, all set within the strange and morbid landscape that Nadar photographed.[72] In these images, the brilliancy of the electric light recedes, the technical requirements of creating such images smoothed over in the translation to newspaper illustration. The accompanying captions and texts, however, do not fail to describe these images as ones made by electric light. Though photography had yet to be materially incorporated into the illustrated newspaper in the 1860s, this chain of mediations—electric light, photographic negative, photographic

4.16 Félix Nadar, *Catacombes de Paris: Crypte n°9* (Catacombs of Paris: Crypt no. 9), 1861. Collection of the Département des Estampes et de la photographie, Bibliothèque nationale de France, Paris.

print, engraving, illustrated weekly—was still tied to its source (a "new" application of photographic technology) by the words "by electric light."

As if underscoring the dual subject matter of the series—the new architecture of the renovated catacombs and the expansive glare of electric light—a retrospectively annotated print shows us a disorderly pile of bones with the caption "electric light" in capital red letters and "the catacombs of Paris" in slightly smaller black script (figure 4.16). Reminiscent of the attestations to priority used on the various prints that have come to stand in for the first aerial photograph and the labels assigned to the images of the caissons described earlier in this chapter, these annotations reveal what Nadar, at least retrospectively, believed to be the true subject of these prints. This double motif was also noted by journalists remarking upon the difficulties encountered throughout the photographer's journey into the underground. "Half asphyxiated by the noxious gases from the electric battery under those suffocating vaults," writes a journalist in *Le moniteur universel*, "Monsieur Nadar has already managed to assemble about twenty perfectly successful negatives."[73] In a similar though more suggestive tone, the *British Journal of Photography* recounted how Nadar had "electrified the mortal remains of the past generation in a remarkable manner."[74] In each instance, the visual or journalistic narratives emphasize the production of the photographs as much as the site that they picture, if not more so. Likewise, when two of the prints of the catacombs were exhibited at the International Exhibition of 1862 in London, they were shown alongside other examples of Nadar's experimentation with the application of electric light to photography and described as "two photographs made with the help of the same light in the catacombs of Paris."[75] Stories of the difficulty of the journey underground, descriptions of the technical feat exemplified in the production of the images, and the allegorical power of electric illumination demarcate the images of the catacombs as a first.

Like the firsts described in other chapters, the first photographs of the Parisian catacombs are better understood as an event or a process rather than a single iconic image—a characteristic of the first that Nadar implicitly acknowledges in his different accounts of the production of photographic images. Firstness is articulated in a processual manner across different forms of media. Tracking this daisy chain of mediations used to construct the narratives of priority around these images also reveals its own infrastructure. Published texts (often in iterative forms), manuscripts, photographic albums, captions, and newspaper reproductions of photographs, among other media, are marshaled to produce an image, or set of images, as "first." To understand these projects in this way is to sometimes push the photograph to the side, to imagine its movement, to listen for it in the body of texts, to see its echoes on the pages of newspapers, and to examine its iteration in different photographic techniques.

Underwater at Marseille and underground in Paris, infrastructure is front and center. But the techniques by which these photographs were produced are also in focus. It is by way of this iconographic union (between caisson and wired light, sewer and battery wire, catacomb and carbon arc lamp) that the image of infrastructure reveals itself. As I've argued here, these images form part of the infrastructure undergirding the invention of Nadar as a figure of historical significance. The production of these images, and their continued circulation in histories of both Nadar and French culture, demonstrates the entangled, contested legacies of both man and nation. It is particularly fitting that some of Nadar's last images picture the construction of a port—a site of exchange and mobility. That it is Marseille's port, infrastructure key to France's colonial project and a point of France's connection with a global economy, should, I argue, prompt the rereading of all his images with this infrastructural frame in mind. In the next chapter, I examine how all these stories come together in Nadar's idiosyncratic memoir-qua-media-historical tome *When I Was a Photographer*. Together, these firsts make history.

5

WHEN I WAS A PHOTOGRAPHER, OR THE HISTORY OF PHOTOGRAPHY IN PHOTOGRAPHIC FIRSTS

Photography itself was once a first. This invention is the precondition for the firsts described across the previous four chapters of this book. Inflammatorily describing the introduction of this medium to our world, Nadar writes: "Exploding unexpectedly, totally unexpectedly, surpassing all possible expectations, diverting everything that we thought we knew and even what could be hypothesized, the new discovery indeed appeared as, and still is, the most extraordinary in the constellation of inventions that already have made our still unfinished century—in absence of other virtues—the greatest of the scientific centuries."[1] Photography's invention was, for Nadar, unexpected—a surprise. This revelation was therefore unlike the other firsts described across this book, which, to varying degrees, were all positioned as the inevitable fulfillment of a technical idea. In fact, in Nadar's formulation, photography modified this equation entirely. Transforming "everything that we thought we knew and even what could be hypothesized," photography is an allegorical technology: both a chemical process by which other technical hypotheses could be realized and a medium whose history could be extrapolated to explain invention writ large. Surveying Nadar's idiosyncratic book *Quand j'étais photographe*, in this chapter, I argue that, for Nadar, photography was a technical trope representing the transformation of historical experience.[2] The semi-autobiographical collection of short, distinct chapter-length texts puts Nadar's firsts in constellation with many of the other inventions of the nineteenth century, aligning photographic novelty with other large-scale social, political, and technological transformations.

Originally published as a series of articles in his son Paul's journal *Paris-Photographe*, *Quand j'étais photographe* recounts many events familiar to historians of photography, alongside some lesser-known stories. But rather than simply repeat the even then ossified list of events leading to the invention of photography and subsequent innovations, the book uses a series of stories, reminiscences, and tall tales to describe how photography transformed everyday (and not so everyday) experiences. Photography, Nadar suggests, is a key example of the influence of technology on modern life and suggestive of the possibility of other new technologies having a similar impact. It is this framework that makes the book such a compelling historical document. It may not be a particularly accurate account of photography's history, but, in a close to final act of invention, Nadar models just how important media representations were (and are) to the rhetoric of innovation.

Throughout *Quand j'étais photographe*, Nadar describes the relationship between media representations and the individuals and technologies that formed the cultural surround of photography's "first century," painting a dynamic picture of when photography was new.[3] In what follows, I examine a handful of chapters from *Quand j'étais photographe* to examine how Nadar narrates the story of photography's "first" century. While I draw attention to only five of the chapters here, the reader will note that frequent reference has already been made to *Quand j'étais photographe* across this book. In many ways, this small volume from the fin de siècle has been central to the invention of Nadar. To complete my study of this invention, this chapter sifts through the narratives that Nadar wrote about inventions other than his own. These texts, I argue, demonstrate how Nadar imagined his own firsts within that "constellation of inventions" that characterized what was, in Nadar's words, "the greatest scientific century." For Nadar, photography was a model for how inventions become historical, moving from the "universal stupefaction" experienced at photography's introduction toward speculations on its many novel applications and, finally, arriving at a record of the medium's experimental legacy, all written down as "history."[4] This process—from new media to historical subject, from idea to fact—parallels the production of the photographic first. But as I've demonstrated over the last few chapters, firsts are not facts, however charismatic their reputation might have become.

PHOTOGRAPHY AS NEW MEDIA

Technical novelty is often registered in the awe, surprise, or fear experienced by the user or audience. Such intense emotions are frequently recorded in reports of the introduction of new media.[5] The first chapter of Nadar's book, "Balzac and the Daguerreotype," describes just such an encounter, recounting the French realist writer Honoré de Balzac's reported fear of being photographed.[6] But more than esoteric superstition—or, as Nadar puts it, "a vague apprehension of the Daguerreian operation"— Nadar's written account of Balzac's photographic reticence is also an example of what anthropologist Michael Taussig has called the "mimesis of mimesis."[7] Taussig, describing a scene from filmmaker Robert Flaherty's 1922 film *Nanook of the North* where an Inuk man, Allakariallak (the titular Nanook), bites a phonograph record, suggests that accounts of the "alleged primitivism" of an individual encountering a novel technology for the first time are just as much about those doing the accounting. Or, as Christopher Pinney has put it, summing up similarly anthropological accounts of "primitive" enchantment with modern technologies of reproduction, "the native's 'difficulty' with photography provides the alibi for the modern's own desire to find in the photographer the descendants of 'augurs and haruspices.' "[8] Indeed, for Nadar, the story of Balzac and the daguerreotype was one of cyclical enchantment and disenchantment, indexing Nadar's own, almost ethnographic, desire to articulate the power of the new (or the first). As with other accounts in this book, my interest here lies less in assessing whether Nadar's story about Balzac is "true" than in accounting for what the story allows Nadar to say about the history of photography as a novel technology.

Much of *Quand j'étais photographe* is devoted to articulating Nadar's own theory of technicity. In our photographer's estimation, the history of photography offered readers both an example of a generalized affective response to the introduction of a novel technology *and* the story of the widespread adoption of that same technology. Writing in the last decade of the nineteenth century, more than fifty years after the announcement of the French state's purchase of the daguerreotype process in 1839, Nadar suggests that readers would do well to remember the "universal stupefaction" experienced in response to the introduction of photography.[9] In his estimation, a recurring disbelief in the possibility of new technologies causes individuals to react in fear rather than appreciation when confronted by the new. "By nature," Nadar writes, "we are hostile to everything that disconcerts our received ideas and disturbs our habits."[10] This is a problem. When individuals with power to facilitate the development of new technologies—for example, members of the powerful French Académie—fail to believe in the possibility of the new ideas presented to them, technology (read: society) is doomed to stagnate. The phonograph, Nadar reminds us, had just recently been dismissed as a "ventriloquist hoax" by a "member of the Institute."[11] From the vantage point of the 1890s, Nadar's point here is twofold: Firstly, it is human nature to be fearful of the new, but, secondly, this alarm is quickly supplanted by widespread acceptance of the new technology. Photography's history, Nadar suggests, offers a useful model for understanding the lifecycle of novel technologies. It is from society that technology emerges; therefore, a culture that recognizes how "germinating ideas" are realized will be all the better equipped to capitalize on the productive forces and cultural expressions enabled by these new technologies.[12]

But, as Nadar queries, was "Balzac's terror before the Daguerreotype ... sincere or simulated?"[13] In the text, the author's "terror" of photography is attributed to his (mis)understanding of photography's material ontology. Ventriloquizing Balzac, Nadar writes:

> Each body in nature is composed of a series of specters, in infinitely superimposed layers, foliated into infinitesimal pellicules, in all directions in which the optic perceives this body.
>
> Since man is unable to create—that is to constitute from an apparition, from the impalpable, a solid thing, or to make a *thing* out of *nothing*—, every Daguerreian operation would catch, detach, and retain, by fixing onto itself, one of the layers of the photographed body. It follows that for that body, and with every repeated operation, there was an evident loss of one of its specters, which is to say, of a part of its constitutive essence.[14]

According to Nadar, Balzac believed that a photograph was a material remnant of that which had been photographed, the resulting image akin to a spectral skin peeled off its subject. Balzac's purported philosophy of photography is likely derived from the Epicurean Roman poet Lucretius's account of the nature of images.[15] For Lucretius, reflections on the surface of water or a mirror—in antiquity, made of metal, not unlike the surface of a daguerreotype—are a model for understanding the atomistic universe and the nature of vision.

Balzac's "theory" of photography as a skinning of the filmic layers of the material world also helps Nadar to articulate an epistemological point: "Man is unable to create," nor to "make a *thing* out of *nothing*." Balzac himself put it in these terms in his 1847 novel *Cousin Pons*. Embedded in a philosophical interlude occasioned by the plotting landlady Mme Cibot's visit to a fortune teller, Balzac describes the unlikeliness of photography's invention:

> The steam engine was condemned as absurd, aerial navigation is still said to be absurd, so in their time were the inventions of gunpowder, printing, spectacles, engraving, and that latest great discovery of all—the daguerreotype. If any man had come to Napoleon to tell him that a building or figure is at all times and in all places represented by an image in the atmosphere, that every existing object has a spectral intangible double which may become visible, the Emperor would have sent his informant to Charenton for a lunatic.[16]

These inventions, described as unfamiliar or even unlikely, are like "image[s] in the atmosphere." Several pages later, Balzac explicitly aligns the predictive power of fortune telling with the creative power of invention, suggesting that both are parallel modes for thinking about the future: "Certain beings have the power of discerning the future in its germ-form," just like "the great inventor sees a glimpse of the industry latent in in his invention, or a science in something that happens every day unnoticed by ordinary eyes."[17] The acts of inventing and telling the future are cut from the same cloth, as

> the world of ideas is cut out, so to speak, on the pattern of the physical world ... As, for instance, a corporeal body actually projects an image upon the atmosphere—a spectral double detected and recorded by the daguerreotype; so also ideas, having a real and effective existence, leave an impression, as it were, upon the atmosphere of the spiritual world; they likewise produce effects, and exist spectrally ... ,

and certain human beings are endowed with the faculty of discerning these "forms" or traces of ideas.[18]

Balzac parallels the photographic image and the latent idea. This concept greatly appealed to Nadar—recall his role as an "archivist" of nascent aerial ideas described in chapter 1—as he certainly imagined himself capable of "discerning" these "traces of ideas." As a photographer, and the originator of many photographic firsts, Nadar positioned himself as an inventor (catching "a glimpse of the industry latent in his invention") and an impassioned advocate for a society better attuned to welcoming new technical ideas. The story of Balzac's understanding of photography thus offers Nadar a chance to argue that, to better ready oneself to receive these ideas from the atmosphere, one had to tame the "universal stupefaction" that was the commonly held response to the introduction of new technologies.

This stupor could be overcome. In fact, Nadar had a daguerreotype of Balzac taken by Louis-Auguste Bisson in his personal collection (a version of figure 5.1). A photomechanical reproduction of the daguerreotype was published in the same issue of *Paris-Photographe* as Nadar's essay on Balzac (figure 5.2). In the pages of Nadar's essay and in his personal collection, Balzac had overcome his reputed "fear," and his likeness had been permanently etched on the surface of the daguerreotype. In print and in theory, "Balzac and the Daguerreotype" describes the integration of new media into an existing ecology of representation.[19] Balzac's rapprochement with the daguerreotype was made material by the object in Nadar's own collection (and its reproduction in his son's journal), a process of familiarization that could stand in for a larger societal transition from awesome reception to widespread acceptance of new technological ideas. With anthropological specificity, Nadar attends to the duality of enchantment and disenchantment accompanying the introduction and adoption of new technologies. But even more importantly for my argument, in doing so, he reveals the relationship between the affective experience of the new and its historical mediation—Balzac's Daguerreian apprehensions are resolved in the form of the daguerreotype portrait in Nadar's own collection just as Nadar's photographic firsts are concretized via the media constellations I've been charting across this book.

This shift—from novelty to acculturation, from first to fact—was, for Nadar, indicative of a transformation of historical experience initiated by a period of unparalleled technological change. As individuals witnessed

5.1 (*above left*) Louis-Auguste Bisson, *Honoré de Balzac*, daguerreotype, 1842. Collection of Paris Musées and Maison de Balzac.

5.2 (*above right*) Dujardin (héliogravure), Chardon-Wittman (printing), Photomechanical reproduction of a Louis Bisson daguerreotype of Honoré de Balzac, *Paris-Photographe* 1, no. 2 (1891): n.p.

the introduction and popularization of numerous new media and technologies across the nineteenth century, there was an overwhelming sense that the "proliferation of germinating ideas" unleashed in the wake of the "glorious haste of photography's birth" was speeding up the process of technological development and its impact on social life.[20] Writing in the early 1890s, Nadar deployed Balzac's reported initial mistrust and later acquiescence to the daguerreotype as an allegory of larger significance for understanding the history of invention. By recounting, again and again, how "ideas" became material "facts," Nadar worked to assimilate the concept of technological development as a natural social good—often placing the photographic first as an emblem of this process. In the following chapters of *Quand j'étais photographe*, Nadar plays on the rhetorical and material slipperiness of invention to craft a curious portrait of photography's "first" fifty years.

SPECULATIVE APPLICATIONS

Specters of other photographic firsts lay in wait. The second chapter of *Quand j'étais photographe*, "Gazebon Avenged," first published in *Paris-Photographe* in June 1891, considers the possibility of the long-distance transmission of photographs. Like all photographic firsts, this supposed invention is brought to our attention via a series of representations. The chapter opens with a transcription of a letter dated August 27, 1856, that Nadar reportedly received at his studio in Paris from a certain Gazebon, the proprietor of a café in the city of Pau in southwestern France. In the letter, Gazebon describes how a "dramatic artist" named M. Mauclerc had shown him a daguerreotype portrait that Mauclerc described as having been made by Nadar in Paris while he was in Eaux-Bonnes (also in southwestern France) "by means of the electric process."[21] Gazebon requests that Nadar make a portrait of him using the same process ("in colour," no less.)[22] Seeing these two names together—Gazebon and Mauclerc—Nadar recalls that this was not the first time he had received a letter from Gazebon, recounting how, two years prior, Gazebon had written to him regarding a clock that Mauclerc had assured him was very valuable, adding that the only other such specimen was in Nadar's own personal collection. Both times Nadar declined to reply to the letter, "But this letter begged to be kept as a specimen, and as a collector pins a rare butterfly, I gave it a place in a special box."[23] This lepidopterist urge proved useful when, years later, Nadar could call on this letter as yet another specter of an idea, off in the atmosphere.

 Nadar sets the scene. One evening after a dinner at home, Nadar is sitting conversing with a friend when an unannounced visitor arrives. After a series of apologies, the young man explains his tenuous connection to Nadar: His mother had worked for Nadar's mother in Lyon, and he himself had worked for the son of a friend of Nadar's. Mention of his previous employment gives the young man a chance to list his professional qualifications as an electrician, noting, as Nadar was often wont to do, a string of famous inventors and inventions with which he was associated. Laying out the technical context for his proposal of a new photographic method, he refers to work on George Leclanché's electric batteries, Gustave Trouvé's dual-motor electric velocipede, Paul-Gustave Froment's clocks, Marcel Deprez's generator and energy transmission, Clément Ader's telephone, and Giovanni Caselli's autographic telegraphy.[24] After describing his professional proximity to this litany of electrical innovations, he begs

Nadar and his friend "not to do me the honor of taking me for an inventor. I am only a young man, and it is not at all a discovery that I claim to bring to you. It is only a simple find, a change, a coincidence in the laboratory."[25] After this lengthy prelude, the young man states that he has invented a system for photography at a distance. The mysterious invention is put in similarly awesome terms as Nadar's evocation of Balzac's theory of the daguerreotype. He cautions that "what I have come to reveal to you is so . . . extraordinary, so beyond, even for you, what is recognized as accepted, classified, catalogued" that Nadar and his companion must "kindly agree not to judge [him] right way as crazy or impudent."[26]

But what, exactly, had this young man invented? He claimed that his development of a system for the long-distance transmission of photographs was motivated by "the latest published experiment on photophony," saying to himself: "If the results obtained by Mr. Graham Bell and Mr. Sumner Tainter have established that all bodies can emit *sound* under the action of *light*, why would we refuse to accept from light itself what light can offer us?"[27] In photographic terms, he asks: Why shouldn't it be possible for the action of photographing a model in the same room to be "reproduced over greater distances?"[28] Finally, the young man arrives at the substance of his request: that Nadar permit him to test his hypothesis regarding the possibility of taking photographs at a distance in Nadar's own studio. While Nadar reports being visibly unfazed by this hypothesis and proposition—he is no Balzac—his friend is excited by the young man's request, clarifying: "So, you say that, across all distances and beyond your field of vision, you hope to take photographs?"[29] The young man confidently replies: "I do not hope to take them, sir; I take them."[30]

The young man replicates several of Nadar's own favorite strategies for claiming a first after the fact: comparisons to other important inventions and published newspaper reports describing a public experiment. In response, Nadar and his friend are "dumbfounded," but even this reaction is couched within a larger technological narrative.[31] Just the day before, they had visited the "exhibition of electricity," almost certainly referring to the Exposition internationale d'électricité held in the late summer and fall of 1881.[32] As discussed in chapter 2, the exhibition included demonstrations of the seemingly endless applications of electricity, including an electrical tramcar and electric lamps, alongside electric inventions by several of the electrical engineers referenced by the young electrician in Nadar's story—including Marcel Deprez's demonstration of the long-distance transmission of power and Giovanni's Caselli's "autographic telegraphy," or pantelegraph.³³ Primed by the panoply of inventions the pair had witnessed at the exhibition, he and his friend were prepared to give the young electrician a chance.

This belief in the power of invention was, the young electrician reminds them, Nadar's hallmark. "Will you permit me, Mr. Nadar, to express my surprise at having encountered such resistance in a man who—the first!—thirty years before anyone else dreamt of it, predicted, explained, and even baptized the *Phonograph*?" Was it not Nadar who "took the first underground photograph by artificial lights and also the first photograph from the basket of an aerostat?"³⁴ Nadar, the young electrician argued, should be almost duty bound to support those who dared to dream of new photographic applications, given his own "firsts." There is no mention of any specialty assessment of the possibility of photography at a distance, nor a summary of the state of the exact technologies needed to realize such a feat. Instead, as in Nadar's other firsts, talking about, writing on, and representing technical novelty were imagined to be the motive force behind the realization of such ideas. Nadar, apparently overcome by the possibility of invention, is called back to reality by the substance of the young man's request: money for supplies. As the friends debate whether they have been hoodwinked, Nadar concludes that, no matter the outcome, the young man's performance—not unlike a condensed version of Nadar's own accounts, records, and archives of his photographic firsts—was worth the money.

The young electrician is never to return, but this isn't the end of the story. Nadar ends the chapter with a quote and two postscripts. The quote is from Jean-Baptiste Biot: "*There is nothing easier than what was done yesterday; nothing more difficult than what will be done tomorrow.*"³⁵ This quote concludes Nadar's roundabout way of stating that to assume that something can't be done is as superstitious as assuming it can be done. He remains hopefully agnostic about whether the proposal proffered by the young electrician will ever be realized. This ambiguity is interrupted by the two postscripts, an epistolary gesture that gives a sense of the chapter unfolding in linear time. The first notes that the young electrician's request that the long-distance photographs be in color had been realized by Nadar's "eminent correspondent and friend," Raphael Eduard Liesegang, a German chemist and photographer.³⁶ The second postscript, "from this very morning," notes the success of Guglielmo Marconi's wireless telegraphy." These twinned events provide the occasion for Nadar to end optimistically: "What can we not dream of!"³⁷ This story demarcates another stage in photography's life cycle, departing from the story of photography's reception offered in "Balzac and the Daguerreotype" toward a future in which photography's endless applications had simply to be imagined to be realized.

EXPERIMENTAL HISTORY

Narratives of scientific developments become histories of invention when written down. The chapter "Doctor van Monckhoven" opens with an ardent ode to French industrial chemistry at the beginning of the nineteenth century, a period that saw a profound shift in the nature of scientific instrumentation and knowledge of the chemical transformation of matter.[38] Named for the Belgian chemist and photographic researcher Désiré van Monckhoven, the chapter was first published in *Paris-Photographe* in July 1892.[39] The text begins with a characteristically passionate reference to a text that "can elevate the soul and strengthen the heart," referring to a report presented to the Convention nationale by the French chemist Antoine François, Comte de Fourcroy, in 1795.[40] Fourcroy's report eulogized a number of scientific developments used during the French Revolutionary Wars. The report is summarized over the next few pages of Nadar's text, including a list of several scientific achievements that contributed to France's industrial independence.[41]

Turning to the essay's namesake, Van Monckhoven, Nadar claims that Fourcroy's report has always reminded him of Van Monckhoven and, vice versa, that mention of the Belgian chemist and researcher never failed to remind him of Fourcroy's text "because Van Monckhoven was also one of those scholars from whom one can order a discovery."[42] In Nadar's estimation, Van Monckhoven would have been "the most brilliant" in the canon of eighteenth-century luminaries, such as "Condorcet, Lavoisier, Monge, Chaptal, Vauquelin, Lalande, Fourcroy, Bossut, Darcet, Conté, etc."[43] Nadar inserts a nineteenth-century photographic researcher into the pantheon of scientists commonly evoked in accounts of the origins of the French Enlightenment. This anachronistic addition demonstrates both the efficiency with which this historical narrative had been propagated across the nineteenth century and Nadar's interest in replicating this narrative structure in his account of the history of photography. Nadar also positions Van Monckhoven within a grouping of nineteenth-century scientists like Jules Janssen, Paul and Prosper Henry, and Étienne-Jules Marey.[44] Not incidentally, this list focuses on scientists who were invested in the application of photography in their respective fields. Nadar places experimental photographers within this lineage of chemists, scientists, and philosophers, affirming photography's place in the history of French industrial modernity.

Nadar's emphasis on the experimental method as a motive force in the history of photography has its roots in the philosophy of history at the turn of the nineteenth century. Scientific innovation was understood to be a central component in the process of historical change. But was such change to be understood as cyclical, repetitive, progressive, or teleological?[45] So-called progress as we know it is not a self-evident concept. Reinhart Koselleck has described the new model of history that slowly emerged over the course of this period as one in which "the experience of the past and the expectation of the future moved apart; they were progressively dismantled, and this difference was finally conceptualized by a common word, 'progress.'"[46] However, as Koselleck outlines, this was by no means a unified process, and the project of the constitution of "progress itself" as a "collective singular" was an ongoing rhetorical project.[47] The photographic manual—as embodied by Van Monckhoven's *Traité*—is deployed by Nadar as an example of how this process plays out in practice. Moving between Fourcroy's report, Van Monckhoven's *Traité*, and Nadar's own text, we can see how invention narratives (or firsts) are the building blocks of these grand narratives.

Van Monckhoven's place in history was, in Nadar's estimation, as much to be credited to his method as to the outcome of his experiments. It was Van Monckhoven's devotion to continued experimentation that distinguished him as a photographic researcher and a worthy historical subject. After listing the chemist's accomplishments in photographic research, Nadar places Van Monckhoven as a direct heir to the "Niépces, the Talbots, the Poitevins"—all photographers known for their experimentation.[48] Van Monckhoven had published a number of different editions of his book, each time updating the text to describe new photographic processes and applications.[49] Nadar suggests that these successive iterations are a testament to Van Monckhoven's commitment to the public dissemination of experimental results, emphasizing that the publication is different from "those editions of recent invention, published in fantastic quantities, but in reality only different on the covers."[50] To Nadar's mind, with each edition Van Monckhoven was contributing to the field rather than simply capitalizing on the growing number of photographers to whom he might sell copies. Nadar mobilizes his veneration of Van Monckhoven's experimental and publishing practices to demonstrate the inextricability of experimental narratives from origin myths. Van Monckhoven's innovation is evident, Nadar argues, because we can see a progressive development and the addition of novel photographic techniques, processes, and applications across the publication history of the book. This observation highlights Nadar's evergreen interest in the concept of invention: "Everywhere and in every new science, doing is as necessary as knowing. From invention and theory, Van Monckhoven proceeds everywhere and immediately to practice, to action." As in most photographic manuals in the period, each volume prefaces Van Monckhoven's inventions with an introductory short history of photography. The innovations outlined in the following sections of the book are always already to be understood as part of the future history of the medium. The modern photographer is to be simultaneously a tireless experimenter, a scientific popularizer, and a historian—all roles Nadar claimed for himself with pride.

WHEN WAS NADAR A PHOTOGRAPHER?

The history of photography was being written as it was being made. The chapter titled "The Primitives of Photography," first published in *Paris-Photographe* between August and December of 1892, is perhaps the only text that could be described as a history of photography. But this is not the typical account of names and developments that was repeated across French photographic literature from 1839 onward.[51] Nadar's account is highly personalized, mediated by his own memories and written from his (self-appointed) position as the doyen of French studio photography's "golden era." Nadar suggests that his perspective, in all its idiosyncrasy, registers a shift in historical experience. He writes:

> We are in an era of exasperated curiosity that explores everything, men and things; in the absence of the great history that we no longer know how to produce, we pick up the crumbs of small histories with such zeal that our attention becomes wide-eyed in front of a stamp-collector.[52]

Universal history was on its way out, with the individual on the ascendent. As a photographer, collector, and writer uniquely attuned to the possibility of future interest in the early history of photography, Nadar appoints himself as historian of his first generation—a collector of these "small histories."⁵³ Paralleling Nadar's own advanced years, photography's "primitive," "primordial" era is positioned as distinctively past.⁵⁴ This past was one worth excavating if only to understand the mechanism of its development. For Nadar, the future of photography depended on it.

This "desire for history," to borrow Stephen Bann's words, manifested itself in the texts of early photographic practitioners, Nadar included.⁵⁵ Nadar begins with a list of names intersected by italicized titles: "*daguerreotypists*," "*photographers*," and "*photosculpture*," ending with his own name.⁵⁶ Nadar shows a marked preference for photographers who, like himself, were continually occupied with extending the technical possibilities of the medium. Nadar describes the photographer Gustave Le Gray as an "industrious and remarkably intelligent researcher" at the culmination of his lengthy and emphatically positive account of the photographer's practice.⁵⁷ Recounting the heights of André-Adolphe-Eugène Disdéri's career, Nadar credits Disdéri's later fall from grace to his inattention to photographic progress, which "every day would bring us something to learn from it."⁵⁸ Describing the first Société française de la photographie exhibition in 1855, Nadar notes that while the public was fawning over celebrity portraits, the "specialists" were "examining the indelible proofs of Poitevin, Moitessier, Taupenot, Charles Nègre, Baudrand, and La Blanchère, and the lithographic transpositions of Lemercier—anticipating through the breakthrough of these first paths the limitless immensity of the domain henceforth secured for photography."⁵⁹ Nadar's emphasis on early experimenters with photomechanical reproduction is bookended by references to the possibility of color photography and mention of a demonstration of a novel negative retouching process. Nadar describes the retouching process as "a resource for which everyone had yearned but never known, as everyone had imagined it without naming it."⁶⁰ Nadar underlines the ongoing experimental nature of photography in a manner analogous to the development of electrical media and chemistry discussed above. In this way, he conceptualizes photography's historical development as analogous to the experiment, imagining the photographic proof as the natural conclusion of experiment and of historical change itself. In this formulation, invention (including photographic firsts) becomes the motor of history.

Transformations in historical writing were coeval with a shift in historical experience. These reorientations are the subject of the final chapter of Nadar's volume, entitled "1830 and Thereabouts." Not previously published in *Paris-Photographe* like most of the other chapters, the text was formulated as a kind of miniature bildungsroman of the nineteenth century. A litany of names and objects unfurl across the text to mimic the acceleration of historical experience. Aligning his own biography with that of the nineteenth century—not unlike the iconography of the Chevreul centenary discussed in chapter 3 of this book—Nadar describes the 1830s as a period of adolescence; that is, a period of physical revolution unleashing an accelerated period of growth and an unknown future. Visualizing this unspoken thesis, Nadar cites a scene of a steamboat on the River Thames in Victor Hugo's *Les Misérables*: "It was a piece of mechanism which was not good for much; a sort of plaything, the idle dream of a dream-ridden inventor, a utopia; a steamboat."[61] In Nadar's account, as in much of modern European historiography, these "dream-ridden" inventors are vindicated, their utopias rendered not only real but world altering. Describing the introduction of steam power as a world-historical event, Nadar writes: "As if from an enchanter or from a stagehand, the first whistle of the first locomotive gave the wakeup signal for everything to begin. A whole new world is moved in this universal April, it feels, it stirs, it grows uncontrollably, it stops in order to find where to set foot next: everything is again put into question. Paris, heart and brain, is on fire."[62] Steam power is here analogized as a temporal motor, inaugurating an age of acceleration both in terms of transport and the realization of fantasies (as it would often turn out, often violently and catastrophically) of how technology could transform the physical world.

For Nadar, technological change was in lockstep with social change. In breathless prose, Nadar attempts to sum up the cultural transformation witnessed in his life, including, for example, the discovery of prehistoric dinosaurs, the fall of organized religion, and the rise of utopian socialism.[63] Historical acceleration is aligned with the shifting political situation of the French state and the expansion of geological time. Describing how, to his mind, this temporal experience dissolves political affiliations, Nadar writes: "The ideologues of yesterday, who the days before yesterday were called 'the blues,' are today called the republicans, in order to be called the socialists tomorrow; they will then be called the anarchists, until something better comes along."[64] This hastening was also, for our author, characteristic of transformations of the built environment, for, as Nadar offers, Hector Horeau had "not yet created the architecture of the future in iron and cement."[65] Bringing together the "architecture of the future" in steel and glass, now understood as defining characteristics of architectural modernity, with the transformation of social norms in religion and politics, Nadar associates change with speed and in so doing levels what were inevitably incoherent and uneven "transformations."⁶⁶ Nadar's social class and occupation afforded him the opportunity to experience acceleration primarily as a mode of transformation rather than exploitation. This temporal register is thus not a fact but a subjective experience.

Nadar is, of course, not the only writer to describe the nineteenth century as a period of acceleration.⁶⁷ In the nineteenth century, as in the twenty-first, speed, and the relation of technology to acceleration, was a common metaphor with which to sum up epochal transformations in the experience of temporality. But, as Sarah Sharma has put it, "The temporal is not a general sense of time particular to an epoch of history but a specific experience of time that is structured in specific political and economic contexts."⁶⁸ Sharma, in her poignant critique of what she terms the "speed theorists" of the twenty-first century, argues that without attending to variable and interdependent experiences of temporality, the "subject" of speed theory—that is, the individual experiencing acceleration—risks being conflated with "the same subject who will confirm speedup most readily as *the* reality."⁶⁹ As a "speed theorist" of his own time, Nadar places himself as the subject experiencing the disorientation of what he articulates as an acceleration. In doing so, he universalizes his subject position—as a white, male French citizen—and his occupation—a photographer, an inventor, a writer—as both the experiential subject of this reorientation of temporal experience *and* the agential cause of it. He is both motor and passenger, to keep Hugo's transportation metaphor alive. But unlike Hugo, to name just one author, Nadar does not devote his attention to the political and economic structures subtending his steamboat utopia.

Nadar continues his rambling verbal tour of Paris, in the spirit of the Benjaminian ragpicker, through the flea markets, where, he notes, one can still find first-edition Dürers and Rembrandts and where the museums of Paris (Cluny and Carnavalet) find their wares. In classically Nadarian twists, the dirt of the streets leads into a discussion of the ill health of the Romantic generation, an excuse to mention yet another litany of scientists: Louis Pasteur, Jacques-Joseph Grancher, and Émile Roux.⁷⁰ This tale rambles on through the theater and dinner crowd to discuss changes in newspaper subscription models, noting Émile de Girardin's proposition to sell a newspaper for forty francs a year as opposed to eighty.⁷¹ In a rhetorical tour of Paris, Nadar sums up the spirit of *Quand j'étais photographe*, one in which the development of art and science are understood as nothing less than the development of history itself.

Despite all this, the temporality of Nadar's text is often difficult to parse. While the title of the book signals a definitive pastness, it also offers insight into how Nadar positioned himself in relation to history, the *when* in the title nodding to the fact that every moment would soon be recounted as past, entering (he hoped) the *grand récit* of history. Others, such as Eduardo Cadava, have interpreted the past tense of the title to be an indication of Nadar's investment in the relationship between photography and remembrance, between photography and death. However, I would argue that the past tense instead signals Nadar's self-positioning in relation to history. Very much still alive (and still, I think he would say, a photographer), Nadar was both a protagonist in, and the author of, this wildly idiosyncratic account of photography's "first century." Rather than death, the primary theme here is dynamic change—of revolution, of circulation—again, all motive processes in which Nadar afforded dynamic, agential roles for himself and photography. If there is a death described here, it is obsolescence: of technologies, of political systems, of aristocratic classes. Photography was a cultural practice and a cultural object, an extraordinary invention quickly made every day in the wake of invention after invention. So even if it was an insurance against death, photography was also an offering to the future.

Across *Quand j'étais photographe*, the precise relationship between historical experience and new media is analogized with reference to literary and historical moments, including Balzac's theory of photography, Fourcroy's testament to French national innovation during the Revolutionary Wars, and the development of glass studio architecture prompted by the explosion of photographic studios in mid-century Paris. But, in describing the influence of these inventions, Nadar also notes the nature of the objects, texts, and images that witness these "revolutions" and speak of them to future generations. Nadar's own daguerreotype of Balzac stands in as an object metaphor for photography's enculturation; the letter from Gazebon recalls the idea of photography at a distance, once unfulfilled; Fourcroy's report is a sweeping testament to the power of French invention; and the streets and *objets* of Paris offer an archaeology of the nineteenth century. The book thus offers both a history and a method, a story and a set of objects upon which to pin these tales. But in penning these strange little "old wives' tales," as Rosalind Krauss memorably called them, Nadar drops in his own objects, his own firsts—the first aerial photography, the first photographs by electric light, the first photographs of the Paris sewers and catacombs. In the process, he intermingles cultural and scientific history with tales of his own ingenuity.[72]

Nadar's narratives warrant as many questions as they answer. *Inventing Nadar* is an experiment in seeing what comes from simultaneously taking Nadar seriously as an archival subject and denaturalizing the self-evidence of his claims to priority. Nothing about Nadar, nor about the history of photography, is inevitable or obvious. *Quand j'étais photographe* is extremely instructive in this regard. Not only has it become one of *the* primary sources for accounts of the early history of photography in France but its ideology of invention has been subsumed into accounts of photography as a "new" medium in the nineteenth century. In this book, I've devoted myself to dissecting the "photographic first" to reveal the duality at the heart of every use of the descriptor "first" to describe an object, a technology, or an event. At its core, the word accounts for a desire for a bare technicity—an unmediated relationship between human and technology, untouched by history and context. But look closely and its use as an appellation reveals the processes of mediation necessary for the first to have been visible as such in the first place: interventions made in word and in material that forevermore shape how these stories are told. In the epilogue, I follow these stories into the twentieth century to see where they end up after Nadar's—and the Nadar studio's—death.

EPILOGUE

The History of Photography,
as Told to Me by Nadar

The machinations traced across this book proved very effective. In 1937, in the now canonical Museum of Modern Art (MoMA) exhibition titled *Photography 1839–1937*, Nadar's "first" aerial photograph was included in a section surveying the development of aerial photography.[1] Curator Beaumont Newhall, writing in the catalog accompanying the exhibition, described Nadar's aerial contributions in succinct terms: "Nadar was as famous for his aeronautical experience as for his photography. In 1858, he combined the two, and took the first aerial pictures."[2] The first aerial photograph even warranted mention in the brief biography of Nadar offered in the index: "Born 1820. Began as caricaturist, 1842. Opened photographic studio, 1852. Aerial photographs, 1858. Commander of balloon corps, siege of Paris, 1870. Died 1910."[3] Aerial photography had been given its own section in Newhall's exhibition, and Nadar's "first" was included alongside aerial images by the Air Service of the American Expeditionary Forces, Fairchild Aerial Surveys, V. Hazen, McLaughlin Aerial Surveys, and Albert W. Stevens, as well as other unattributed images, documenting aerial bombardment during the First World War.[4] Nadar's inclusion in the exhibition was thanks, in part, to his son Paul's participation as a lender to the exhibition, but Nadar's aerial photographs remained "first" thanks to the processes of mediation that I have described across this book.

Within the halls of MoMA, Nadar's claim to technical priority was explicitly allied with the artistic imperative of originality—a connection Nadar made repeatedly across his career. This museal canonization represented a dramatic finale in what, by then, had become an eighty-odd year history of the first aerial photograph. In his initial pronouncements, and as was inscribed on his son's enlargement of the image, Nadar had argued that aerial photography would be key to both military surveillance and state surveying. That his image would be included as a predecessor to the twentieth-century capitulation to the aerial view as the lens of state-sponsored violence is both a fitting end to the story of the first aerial photograph and a reminder of the role media origin stories have in naturalizing technological outcomes. All this occurred within an exhibition that has been central to building (and then critiquing) a Modernist history of photography, underlining the utility of following Nadar through the history of photography.

Across the twentieth century—and in between in the collections of major museum collections in North America and Europe—Nadar became central to the emergent canon of photographic history. He was now a "master" of photography's "first century." But, as I've argued across this book, this history (or, more aptly, these histories) has been shaped by the forms and expressions of media that Nadar produced, collected, and circulated during his lifetime. Often in service of articulating the originality of his many and varied projects, these media forms were created and circulated in concert with cultural and institutional processes—the patent office, civic and technologies museums, the scientific press, photographic societies, international exhibitions, among other forms. Nadar was both a self-invention and the product of a specific cultural surround. Nadar's presence and precedence became a precondition for the received history of photography. In other words, Nadar became Nadar to become known to us as Nadar—an invention spanning centuries.

For some readers, it might appear curious that I have not yet addressed the question of *why Nadar*? So far, I've avoided writing this history in the corrective—Nadar was not first, and here's why. Instead, my focus has remained on how and why originality (or firstness) became a value prized in a medium associated with multiplicity and reproduction (even the daguerreotype was prized for its ability to *reproduce* nature, despite the fact that the object itself was singular). But as *Inventing Nadar* has shown, it was only through reproductive media that Nadar was able to make an argument for his singularity—evidence of an emergent celebrity culture dependent on technologies of reproduction and circulation.[5] So why

Nadar indeed? For this author, it is because the materials that defined the photographic first also became the archives upon which several successive generations of North American photography historians based their understanding (if only in part) of photography's first century—histories that for far too long were exceedingly limited in terms of gender and geography. It is very instructive to observe how historians of photography have turned to Nadar (or his son Paul) as authoritative sources on the history of nineteenth-century photography and to ask why Nadar was an interesting subject for theorists of photography at key moments in the development of the history of photography in France and North America. When it comes to Nadar, the history of photography has been constituted by much the same process as the photographic first. That is, it has been refracted through the efforts of a consummate self-promoter living within a cultural milieu primed to recognize individual accomplishment and to preserve those expressions of media for future generations.

To read for the history of photography *as told to me by Nadar* is to examine how Nadar's rhetoric of firstness within the history of a medium of reproduction was particularly useful for curators, critics, and historians eager to affirm the status of photography as a fine-art medium in the twentieth century, the flourish of his signature familiar to those used to looking at paintings. And yet the story of how art history–trained photohistorians or sociologically minded theorists of photography could make dramatic pronouncements about the value of his photographs is a much more material one. The history of photography is shot through with examples of physical encounters between Nadar's images, texts, and archives and European and North American curators, historians, and theorists. This occurred through reading his texts, first-hand experience with his studio (when it was still open under the direction of his son Paul), visits to his archives held in France, and the private and public collections of his prints that were bought and sold across the twentieth century, including to major North American collections. It might seem to be all too obvious a statement to say that researchers engaged with research materials, but if we avoid taking for granted the existence of archives at all—and remember the lengths to which Nadar went to ensure his legacy was preserved—these encounters become even more significant.[6] For example, in that same 1937 catalog noted above, Newhall describes flipping through Nadar's portraits in Paul Nadar's studio. Newhall writes: "To leaf through them is an experience, for they are a pictorial index to four generations of great men."[7] Newhall also references *Quand j'étais photographe* extensively.[8] True or not, Nadar's anecdotal reminiscences (now eyewitness reports) and family collections became the materials with which photography's early history was written.[9]

After his death in 1910, Nadar quickly became a key source for scholars seeking to historicize photography's "first" century. Nadar finds pride of place in the German-born French photographer and writer Gisèle Freund's *La photographie en France: Essai de sociologie et d'esthètique* (1936, later reprised as *Photographie et société*, published in 1974), not only as an exemplary subject of the chapters on Second Empire photography but also as a source on the Parisian photographic culture of the mid-nineteenth century.[10] Even at this early stage of scholarship on Nadar (and on photography more broadly), a dichotomy emerges between the business and art of photography. Though Freund acknowledges that Nadar initially took up photography to make money and to cultivate his growing celebrity, she also describes a split between "entrepreneurial" photographers such as André-Adolphe-Eugène Disdéri and "artistic" photographers like Nadar.[11] And yet Freund's own analysis relies upon the perseverance of the commercial Nadar studio, in the form of Paul Nadar's testimony to his father's career. Freund cites personal interviews with Nadar's son Paul, underscoring the intergenerational management of the studio as key to the perseverance of its legacy.[12] Artistic and commercial legacies are not so easily separatable. Freund's friend, the philosopher and writer Walter Benjamin, was also a close reader of Nadar's writing, citing *Quand j'étais photographe* extensively throughout *The Arcades Project*. In a turn of phrase that Nadar would certainly have enjoyed, Benjamin wrote that Nadar's images of the Paris catacombs were "the first time" that "the lens was deemed capable of making discoveries."[13] Benjamin's description of Nadar's camera as "first" and "capable of making discoveries" is no doubt built upon the discursive armature of the first as laid out by Nadar in texts like *Quand j'étais photographe*. Nadar's accounts of photography stand in as eyewitness testimony for some of photography's early theorists without critical acknowledgement of why it was that these statements were so readily available for their consumption. Selling Nadar was an intergenerational, and thus a historical, enterprise.

COLLECTING NADAR

The 1930s also saw preliminary steps toward the institutionalization of the Nadar studio collections. These collections had remained remarkably intact due to the fact that the studio had been run by three successive generations of the family over almost 100 years (between 1854 and 1948, by father, son, and then Paul's daughter Marthe).[14] Paul Nadar had approached the Administration des beaux-arts in 1933, seeking to donate his studio collection to the French state.[15] However, the process was interrupted by the Second World War and Paul's death in 1939. Years later, in 1949, Julien Cain, the administrator general of the Bibliothèque nationale de France (BNF), brought the collection to the attention of the Direction général de l'architecture, proposing that the collection be bought from Paul Nadar's wife Anne.[16] Letters from late 1949 and early 1950 between Julien Cain and Anne Nadar outline the details of the sale.[17] The BNF would receive all photographic prints, albums, catalogs, ephemera, and archives belonging to Félix and Paul Nadar for the price of 700,000 francs, and the photographic archives would receive all existing glass negatives in exchange for 1,800,000 francs.[18] The 1950 sale secured a large portion of the Nadar studio corpus for public institutions, complementing previous sales of smaller portions of the collection, including the 1897 sale of aerostation- and aviation-related materials to the musée Carnavalet (discussed in detail in chapter 1), the 1907–8 acquisition of 584 drawings related to the production of the *Panthéon Nadar* by the BNF, and the 1943 acquisition of 514 photographic portraits from Marthe Nadar by the BNF.[19] These large collections are complemented by small holdings that were the result of smaller donations that occurred during Nadar's lifetime, including to the Société française de la photographie and the Conservatoire national des arts et métiers in Paris.[20]

Many of the major figures involved in collecting the works of the Nadar studio were also undertaking research on the photographer and his studio. As the prominent French collector of photography André Jammes noted in 1994, the "history of the Nadar holdings is bound up with the evolution of Nadar research."[21] Early monographic research was completed by a constellation of figures in France, including two curators from the BNF, Jean Adhémar and Jean Prinet, and the collector and scholar Michel-François Braive. The 1960s saw the major outputs of this research in the form of a 1965 monographic exhibition at the BNF curated by Jean Adhémar, assisted by Alix Chevallier; a monographic book, *Nadar*, by Jean Prinet and Antoinette Dilasser (1966); and a book by Michel-François Braive entitled *L'age de la photographie*, drawn from Braive's unfinished École du Louvre thesis on nineteenth-century portraiture, which took Nadar as a central figure.[22] This period was also the last era of major acquisitions of original Nadar photographs. Braive's collection, originally acquired from

Marthe Nadar, was sold in two parts: the first to André and Marie-Thérèse Jammes and the second to the American collector Samuel Wagstaff.[23]

If this book has shown anything, it is that Nadar's images—like many images—were forever on the move. By the 1970s, this movement was definitively directed westward across the Atlantic Ocean. Jammes and Wagstaff are arguably responsible for the establishment of Nadar as a key fixture of North American collections of photography. Indeed, Jammes mentored Wagstaff in Paris, advising him on what to buy in Paris and selling him many photographs. Both Wagstaff and Jammes would later sell their collections to the Getty Museum in Los Angeles—with Wagstaff selling his entire collection and Jammes selling a portion. The development and establishment of the Jammes and Wagstaff collections were part of a broader expansion of the market for photographs as distinct original objects with individual authors. The "authenticity" of photographs could now determine their market value based on elements drawn from the art market more broadly, like signatures, condition, and provenance. In the 1970s, the market for photographs reflected the interest in nineteenth-century French photography of these two collectors.[24] This period also saw many books and exhibitions on the Impressionists and the art and visual culture of the Second Empire, which Nadar emerged as representative of in photography.[25] Jammes, in particular, was a key figure in the transatlantic movement of nineteenth-century French photographs, organizing exhibitions like *French Primitive Photography* (a title that borrowed, if not explicitly, from Nadar's essay in *Quand j'étais photographe*), which opened in Philadelphia in 1969, and traveling around the United States.[26]

By the 1970s, Jammes had succeeded in "cultivating a taste for Nadar in the United States," as Isabella Seniuta has put it.[27] Having bought a collection of Nadar's photographs from Michel-François Braive, Jammes sold the doubles from his collection to Wagstaff.[28] Corresponding with Wagstaff in 1976, Jammes writes:

> If I can give you one piece of advice, keep all your Nadars, because there are none anywhere. The dozen that escaped me from Mr. Braive will be enough to establish a very acceptable cost around the world.... I am also struck by the "spectacular breakthrough" that Nadar is making in the United States thanks to you. You Americans are very Victorian, and theologically Victorian. Inhibited by Hill and Cameron, your almost mystical conception of the human face is suddenly

confronted with the realistic and scientific vision of Nadar. [...] It's almost a Marxist shock.[29]

Despite Jammes's elegant efforts to affirm the dichotomy between the "artistic" British calotype and the French "realist and scientific" or even "Marxist" salted paper prints from collodion glass negatives, Nadar's artistry was quickly affirmed in some of the most prominent US artistic institutions via Wagstaff's collection—the occasion for another first. Opening in February 1977, the Metropolitan Museum of Art in New York announced the "first major exhibition in this country of photographs by Nadar."[30] The exhibition, *Nadar, Photographer*, was drawn entirely from Wagstaff's collection.[31] Continuing the "spectacular breakthrough" that Jammes described, the press coverage of the show repeated words Nadar himself would have quite liked. From John Russell in the *New York Times*: "One of the greatest portrait photographers of all time."[32] However laudatory, Russell lamented that the exhibition only featured images from Wagstaff's collection, whereas "two or three phone calls to Paris" would have rustled up some of Nadar's most memorable images: of Charles Baudelaire, the Goncourt brothers, or Honoré Daumier.[33]

Such exhibitions were, however, undoubtedly concerned with more than just offering the best Nadar portraits to the New York public. Exhibition is an excellent way to both build a market for and raise the market value of photographs. Wagstaff stood to financially benefit from the "spectacular breakthrough" of Nadar in the US and the corresponding demand for Nadar photographs. This speculative investment—recall that, in his *Panthéon*, Nadar said that people in 3607 would be searching for a rare and valuable copy of his print—bore fruit when Wagstaff sold his entire photography collection to the J. Paul Getty Museum in 1984 for $5 million, just several years before his death in 1987. His collection became the foundation of the museum's newly formed department of photographs.[34] Nadar's "spectacular breakthrough" on the US scene was complete.

THEORIZING (WITH) NADAR

Nadar's photographs entered North American institutional collections at what was recognized even then as a critical shift in the institutionalization of the study of photography in both museums and universities. Writing in the 1980s, Abigail Solomon-Godeau described photography's gradual accrual of prestige in the academic discipline of art history and within the walls of museums in the context of an "ascendent right-wing politics and an exploding, volatile, and unsatiable art market."[35] This context, she argued, irrevocably shaped precisely what kind of photography would be considered valuable and in what terms: "Photography had to be defined in ways that permitted it to be assimilated within the museological/art-historical model of aesthetic autonomy, the model that, for shorthand, could be termed formalist modernism."[36] At the same time, scholars and critics such as Solomon-Godeau were modeling a different kind of photographic history that considered photography as more than just an image—offering not just a "social" history of photography but also a politically engaged one.[37] Solomon-Godeau again: "I would submit that the history of photography is not the history of remarkable men, much less a succession of remarkable images, but the history of photographic uses."[38] A picture editor turned historian and theorist, Solomon-Godeau was writing against the kind of market for photographs that men like Jammes and Wagstaff were fostering and speculating upon.

Writing about the battle being waged over photography's "museumization" (as Hilton Kramer put it), Andy Grundberg memorably summed up the dispute as one between the "connoisseurs and contextualists."[39] However, Nadar—a photographer who self-positioned himself as both an artist and a business owner—defies easy categorization in either of these frameworks. It is not "out of context" for Nadar's images to be understood as artistic portraits, nor an aestheticizing gesture to consider them objects worthy of collection in a museum, nor unreasonable to think of his images as speculative commodities—all processes foretold by their author, or, should I say, inventor. Nadar is thus both similar and dissimilar to the case of a photographer like Eugène Atget, an "author" who had to be "invented," as Solomon-Godeau put it.[40] While Nadar indeed had to be invented, he initiated the process of canonization himself—most memorably in the form of the photographic first—before handing the baton first to his son, and then to the writers, curators, historians, and theorists of the coming centuries. In the case of Nadar, we might say that originality and genius aren't things "done to" photography by art but things that emerge precisely out of photography's seesaw ontological status between art, science, and, let us never forget, commerce. As the story of Nadar tells us, "originality" and "genius" are culturally specific concepts, circumscribed and shaped by the tales of disputed patents, sales and donations to museums, and the processes self-historicization charted across this book.

But the United States in the second half of the twentieth century was not France in the second half of the nineteenth century. So what *was* Nadar to photographic theorists and historians in the twentieth century—a period in which photography was not only a commodity, a museum object, and the subject of intellectual inquiry but also an old medium among new media? We return to exhibitions. The exhibition *French Primitive Photography* was presented in 1969 at the Philadelphia Museum of Art (co-organized by André Jammes and Aperture publisher Michael Hoffman) as one of a series of exhibitions that began to foster an American academic interest in early French photography.[41] These exhibitions drew from Jammes's own collection, ensuring that the view of nineteenth-century French photography received in the United States was structured by Jammes own thematic categories—themselves based on the language of connoisseurship, as Solomon-Godeau points out.[42] In the introductory catalog essay for the Art Institute of Chicago's exhibition of prints from the Jammes collection *Niépce to Atget: Photography's First Century*, David Travis writes: "If it takes but one genius to make any medium worth study, we might easily say that Atget's genius, Cameron's genius, Nadar's genius, or the genius of that obscure figure, Charles Szathmari, is enough. But to know one of those photographers is just the beginning."[43] This genius focus is, for Solomon-Godeau, exactly the problem. However, as I have traced across this book, art and art history are not the only frameworks of authorship, originality, and genius that photography's history has been embroiled with. In the history of science, technology, and business, concepts such as the individual, the author, the owner, and the patentee provide parallels to the artist and the genius. This model of distributed authorship—patent, apparatus, and image, or of idea, machine, and representation—would also have been particularly interesting for art historians and theorists making sense of contemporary art in the 1970s, including conceptual artists and new media art.[44]

Alongside the acquisition of his photographs by museums in New York City and Los Angeles, Nadar became a recurring character in phototheoretical discourse in the 1970s. During this decade, a number of North American journals published volumes dedicated, for the first time, to the study of photography.[45] Both the September 1976 issue of the art magazine *Artforum* and the summer 1978 issue of the academic journal *October* featured monographic articles on Nadar.[46] The issue of *October* also saw the first partial translation of *Quand j'étais photographe* into English, alongside postmodernist art criticism by authors such as Craig Owens and Douglas

Crimp.⁴⁷ I single out these two articles, and the publications and moment in which they were published, to make a larger point about how Nadar's arguments about photography, history, and technology (not to mention the images entangled in these arguments) have mediated the history and theory of photography. Photographer and critic Max Kozloff, in his *Artforum* article "Nadar and the Republic of Mind," suggests that even Nadar's portraits demand that we "drop these art historical criteria or even stylistic analyses and think about his social psychology."⁴⁸ Similarly, the art historian Rosalind Krauss, in her *October* article "Tracing Nadar," reflects on the renewed significance of Nadar's memoirs for art historians and critics in the 1970s.⁴⁹ Like Kozloff, Krauss argues that what Nadar best communicated were the "psychological" "facts" that characterize the experience of photography as a new media.⁵⁰ Krauss suggests that Nadar's attention to "the conversion of mystery to the commonplace" deserved particular attention at the historical moment she was writing, which, as described above, was seeing a surge in attention to photography by scholars, critics, collectors, and museums.⁵¹ "For at this point, in our turn," writes Krauss,

> we are realizing the immense impact of photography, the way it has shaped our sensibilities without our quite knowing about it, the way, for example, the whole of the visual arts is now engaged in strategies that are deeply structured by the photographic.⁵²

Nadar's reflections, Krauss suggests, are, despite his unreliability as a narrator, useful precisely because they avoid any ontological precision when it comes to defining *what* photography is. This drive to understand or fix photography as a discrete medium is, Krauss analogizes, rather like a recently diagnosed patient trying to pry details about their illness from the doctor. Instead, "Nadar's point is that among other things, photography is a historical phenomenon, and therefore what it *is* is inseparable from what it *was* at specific points in time."⁵³ Photography was and would be multiple.

Steeped in the milieu of conceptual art, where representation and language were front of mind, theorists and critics of photography in the 1970s likely found compelling Nadar's extensive description of photography's first century as one in which photography (and media technologies more broadly) and photographers cocreated new modes of representing social experience rather than offering an unvarnished mirror of reality. It was perhaps this self-reflexive mode that drew so many writers to Nadar, his images, and his writings, as they offered an opportunity to read for what photography was then rather than interpreting what has been left to history now. Nadar remained somehow contemporary, even sixty-odd years after his death.

The debate over photography's status in the 1970s and 1980s was about more than whether nineteenth-century photography should be the subject of aesthetic contemplation or art-historical inquiry. It was precisely photography's ubiquity that prompted many scholars and writers to argue that following photographs as they made their way through institutions could tell us something about the construction of meaning as such. To make photographic history a history of photographic authors rather than a history of photographic subjects was fundamentally to deny the vitality of photography's presence in the world ever since its invention and, even more importantly, to deny the political import of its study. However, in the intervening years between then and now, the history of photography has emerged as much more a field of study than a discipline, overflowing and exceeding its containment within its parent discipline of art history to cover almost every conceivable topic. The relationship between photographic authors and photographic uses remains a tense one. Nadar demonstrates that you can never separate the two. Images are used to construct authorship, whether by their maker (as in the case of Nadar) or by collectors, curators, theorists, and, yes, historians too. By examining the cultural and material history of the photographic first, this book has offered a series of object lessons in the invention of a photographer's legacy—from the museum to the patent office to the history book. Nadar made himself an author by first fashioning himself as an inventor.[54]

EXHIBITING NADAR, NOW AND THEN

By the second half of the twentieth century, Nadar was not only first, he was also *a precedent*—an avant-avant-garde. In a double-page spread in the catalog from an exhibition of Wagstaff's own collection, *A Book of Photographs from the Sam Wagstaff Collection*, an image from Nadar's series of photographs of the Paris catacombs (discussed in chapter 4) is set alongside a Robert Mapplethorpe photograph of a man in bondage gear leaning against a metal staircase in a concrete room.[55] The dramatic juxtaposition is characteristic of Wagstaff's wide-ranging taste and curatorial ethos, but, for my purposes here—exploring how Nadar's pictorial and technical innovation has been constructed and maintained mediatically and transhistorically—the move is unsurprising. Mapplethorpe (Wagstaff's partner) was, for Wagstaff, undeniably part of the photographic avant-garde of the twentieth century. Placed alongside Mapplethorpe, Nadar is canonized as part of Mapplethorpe's artistic genealogy, a forebear of twentieth-century photographic arts. In both images, the subject is double: the libidinal image of a sexual subculture or the repressed image of a city's dead *and* an image of light (electric in the case of Nadar and from above in the case of Mapplethorpe). In each image, the photographer is underground. This move is to some extent characteristic of how Nadar's legacy was being articulated in photographic exhibitions in the United States from the 1960s onward: a "master" of photography's first century, representative of photography's "primitive" time, a "genius" of photography's "golden age," but also inspiration for the photographic avant-garde that would follow ("pioneering" is another word that crops up frequently).[56] Avant-garde is, after all, just another way to say first.

Looking at exhibitions is a useful way to think about how Nadar's place within the history of photography has evolved over time. More recently, the great shift in focus between two major monographic Nadar exhibitions of the late twentieth and early twenty-first centuries evidences the extraordinary influence of the intervening period of scholarship. The 1994 exhibition *Nadar*, held at the Musée d'Orsay and the Metropolitan Museum of Art, foregrounded his early work in photography from approximately 1854 to 1860. Writing in the foreword in the accompanying English-language catalog published by the Metropolitan Museum of Art, the directors of both museums profess the exhibition's commitment to "untangl[ing] the multitude of Nadars and allow[ing] the great portraitist to come forth in all his singular glory."[57] Great care and emphasis are given to the fact that the prints included in the exhibition were all "made by Nadar or under his direct supervision," with the goal of allowing audiences to "now recognize this authentic, personal, and autographic aspect of Nadar's work as readily as they do his famous name."[58] This is, in short, another way to show the "invention" of Nadar, though with the goal of "uncovering" a singular Nadar that would be the Nadar for exhibition-goers in New York and Paris. This Nadar was a photographer and first and foremost a portraitist. This focus is, ironically, made most abundantly clear in an essay by Sylvie Aubenas titled "Beyond the Portrait, Beyond the Artist" that touches upon his non-portrait images. The essay begins with a qualification—"Nadar the photographer, whose career had many vicissitudes and a brief period of true inspiration, was essentially a portraitist"—before going on to describe Nadar's work in all things non-portrait.[59] This characterization is true if one thinks in terms of pure numbers of portraits produced during his lifetime—though many of us would perhaps bristle at the suggestion that what we do most or for money is "who we are." It is instructive to think about how the images and events explored across this book have been characterized as Nadar's "other" interests; as in, his non-portrait interests. As I hope is now abundantly clear, these "other interests" are marginal only if you consider photography to be the only medium, or rather the most important medium, in which Nadar worked. To my mind, it hardly makes sense to judge importance based on volume, especially for an individual who produced so profusely across such a constellation of media. The invention of Nadar as a portraitist also took some work.

Two and a half decades later, a 2018 exhibition titled *Les Nadar: Une légende photographique* opened at the BNF in Paris. This exhibition refused the singular Nadar put forth in most previous exhibitions by refocusing attention on the three Nadars. To do so, the exhibition examined the photographic production of Adrien (Félix's younger brother), Félix, and Paul (Félix's son) together. In addition to addressing pressing questions of authorship, the kind of photographs and documents exhibited shifted greatly between the 1994 and 2018 exhibitions. The 2018 exhibition displayed a greater diversity of archival documents and ephemera, alongside prints with hand-drawn annotations, while the 1994 exhibition favored the so-called "original" prints themselves and included less contextual and material framing. Hosted by the French national library and curated by two of the most prominent experts on the history of French photography, Sylvie Aubenas and Anne Lacoste, the exhibition was evidence of both transformations in photo-historical scholarship and the influence of the exhibition site on the nature of the materials included.[60] The BNF is home to the largest Nadar studio holdings in the world and is thus in a unique place to put on display the rich diversity of media produced by the Nadars over the course of their lives. The great contribution of the 2018 exhibition was that it decentered Nadar from the story of les Nadars and began to examine other thematic currents; for example, the great influence of his son Paul on the stewardship and legacy of the Nadar studio. The exhibition also sought to further contextualize the cultural and scientific milieu in which these three photographers worked, rather than simply the artistic culture from which "Félix" emerged (though it did this well also).[61] As the last (exhibitionary) word on Nadar at the time of writing—Nadar is not singular.

The invention of Nadar continues.

ACKNOWLEDGMENTS

Inventing Nadar is about many things, but perhaps most importantly, it is about the collaborative acts of invention that lie behind all creations—this book is no different. I've been working on this project in one way or another since 2013 and through that time have been privileged enough to enjoy a great wealth of intellectual, creative, and personal relationships that have supported me through to finishing this manuscript.

I had the great fortune to study in the Art History department at the University of Toronto, where I worked with a trio of brilliant and creative scholars—Jordan Bear, James Cahill, and Alison Syme—whose intellectual influence on this project is immeasurable. Richard Taws has remained an engaged interlocutor since my time in the History of Art department at University College London, when I first began to engage with Nadar. I (re)conceived of this project during my time as an international fellow at the Kulturwissenschaftliches Institut Essen. Thank you to Julika Griem and the wonderfully interdisciplinary team of scholars there for a very intellectually spacious summer in Essen. I wrote the bulk of this manuscript as a Social Sciences and Humanities Research Council of Canada Postdoctoral Fellow in the Department of Art History and Communication Studies at McGill University. My gratitude goes to the late Jonathan Sterne for taking in a wayward art historian, for suggesting that this project had a home at Duke, and for his bottomless intellectual generosity. Thank you too to all the members of CATDAWG (Culture and Technology Discussion and Working Group) for welcoming me into the fold here in Montreal.

This book found a happy home at Duke University Press thanks to Ken Wissoker's enthusiasm for this project. My gratitude to Kate Mullen for shepherding this through to production and to Lisa Lawley and the entire Duke University Press team for their care and support for intellectual work amidst profound political uncertainty. I am grateful to the two anonymous reviewers for their generous engagement with this manuscript and for their enthusiasm for my ideas—their engagement has made this book immeasurably better. All errors remain my own.

This work would have been impossible without the support and assistance of librarians, archivists, and curators at institutions in France and in Canada. I wish to thank in particular the staff at the Département des Estampes et de la photographie, the Département des manuscrits, and the multiple libraries of the Bibliothèque nationale de France; the staff of the Archives nationales; the staff of the musée Carnavalet for facilitating access to materials amidst a large-scale renovation of the museum; the staff of the Société française de la photographie; the staff of the musée des Arts et Métiers; the staff of the Bibliothèque historique de la ville de Paris; the staff of the Mediathèque de l'architecture et du patrimoine; the staff of the archives of the Institut national de la propriété industrielle; and the staff of the archives and library of the École nationale des ponts et chaussées.

This work was made possible with the generous support of the Publication Grant Fund of the Leonard A. Lauder Research Center for Modern Art at the Metropolitan Museum of Art. Financial support also came from a publication grant from the Photography Network, whose work fostering a network of scholars thinking deeply about photography is ever more important. Additional research funding came from the Social Sciences and Humanities Research Council of Canada.

An earlier version of a portion of chapter 1 was published in *Grey Room*, and a selection of material from chapter 2 was previously published in a volume titled *Le Laboratoire*, edited by Martha Langford and Zoë Tousignant and jointly published by Artexte and Formes actuelles de l'expérience photographique: épistémologies, pratiques, histoires (FAEP).

Friendship and collegiality make intellectual life possible and all the more enjoyable. Thank you to Amy Wallace and Michaela Rife for their friendship in grad school and beyond. Thank you to Marina Dumont-

Gauthier for carrying a very heavy catalog on a transatlantic flight and documenting an exhibition I couldn't see in person. Thank you to Beth Matthison and Katie Addleman-Frankel for the company in Paris. Thank you to Kathrin Yacavone for reading part of this manuscript and sharing in my enthusiasm for tracing Balzac's ghosts. Thank you to my aerial colleagues Nicholas Robbins and Matthew Hunter for their collaborations, intellectual and otherwise, that shaped the direction of chapter 1 of this book. Thank you to my writing group—Christy Anderson, Camille Bégin, Dragana Obradovic, and Danielle Taschereau-Mamers—for their company amidst the ups and downs of it all. Thank you to Hilary Bergen for her writerly solidarity. Thank you to Rhiannon Vogl for bolstering our shared project of keeping in sound body and mind in wild times. Thank you to Zoë de Luca for everything—may we never get to the bottom of things. Thank you to Akshaya Tankha for an unparalleled intellectual friendship. To my learned friends—Caycie Soke, Katie Kidder, and Jane Harrington—our friendship is my greatest accomplishment.

This book is for my family, each of whom shows me, in their own way, that making and learning are among the best ways to engage with our shared world—Kathy Glinski, Paul Doucet, Jacob Doucet, Hannah Doucet, Olivia Doucet, Maia Doucet, Ben Borys, and Mick Brambilla. Last but most certainly not least, it is for Curtis McCord—for keeping me fed and for your endless love and curiosity.

NOTES

INTRODUCTION: "WHO DO YOU THINK IS THE GREATEST PHOTOGRAPHER IN THE WORLD?"

1. For example, Walter Benjamin said of Nadar's photographs of underground Paris that they were "the first time the lens was deemed capable of making discoveries." He further described Nadar as a "master" of photography's early period. Roland Barthes also spoke in hyperbolic terms (which I have borrowed here as a title) regarding Nadar: "Who do you think is the greatest photographer in the world?" "Nadar," answers Barthes. Benjamin, "Paris, Capital of the Nineteenth Century," 6; Benjamin, "Small History of Photography," 240; Barthes, *Camera Lucida*, 68.

2. A 2012 exhibition curated by Luce Lebart titled *Un laboratoire des premières fois* (A laboratory of firsts) and displayed at the Rencontre d'Arles brought together examples of such *"premiers essais"* from the collections of the Société française de photographie, including several of Nadar's photographs.

3. As Michel-Rolph Trouillot has noted, "In its vernacular use, history means both the facts of the matter and a narrative of those facts, both 'what happened' and 'that which is said to have happened.'" Trouillot, *Silencing the Past*, 2.

4   Christopher Pinney has similarly described images as "unpredictable 'compressed performances' caught up in recursive trajectories of repetition and pastiche whose dense complexity makes them resistant to any particular moment." Pinney, "Things Happen," 266.

5   Benjamin, "Paris, Capital of the Nineteenth Century."

6   Freund's doctoral dissertation, "La photographie en France au dix-neuvième siècle: Essai de sociologie et d'esthétique," broadly considered to be the first doctoral dissertation to focus on the history of photography in France, was published as a book by the influential author and publisher Adrienne Monnier in 1936. As Eduardo Cadava has noted, in the afterword to Benjamin's *Berlin Childhood Around 1900*, Theodor Adorno compellingly analogized Benjamin the author with Nadar the photographer. Freund, *La photographie en France*; Benjamin, *Arcades Project*, 6, 90–91, 309, 673–74, 681; Theodor Adorno cited in Cadava, *Words of Light*, xxi.

7   Chevallier, *Nadar*; Prinet and Dilasser, *Nadar*; Braive, *L'age de la photographie*. On the collection history of the Nadar studio, see Jammes, "Nadar Studio," 116. On Nadar in North American museum collections, see Hellman, "Nadar dans les collections Nord-Américains."

8   See, for example, Begley, *Great Nadar*; Greaves, *Nadar*; Gosling, *Nadar*; Stiegler, *Nadar: Bilder der Moderne*.

9   These theoretical trajectories will be elaborated on in the epilogue to this book. See the summer 1978 issue of *October* and the September 1976 edition of *Artforum*. To these we might also add the spring 1970 issue of *Aperture*, "French Primitive Photography," a title that partially borrows from the title of Nadar's essay "Les primitifs de photographie" in *Quand j'étais photographe*. See also Solomon-Godeau, *Photography at the Dock*, xxii; Sekula, "Neither Metaphysics nor Positivism"; Baldwin and Keller, *Nadar–Warhol, Paris–New York*.

10 Daston, "History of Science."

11 On the dangers of reparative history replicating colonial metaphors, see Loveday, "Pioneer Paradigm."

12 On Nadar's self-portraits, see Lerner, *Experimental Self-Portraits*, 32–58.

13 On Nadar's signatures, see Lerner, "Nadar's Signatures."

14 Where possible, I have elected to include quotations from existing English translations of primary sources. Where this wasn't possible, translations from the French are my own. Nadar initially withdrew permission for his brother to use the name in 1855, an act which was

no doubt prompted by his establishment of his own photographic studio that same year. In response to Adrien continuing to sign "Nadar *jeune*" on his photographs, Nadar decided to bring a suit to the tribunal. After several judgements (on February 28 and April 22, 1856), the definitive ruling was made in favor of Nadar on December 12, 1857.

15 "Exposé de motifs pour la revendication de la propriété exclusive du pseudonyme Nadar. Mémoire adressé à MM. les membres du tribunal de Commerce de la Seine," April 23 1856, Fonds Nadar, YB3–2340, Box 9, Département des estampes et de la photographie, Bibliothèque nationale de France, Paris.

16 "Mémoire adressé à MM. les membres du tribunal de Commerce de la Seine," April 23, 1856, Fonds Nadar, YB3–2340, Box 9, 3.

17 As Jillian Lerner notes, we can see his name in bold capital letters in the title at the top, in smaller capital letters at the bottom of the dedication, and in the form of his signature at the bottom left. See Lerner, "Nadar's Signatures," 110.

18 "Lettre de G. Félix Tournachon (Nadar) à Paul Tournachons" (Letter from G. Félix Tournachon (Nadar) to Paul Tournachon), December 17, 1903, Papiers Nadar, NAF24992, f. 80; See also Lacoste, "Paul Nadar, l'ère de la modernité," 49; Aubenas, Lacoste, and Roubert, "La maison n'a pas de succursale."

19 See Batchen, *Burning with Desire*; Brunet, "Literary Prehistory of Photography"; Brunet, *La naissance de l'idée*; Gasser, "Histories of Photography 1839–1939"; McCauley, "Writing Photography's History Before Newhall"; Siegel, *First Exposures*; Sheehan and Zervigon, *Photography and Its Origins*.

20 Siegel, "Afterword," 404.

21 For a summary of recent approaches to the problem of media specificity in the nineteenth century, see Leonardi and Natale, *Photography and Other Media*.

22 On the evolution of the term *new media*, see Chun, "Did Somebody Say New Media?" Importantly for my argument here, Chun notes that, in the 1990s, the literature on new media often sidelined the commercial history of new media. On the newness of old media, see Gitelman and Pingree, *New Media, 1740–1915*; Gitelman, *Always Already New*; Marvin, *When Old Technologies Were New*; Thorburn and Jenkins, *Rethinking Media Change*.

23 Gitelman, *Scripts, Grooves, and Writing Machines*, 8.

24 Gitelman, *Always Already New*, 26.

25 Chun, "Did Somebody Say New Media?," 3.

26 English translations from *Quand j'étais photographe* are from the 2015 translation by Eduardo Cadava and Liana Theodoratou. Nadar, *When I Was a Photographer*, 1–2.

27 Chun, "Did Somebody Say New Media?," 3. In this regard, newness and firstness are inextricable from colonial modernity. On the use of history writing and the practice of "firsting" as a tool of colonization, see O'Brien, *Firsting and Lasting*.

28 See, in particular, McCauley, *Industrial Madness*; Nesbit, *Atget's Seven Albums*.

29 Solomon-Godeau, "Calotypomania," 4.

30 Edwards, "Why Pictures?"

31 Edwards, "Why Pictures?," 1. I have written about the history of photography as a history of invention elsewhere; see Doucet, "In History, the Future."

32 Edwards, "Why Pictures?," 1.

33 In asking these questions, I am indebted to the work done by scholars such as Elizabeth Anne McCauley and Molly Nesbit. See McCauley, *Industrial Madness*; Nesbit, *Atget's Seven Albums*.

34 The premise that archives shape how history is written has been the subject of much scholarship by both historians and archivists. For a summary of some of this work (with particular attention to the practices of working archivists), see Schwartz, "Archives, Records, and Power." See also Azoulay, *Potential History*.

CHAPTER 1. COLLECTING THE IDEAS IN THE AIR: THE FIRST AERIAL PHOTOGRAPH

1 See, for example, Cosgrove and Fox, *Photography and Flight*, 24–25; Dorrian and Poussin, *Seeing from Above*; Gervais, "Experimentations photographiques," 52; Gervais, "Un basculement du regard"; Haffner, *View from Above*; Kaplan, *Aerial Aftermaths*; Newhall, *Airborne Camera*, 19–21, 30–33; Parks and Kaplan, *Drone Warfare*.

2 On the technical improbability of this early print, see Gervais, "Experimentations photographiques," 52.

3 Schaffer, "Making up Discovery," 16.

4 *Online Etymology Dictionary*, "Symptom," by Harper Douglas, accessed October 28, 2024, https://www.etymonline.com/word/symptom.

5 See, for example, Begley, *Great Nadar*, 1–8, 123–52; Braive, "Nadar premier voyageur dans la lune," 2–3.

6   Nadar's account of his experiments with aerial photography is reprised numerous times, first in 1864 in *Mémoires du Géant*, then in a multipart essay published in his son Paul's journal *Paris-Photographe*, and finally as a chapter in his book *Quand j'étais photographe* in 1900.

7   There are other compelling episodes in the history of Nadar's aerial exploits that are not within the scope of this chapter. For example, during the 1870 siege of Paris, Nadar proposed reviving the military ballooning corps that had been established in 1792. While the decree was not approved, Nadar briefly maintained an observation balloon at the place d'Auteuil. Nadar kept careful records of his correspondence with officials during this period, and his archives on this subject held at the Bibliothèque historique de la ville de Paris represent a compelling snapshot of institutional debate on the utility of balloons in 1870. See Collection Nadar, "Correspondance addressée à Nadar et documents relatifs au role des ballons pendants le Siège de Paris de 1870," MS-NA-471, Bibliothèque historique de la Ville de Paris, Paris, France; Nadar, *Les ballons en 1870 C.E.*; Nadar, "Obsidional Photography." See also Jestaz, "Soutenir l'aérostation." My thanks to Juliette Jestaz for sharing her research on this collection with me.

8   Think, for example, of the "first" full-color images from the James Webb Space Telescope, "launched" to the public in a particularly dramatic way in the summer of 2022. The images were neither the "first" produced with the telescope nor exactly photographs, but they were, and are, repeatedly described as such by NASA, European and Canadian space agencies, and the media.

9   Nadar, *Mémoires du Géant*, 51–52.

10 On the role of anonymous inventors in photo-historical writing, see Gunthert, "L'inventeur inconnu."

11 Andraud, *Exposition universelle de 1855*, 98.

12 Andraud, *Exposition universelle de 1855*, 98.

13 Nadar, *Mémoires du Géant*, 51–52.

14 Nadar, *When I Was a Photographer*, 71; *Mémoires du Géant*, 63.

15 To name just three other rather famous nineteenth-century figures who penned similar phrases, Oliver Wendell Holmes, upon being accused of poetic plagiarism, stated that "the spores of a great many ideas are floating in the atmosphere"; Charles Babbage, in his 1838 work of natural theology, described the atmosphere as a repository of all human knowledge, past and present, writing that "the air itself is one vast library on whose pages are for ever written all that man has ever said or woman whispered"; and Thomas Jefferson described ideas as "like fire, expansible over space, without lessening their density

16   in any point, and like the air in which we breathe, move, and have our physical being, incapable of confinement or exclusive appropriation." Oliver Wendell Holmes cited in Wentersdorf, "Underground Workshop," 8; Babbage, *Ninth Bridgewater Treatise*, 112; Thomas Jefferson cited in Pottage and Sherman, *Figures of Invention*, 4.

16 Félix Tournachon, "Un système de photographie aérostatique" (A system of aerostatic photography), patent no. 38509, 1858, Institut national de propriété industrielle, Paris, France.

17 The scholarship on the evolution of modern patent law is voluminous. For this summary, I have relied primarily on the account outlined in Biagioli, "Patent Republic." The relationship between the patent process and Nadar's claims to priority will be the subject of chapter 2.

18 Pottage and Sherman, *Figures of Invention*, 12.

19 Pottage and Sherman, *Figures of Invention*, 2.

20 As Pottage and Sherman note, in patent law from the twentieth century onward, this kind of abstract theorization on what an idea is has been replaced with a debate over "what kinds of rights one can have in ideas." Pottage and Sherman, *Figures of Invention*, 4.

21 We see this even today, with Silicon Valley tech founders regularly citing (often dystopian) science fiction as the origin of their ideas. Even though these figures would rarely pass up an opportunity to acclaim their own inventive prowess, there is evidently cultural currency in their professed goal of "realizing" existing ideas expressed in art and literature of the past.

22 Nadar, *Mémoires du Géant*, 55.

23 Nadar, *Mémoires du Géant*, 58.

24 Andraud, *Exposition universelle de 1855*, 136.

25 Heatherton, "Making the First International," 295.

26 Nadar, *Mémoires du Géant*, 93.

27 Aubenas, "Beyond the Portrait," 101.

28 As de La Landelle put it, aluminum "appeared to have been discovered for the construction of aéronefs." Quoted in Bardon, *Gustave de Ponton d'Amécourt*, 82.

29 Ponton d'Amécourt employed a variety of figures to build different parts of the model. He engaged the talents of M. Joseph, a skilled clockmaker from Amiens, who, after making ten different models, finally succeeded in building one that could stay airborne, if only momentarily. On June 3, 1862, the helicopters were presented publicly to much excitement. The pair would later coin the word *aviation* (the field of research into machines heavier than air) and the verb *avier* to describe the new sensation

of flying. De La Landelle cited in Bardon, *Gustave de Ponton d'Amécourt*, 84. See Roger Musnik, "Gabriel de La Landelle (1812–1886)," *le BloGallica* (blog), *Gallica*, November 3, 2014, https://gallica.bnf.fr/accueil/fr/html/gabriel-de-la-landelle-1812-1886. See also, *Trésor de la langue Française informatisé*, "Aviation," ATILF (CNRS/Université de Lorraine), accessed July 13, 2018, http://www.atilf.fr/tlfi.

30 Ponton d'Amécourt was evidently frustrated by this, noting that it was as if the promoters (Nadar and de La Landelle) were taking the place of the inventor (Ponton d'Amécourt): "It would be ungrateful for me to suggest the MM. Nadar and de La Landelle took the merit of the idea they sponsored for themselves; they were kind enough to put me in the same rank as them: but despite their good will, their resounding personalities left the inventor holding the patents far behind them in public notoriety." Ponton d'Amécourt, *Collection de mémoires*, vii.

31 Nadar, *Mémoires du Géant*, 121.

32 Nadar, *Mémoires du Géant*, 135.

33 De La Landelle, *Dans les airs*, 182.

34 In the full series, there are four different helicopter models pictured. See "Receuil. Photographies Positives. Oeuvre de Félix Nadar," FOL-EO-15(11), Département des Estampes et de la photographie, Bibliothèque nationale de France, Paris.

35 Nadar, "Manifeste de l'autolocomotion aérienne," *La presse*, August 7, 1863; Nadar, "Manifeste de l'autolocomotion aérienne," *L'aéronaute*, numéro spécimen, [1863], 1–4.

36 Nadar, *When I Was a Photographer*, 156.

37 Eric Kluitenberg has described the importance of such an "imaginary" register in media history. See Kluitenberg, *Book of Imaginary Media*, 9.

38 The inclusion of Doré's frontispiece as an illustration in Christopher Hatton Turnor's 1865 *Astra Castra: Experiments and Adventures in the Atmosphere* with the simple caption "M. Nadar's ideas" would seem to indicate that the assignation of the helicopter as part of Nadar's personal legacy had also extended beyond France. Turnor, *Astra Castra*, 341.

39 Nadar, *When I Was a Photographer*, 142.

40 See Nadar, *Le droit au vol*. I cite here from the 1866 English translation (Nadar, *Right to Fly*). Nadar's monumental balloon project builds on a tradition of public and "spectacular" demonstrations of scientific principles and physical phenomena. See, for example, Raichvarg and Jacques, *Savants et ignorants*; Tresch, "Prophet and the Pendulum."

41 "Falsifying" refers to Karl Popper's theory of falsification. See Godfrey-Smith, *Theory and Reality*, 57–63; See also Popper, *Logic of Scientific Discovery*, 57–63.

42 For Nadar's statistical claims regarding facture, see Begley, *Great Nadar*, 138. Nadar would later win a legal dispute with the Godard brothers in which the latter alleged (accurately, it would seem) that Nadar had not paid them in full for the cost of the balloon. Nadar argued (successfully, if cynically) that technical issues had precipitated the crashes, and that these defects rendered his agreement with the brothers void.

43 Begley, *Great Nadar*, 141.

44 Nadar describes these ascensions (and crashes) at length in *Mémoires du Géant*, (see pp. 231–354).

45 After repairing the destroyed balloon, Nadar also displayed the *Géant* (inflated but not in flight) at the Crystal Palace in London later in 1863, thus further emphasizing the static quality of ballooning technology. Shorter ascensions were also undertaken in Brussels (1864) and Amsterdam and Lyon (1865).

46 See Sinnema, "Representing the Railway." On the media production emanating from the *Géant* project, see Doucet, "Idea Was in the Air."

47 Jestaz, "Soutenir l'aérostation."

48 Nadar, *Mémoires du Géant*, i–xx.

49 Nadar, *Mémoires du Géant*, iii.

50 Nadar, *Mémoires du Géant*, iv. A large portion of the Tissandier collection is now in the collection of the Library of Congress in Washington, DC. For a detailed description of other collections of aviation material, see Jestaz, "Soutenir l'aérostation."

51 A letter from Nadar cited by A. L., "La collection de Nadar."

52 "Arrêtes préfectoraux pris en execution des deliberations du Conseil municipal," *Bulletin municipal officiel de la ville de Paris*, no. 333 (December 10, 1897): 9. The catalog accompanying the sale indicated the total number of items as 5,493, though this total may not have been completely accurate. For an exact breakdown of the objects in the collection, see Jestaz, "Soutenir l'aérostation."

53 Newspaper clipping, *Le Figaro*, May 17, 1895, 1, Fonds Nadar, YB3-2340, Box 17.

54 Henriot, "Catalogue des publications," 21. Catherine E. Clark has described the early institutional and collecting histories of these institutions in light of the changes wrought upon the urban fabric by the renovation of Paris directed by the Baron Haussmann, signaling

55 "Chez Nadar. Une collection aérostatique—Don à la Ville de Paris," *La patrie*, September 20, 1897, 1, newspaper clipping, Fonds Nadar, YB3-2340, Box 17. In a letter from June 1897, Laussedat expressed his dismay that Nadar had offered his collections to the city of Paris. Laussedat had also been in correspondence with Nadar to collect several other photographic prints for the CNAM collections. See "Lettre de Colonel Laussedat à Félix Nadar" (Letter from Colonel Laussedat to Felix Nadar), June 14, 1897, Fonds Nadar, NAF24275, f. 399.

56 The catalog organizes the collection into material type and chronological subject order (not publication date). Subsections include books and pamphlets, periodicals, autographs (letters alphabetically organized into historical, modern, and contemporary sections), manuscripts (organized alphabetically by subject), translations of his own books, maps and plans, prints, photographs, drawings, paintings, faïence, and commemorative medals. See Félix Nadar, "Collection Nadar: Documents relatifs à l'histoire de la navigation aérienne," unpublished manuscript, n.d., A0788, musée Carnavalet, Paris, France.

57 Nadar divided the visual material in his collection into the following categories: military aerostation, posters, caricatures, public events, *Géant*, history of aerostation, music, portraits of aeronauts, the siege of Paris, technology, and panoramic views. To the collections held at the musée Carnavalet and the Bibliothèque historique de la ville de Paris we can add the numerous boxes of archival materials dedicated to aerial-related material conserved at the Département des estampes et photographies and the Département des manuscrits at the Bibliothèque nationale. See primarily Fonds Nadar, YB3-2340, Boxes 12–19, and EO-15(11)-FOL. The presence of photographs of Leonardo da Vinci's sketches of flying machines among Nadar's papers indicates his attention to the longer visual histories of the legacy of such technological prophecies. See photograph of drawing by Leonardo da Vinci, Fonds Nadar, YB3-2340, Box 17. Prints of these images can also be found in the Nadar collection at the musée Carnavalet.

58 Nadar, "Manifeste de l'autolocomotion aérienne."

59 As Jestaz notes, in acknowledgment of the publicity efforts of the Société, many inventors sent their ideas (in the form of letters, sketches, and the like) to Nadar and the Société. These documents can be found within the manuscripts held at the Bibliothèque historique de la ville de Paris. Jestaz, "Soutenir l'aérostation."

that the museum and the library were explicitly created to preserve documents "of objects and buildings that might soon be gone." See Clark, *Paris and the Cliché of History*, 2.

60   Jacques Babinet, "Introduction," in Nadar, *Mémoires du Géant*, unpaginated.
61   Nadar, *Right to Fly*, 9–10.
62   De La Landelle, *Société d'encouragement*, 5.
63   On the genealogy of invention in the arts and the sciences, see Godin, "Innovation."
64   Loh, *Titian Remade*, 21–22, 41–42.
65   Schaffer, "Scientific Discoveries," 397.
66   Schaffer, "Scientific Discoveries," 397.
67   See under *"inventer"* in *Dictionnaire de l'Academie française*, 6th ed., vol. 2 (Paris: Imprimerie et Librairie de Firmin Didot Frères, 1835); *Dictionnaire de l'Academie française*, 7th ed., vol. 2 (Paris: Librairie de Firmin-Didot, 1878). This understanding of the word is addressed in Nadar's photo-interview with the scientist Michel-Eugène Chevreul, discussed in chapter 3 of this book. See Félix Nadar, "L'art de vivre cent ans," manuscript, NAF13828, f. 24, Département des manuscrits, Bibliothèque nationale de France, Paris.
68   Nadar's more familiar domain of photography was another site for transhistorical collaboration in technological innovation, usually framed in the history of photography as the "prehistory" or "origins" of photography. See Batchen, *Burning with Desire*; Thélot, *Les inventions littéraires*; Eigen, "On Purple."
69   De La Landelle, *Aviation*, 96–97. In the first edition of this book, de La Landelle misremembers the location as the British Museum, but in a subsequent edition from the same year, he notes that correspondence with M. Lucien Serres reminded him that the *"hélice"* in question was from the Patent Museum and had been patented by Henry Bright in 1859. Bright's model was not, as with Ponton d'Amécourt's, a device powered by a motor but rather a *"projet en relief"* for a propeller-assisted parachute. Serres's correction appears to have been motivated by his own research into the possibility of propeller-powered flight, which he believed had been developed before Ponton d'Amécourt's model. What this episode reveals more than anything is the flurry of models for heavier-than-air flight that were being developed during this period and the immense importance attached to publicity by those involved. See also Hayward, "Review of Helicopter Patents."
70   De La Landelle, *Aviation*, 96–97.
71   De La Landelle, *Aviation*, 119.
72   De La Landelle, *Aviation*, 97.

73 Nadar, "La première épreuve de photographie aérostatique (suite)," pt. 3, 248. Nadar's 1893 version of the essay on aerial photography unfolded in three parts, of which this was the third. For the first two, see Nadar, "La première épreuve de photographie aérostatique," pt. 1; Nadar, "La première épreuve de photographie aérostatique (suite)," pt. 2.

74 Nadar, "La première épreuve de photographie aérostatique (suite)," pt. 3, 248n. Curiously, the footnote is signed with Nadar's signature "N," another authorial flourish to round out this story.

75 Nadar, "La première épreuve de photographie aérostatique (suite)," pt. 3, 248n.

76 For a summary of the state of the field published by one of Nadar's contemporaries (and fellow collectors), see Tissandier, *La photographie en ballon*.

77 Exposition internationale, *Exposition universelle internationale*, 66.

78 This print indicates that Nadar used a multi-lens carte-de-visite camera for this series.

79 Tissandier, *La photographie en ballon*, 1–6.

80 "Lettre de Colonel Laussedat à Félix Nadar" (Letter from Colonel Laussedat to Félix Nadar), June 30, 1893, Fonds Nadar, NAF24275, f. 383.

81 "Lettre de Colonel Laussedat à Félix Nadar" (Letter from Colonel Laussedat to Félix Nadar), July 8, 1893, Fonds Nadar, NAF24275, f. 385.

82 Kaplan, *Aerial Aftermaths*, 8–15.

83 This is an issue that was explored by Allan Sekula in his 1975 article on World War II aerial reconnaissance photography and Edward Steichen. See Sekula, "Instrumental Image."

CHAPTER 2. PATENT PRIORITIES: THE FIRST PHOTOGRAPH BY ELECTRIC LIGHT

1 Critical to my understanding of this material is the reconfiguration of the archive by the processes of historical inquiry. Much of the material pertinent to this chapter is compiled in a folio in the collection of the Département des Estampes et de la photographie at the Bibliothèque nationale de France bearing the label: "Création de la Photographie à l'électricité (inédit) / Ces 32 Documents ont été communiqués au Grand Chroniqeur Historien (en 1913) Frédéric Masson qui devais avant la guerre de 1914 faire la biographie de Nadar, Pierre Petit ainsi que celles de leurs contemporains." See "Chemise contenant le dossier sur la

lumière électrique," n.d., manuscript, BOITE FOL A-EO-15, Département des Estampes et de la photographie, Bibliothèque nationale de France, Paris. Frédéric Masson was a French historian who wrote primarily about the Napoleonic era but was, according to this note, reportedly working on a text on nineteenth-century French photographers before his death in 1923.

2   Pottage and Sherman, *Figures of Invention*, 8–9.

3   This chapter joins a broader recent current of investigations into the economic history of photography. See, for example, Edwards, "Why Pictures?," and the special issue of the French journal *Photographica* titled "Prix, coût, valeur: Pour une histoire économique de la photographie," edited by Éléonore Challine and Paul-Louis Roubert.

4   This approach is informed by scholars such as cultural historian Jane M. Gaines and Marxist legal theorist Bernard Edelman, who consider legal disputes (and the laws that structure them) as cultural texts. It should be noted that both texts deal primarily with the notion of copyright or the "author's rights" with regard to the legal (not to mention aesthetic, economic, and philosophical) problem of cultural representations as property forms. In my case here, I consider instead the problem of intellectual property in the form of patented photographic processes. See Gaines, *Contested Culture*; Edelman, *Ownership of the Image*. On the history of the *droits d'auteur* in France, see Nesbit, "What Was an Author?"

5   "Prior examination" refers to the part of the patent process that considers the relative novelty of a given invention, determined based on the existence of "prior art" and/or the "non-obviousness" of the invention under consideration.

6   See Baudry, "Examining Inventions." Despite the technical absence of the requirement of "prior examination" to determine the utility and novelty of invention, Baudry argues convincingly that the Comité consultatif des arts et manufactures effectively functioned as an examining body. The committee was primarily concerned with the novelty of inventions and "shaping a normative language for the description of technical artifacts." What this process indicates, Baudry argues, is that "both the inventor and the public were figures whose role had to be actively construed." Baudry, "Examining Inventions," 75.

7   A significant revision of the patent process also occurred in 1844, including a ban on the importation of patents (though noncitizens were authorized to take out French patents under the same conditions as citizens) and the introduction of the Brevété sans garantie du gouvernement (SGDG). These reforms occurred within a broader reconfiguration of France's (and Europe's) industrial economy. Con-

sidering the challenges presented, for example, by British competition, French industry was understood to be in need of legal protection. Galvez-Behar, "Patent System."

8   Mario Biagioli has argued that the move from patents as privileges under the ancien régime toward the Republican construction of patents as intellectual property "parallels the demise of political absolutism." Jérôme Baudry, however, distinguishes between the liberal democratic theory behind intellectual property law in France and the realities of practice during the first fifty years of French patent law post-1791, which, as he notes, remained highly influenced by the elite bodies of scientific authority. See Biagioli, "Patent Republic"; Baudry, "Examining Inventions."

9   See Crosland, "Popular Science and the Arts"; Béguet, *La science pour tous*.

10 Cercle de la presse scientifique quoted in Crosland, "Popular Science and the Arts," 306. On the foundation of this group within the broader context of "scientific popularization," see Nieto-Galan, *Science in the Public Sphere*, 99.

11 Cercle de la presse scientifique, "Convocation à une réunion au Cercle de la presse scientifique le 23 avril 1859," A-EO-15, Département des Estampes et de la photographie, Bibliothèque nationale de France, Paris.

12 See Doucet, "Excessive Luminosity."

13 This annotation was presumably added upon the presentation of the prints to the Société française de la photographie in 1861.

14 Recall Simon Schaffer's description of the moment of discovery as a moment when an idea is fixed as fact and ascribed an author. See Schaffer, "Scientific Discoveries," 397.

15 Nadar, *When I Was a Photographer*, 85.

16 If we take this to be true, nineteenth-century debates over whether photography was an "art" take on an entirely bourgeois, commercial character. Photography needed to be an art so that photographers could become authors/owners. See Edelman, *Ownership of the Image*, 39–51.

17 The secondary literature occasionally implies that Nadar had patented a method for taking photographs by electric light. See, for example, Rice, *Parisian Views*, 157. For the patent text, see Félix Tournachon, "Le tirage des épreuves photographiques positives par la lumière électrique ou la lumière du gaz" (The printing of positive photographic prints by electric or gas light), patent no. 48442, 1861, Institut national de propriété industrielle, Paris, France.

NOTES TO CHAPTER 2   169

18 The printing process Nadar was referring to was developed by Albert Moitessier and presented to the Académie des sciences in 1855. Moitessier obtained positive prints on collodion plates from negative prints, which were later transferred to paper. Moitessier argued that this process was superior to printing directly on paper because of the sensitivity of the collodion surface and the delicate gradation of tones. This process also enabled enlargements of the negative in which the resulting prints were much sharper than those that had been printed on paper. See Moitessier, "Procédé pour obtenir des épreuves positives," 120–21.

19 Félix Tournachon, "Le tirage des épreuves photographiques positives par la lumière électrique ou la lumière du gaz" [The printing of positive photographic prints by electric or gas light], patent no. 48442, 1861, Institut national de propriété industrielle, Paris, France, 7.

20 See, for example, the print in the collections of the Bibliothèque nationale de France: Félix Nadar, "La main du banquier D*** (étude chirographique), tirée en une heure à la lumière électrique" [The hand of banker D***, printed in one hour by electric light], uncropped, full-plate albumen paper print from a collodion glass negative, 24.2 × 20.1 cm., EO-15 (22)-PET FOL, Département des Estampes et de la photographie, Bibliothèque nationale de France, Paris.

21 Nadar, *When I Was a Photographer*, 85–86.

22 On the photographic flash, see Dinkar, "Pyrotechnics and Photography"; Flint, *Flash!*

23 Davy cited in Ayrton, *Electric Arc*, 24.

24 *Arc* was a term given to the phenomenon years later in 1820 after François Arago and Davy's effectively simultaneous predictions and experiments with the effect of magnets on the arc. Demonstrating the effects but not the cause of what would later be defined by Michael Faraday as electromagnetic induction, Arago had demonstrated the phenomenon in an experiment in 1824 in which a rotating copper disc caused a magnetic needle situated above it to spin.

25 Serrin received international acclaim after exhibiting the lamp at the 1862 exhibition in London. See Iselin and Le Neve Foster, *Reports by the Juries*, 100.

26 Experimenting with the electric arc himself, Bunsen had also invented a comparative photometer, designed to measure the intensity of light produced by the arc lamp in combination with his batteries. Stock, "Bunsen's Batteries," 101.

27 Nadar, *When I Was a Photographer*, 85.

28 Nadar, *When I Was a Photographer*, 84.

29 Nadar, *When I Was a Photographer*, 84.

30 I am grateful to Amy Wallace, whose work on the evolution of glass painting studios in the nineteenth century prompted me to rethink this trajectory. See Wallace, "Studios of Nature."

31 In a speculative yet never realized collaboration with the wealthy financiers the Pereire brothers, whose company had been contracted for the construction of the Grand hôtel du Louvre, Nadar had developed the idea of constructing a photography studio on the roof of the hotel to capture the full range of sunlight available in central Paris in 1856. As Anne McCauley has noted, the Pereire family were frequent investors in Nadar's photographic businesses. McCauley, *Industrial Madness*, 62. The relationship between the Saint-Simonian financiers and the photographer is also noted by Helen Davies in *Emile and Isaac Pereire*, 3, 157–58.

32 McCauley, *Industrial Madness*, 68.

33 McCauley, *Industrial Madness*, 68.

34 Van der Weyde, "Photography by Electric Light," 370; See also Van der Weyden, "Henry van der Weyde," 72n11.

35 Van der Weyde, "Photography by Electric light," 371.

36 Van der Weyde, "Photography by Electric Light," 370; See also Van der Weyden, "Henry van der Weyde," 72n11.

37 Van der Weyde, "Photography by the Electric Light," 370.

38 A Grove battery is a type of early battery designed by William Robert Grove. A Fresnel lens is a type of compact lens designed originally for use in lighthouses by Augustin-Jean Fresnel. In 1877, Van der Weyde introduced a gas-powered dynamo into his studio installation, and in 1891, he was connected to the Pall Mall Electric Light Company. Van der Weyden, "Henry van der Weyde," 69.

39 Van der Weyde, "Photography by Electric Light," 371.

40 Henry van der Weyde, "An improvement in illuminating objects to be photographed, and the interior of public and other buildings," February 2, 1878, patent no. 446, provisional patent specification, British Library, London, United Kingdom. The copy of the patent specification held in the patent collection of the British Library specifies that this specification was rendered void because the patentee had "neglected to file a Specification in pursuance of the conditions of the Letters Patent." This patent filing followed a previous filing by the photographer in 1876 that outlines a system for using a "water lens" to diffuse direct light (whether natural or artificial) for the purpose of photographic portraiture. See Henry van der Weyde, "An

41 improvement in photography and apparatus used therein," April 25, 1876, patent no. 1747, British Library, London.

41 Tissandier, *La photographie*, 182.

42 Wilson, "Day Among the Paris Studios," 411.

43 For example, the minerals used to create light-sensitive photographic materials. See Angus, *Camera Geologica*.

44 The Exposition internationale d'électricité was the largest in a wave of specialist electricity exhibitions in the 1880s, developing in part from the great interest in the use of electricity at the Paris Exposition universelle in 1878. On the influence of such electric exhibitions in visual culture, see Clayson, "Bright Lights, Brilliant Wit."

45 See Ministère des Postes et Télégraphes, *Exposition internationale d'électricité Paris 1881*.

46 As Kate Flint has noted, several early photographers experimented with the application of electric light to photography, such as Léon Foucault's use of the carbon arc lamp to take daguerreotypes of medical specimens and Sergei Lvovich Levitsky's portraits by battery-powered arc lamp. Flint, *Flash!*, 34. See also Eder, *History of Photography*, 528–30.

47 Edelman, however, doesn't examine the role of patents in structuring the photographer as inventor-capitalist prior to becoming an artist-creator. Edelman, *Ownership of the Image*, 49.

48 "Cours de cassation," 42. Antoine Lumière had purchased the patent rights for the Rhône region from Liébert and later received compensation for his fees in light of the voiding of Liébert's patent. See "Art. 3655"; "Lettre de Antoine Lumière à Pierre Petit" (Letter from Antoine Lumière to Pierre Petit), January 11, 1885, EO-15-A, Département des estampes et photographie, Bibliothèque nationale de France, Paris.

49 See, for example, "Un système de photographie aérostatique" (A system of aerostatic photography), patent no. 38509, 1858, Institut national de propriété industrielle, Paris, France.

50 Pottage and Sherman, *Figures of Invention*, 25.

51 "Lettre de Félix Nadar à un confrère" (Letter from Félix Nadar to a colleague), May 10, 1882, EO-15-A.

52 "Lettre de Félix Nadar à un confrère" (Letter from Félix Nadar to a colleague), May 10, 1882, EO-15-A. Several other letters from other photographers also supported Petit's claim, including a letter from G. Fangui declaring that he had also photographed with electric light in April 1877. See EO-15-A, Département des Estampes et de la photographie, Bibliothèque nationale de France, Paris.

53 "Lettre de Félix Nadar à Pierre Petit" (Letter from Félix Nadar to Pierre Petit), November 22, 1884, EO-15-A. A later letter confirms exactly what he was sending; see "Lettre de Félix Nadar à un confrère" (Letter from Félix Nadar to a colleague), January 27, 1885, EO-15-A.

54 La loi du 5 juillet 1844, cited in Emptoz, Marchal, and Hangard, *Aux sources de la propriété industrielle*, 46.

CHAPTER 3. SOUND REPRODUCTIONS: THE FIRST PHOTOGRAPHIC INTERVIEW

1 Thomas Grimm cited in Gervais, "Interview of Chevreul, France, 1886," 35; See also Grimm, "L'art de vivre cent ans."

2 The interview had originally been intended for publication in *L'illustration*, a publication that would later become known for its photographic reproductions.

3 In keeping with other chapters, I refer to Félix Nadar by the shorthand Nadar and his son Paul Nadar by his first name.

4 Félix Nadar, "L'art de vivre cent ans," manuscript, NAF13828, f. 18, Département des manuscrits, Bibliothèque nationale de France, Paris.

5 Geneviève Reynes has, for example, painstakingly corroborated various incomplete and iterative captions on different prints with the manuscript of the interviews. Reynes, "Chevreul interviewé part Nadar." See also Michèle Auer's annotated republication of the interview: *Paul Nadar: Le premier interview photographique*.

6 The editors describe the interview as transforming the "reader" into a "spectator." See the extended note from the editors in the manuscript version of the interview: "Avis des Editeurs," in "L'art de vivre cent ans," f. 18. Vanessa Schwartz has framed this phenomenon as part of a broader culture of spectacle and "one of the means in which a mass culture and a new urban crowd became a society of spectators." Schwartz, *Spectacular Realities*, 2.

7 Sterne, *Audible Past*, 220.

8 La redaction du Journal illustré, "Avis des editeurs" in "L'art de vivre cent ans," f. 11.

9 For example, Elizabeth Edwards has described photographs as "material performances that enact a complex range of historiographical desires." Edwards, "Photography and the Material Performance," 130–31. See also Edwards, *Camera as Historian*.

10 Given that the previous Exposition universelle had taken place in 1878 and that the exhibitions were planned every eleven years, 1889 was the logical date for next French exhibition. In addition to the

political instability of the still nascent Third Republic, many feared that the simultaneous celebration of French Republicanism sure to be embodied in the 1789 centennial would deter other European nations still under monarchical rule from sponsoring national entries. See Nelms, *Third Republic*, 48.

11 See Bann, *Clothing of Clio*; Keylor, *Academy and Community*.

12 Carey, "Problem of Journalism History."

13 Carey borrows from the historian Herbert Butterfield to describe (and argue against) the "Whig interpretation of journalism history" that "views journalism history as the slow, steady expansion of freedom and knowledge from the political press to the commercial press, the setbacks into sensationalism and yellow journalism, the forward thrust into muckraking and social responsibility." Carey, "Problem of Journalism History," 88.

14 Hugo, *Les Misérables*, 102.

15 A longer publication of the complete transcripts of the interviews was planned but never realized, remaining in manuscript form. Based on three separate interview sessions, the manuscript version of the interview text includes a slightly longer introduction by the newspaper editors and five thematic sections: human longevity, photography, ballooning and aviation, Chevreul's color theory, and a general concluding conversation. In addition to these five edited "conversations," a section entitled "The Spirit of Chevreul" was added between the editor's introduction and the first interview, a biography of M. Chevreul was inserted directly after the final chapter, and a letter from the author Gustave Grignon responding to the text was included at the very end of the manuscript. In contrast, the newspaper version of the interview included an editorial introduction and shortened sections subtitled "The Secret of Longevity," "Regiments," "Hygiene and Philosophy," "Photography," "Balloon," "Colors," and "Spiritualism"—categories that reflected Nadar's interests as much as they did Chevreul's. See Nadar, "L'art de vivre cent ans," NAF13828; Nadar and Nadar, "L'art de vivre cent ans."

16 Nadar and Nadar, "L'art de vivre cent ans," 284.

17 La redaction du Journal illustré, "Le journal illustré à ses lecteurs," 282.

18 Nadar, "L'art de vivre cent ans," f. 18.

19 La redaction du Journal illustré, "Le journal illustré à ses lecteurs," 282. As Jean-Marie Seillan notes in his survey of the history of the interview in modern French journalism, it is difficult to date the origin of the practice, in part because the method predates the im-

20 Delporte, *Les journalistes en France*, 60–77.
21 Schudson, "Question Authority," 565.
22 Schudson, "Question Authority," 565.
23 Schudson, "Question Authority," 566. Seillan also notes this problem in the case of the history of the interview in France, noting that journalists recorded the interview either by simple notetaking or shorthand (stenography) to be reinterpreted later for publication. Seillan, "L'interview," 1029.
24 Recounting the origins of the project in a lecture at the École des hautes études commerciales, later published in *Paris-Photographe* in July 1892, Paul Nadar notes that, though he had the idea to combine photography and phonography, the commercial unavailability of such a device presented a barrier to its use. There was no mention of how a phonograph recording would have been presented in a newspaper. The twinning of new media technologies with new forms of reportage was characteristic of many new publications that emerged during this period. For example, Christian Delporte notes that *Le Matin* (founded by Alfred Edwards in 1884) was explicitly described as a daily providing "telegraphic information." P. Nadar, "Progrès et applications de la photographie (suite)," 284–86; Delporte, *Les journalistes en France*, 63.
25 Pierce, "Writing at the Speed of Sound."
26 Pierce, "Writing at the Speed of Sound," 139.
27 Nadar, "L'art de vivre cent ans," f. 13.
28 Nadar, "Les histoires du mois." The term was used by Nadar in this article in 1856 and summarized again in *Mémoires du Géant* in 1864. See Nadar, *Mémoirs du Géant*, 271–73.
29 Nadar, "Les histoires du mois." Though the "acoustic daguerreotype" isn't originally described in terms of its implications for journalism, it does appear just before several of Nadar's meandering paragraphs on the profusion of newspapers being published at the time.
30 Nadar is not alone in his descriptions of the possibility of using sound recording to "daguerreotype" sound. For example, Édouard-Léon Scott de Martinville's phonoautograph was also described in photographic terms. See Feaster, "Daguerreotyping the Voice," 18–23.
31 Nadar, *Mémoires du Géant*, 272.

32   Lissajous's system visualized sound waves by directing light (in this case, candlelight) at a mirror attached to a vibrating metal tuning fork and projecting it onto a wall. The high-speed vibration of the light reflecting off the mirror affixed to the tuning fork appeared as a line on the wall. When bouncing light off multiple tuning forks at the same time, the respective "lines" of sound created by the vibrations appeared perpendicular to each other, creating a pattern (known as a Lissajous curve) reflecting the ratio between the varying frequencies of the tuning forks. See Lissajous, "Mémoire sur l'étude."

33   This distinction is also critiqued by Jonathan Sterne, who argues that the definition of sound reproduction technologies as a division of sound from the copy assumes that "at some point prior to the invention of sound reproduction technologies, the body was whole, undamaged, and phenomenologically coherent." Likewise, here, "history" is imagined as something that could be grasped and copied—even occasionally in the prose of recent scholars who acknowledge the Nadars' efforts as a kind of failure of reproduction. Sterne, *Audible Past*, 21.

34   Léon Scott cited in Sterne, *Audible Past*, 45.

35   Ader's prototype was developed in 1881, but a subscription service allowed listeners, perhaps most infamously the writer Marcel Proust, to listen to live theater broadcasts via the services of the Compagnie du théatrophône from 1890 to 1931. As Melissa van Drie notes, Ader's system was commercialized by two engineers named Belisaire Marinovitch and Geza Szarvady, proprietors of the Compagnie du théatrophône who managed to arrange for their exclusive right to broadcast performances of the opera in advance of Ader. See Van Drie, "Hearing Through the Théâtrophone."

36   See Collins, "Theatrophone"; Gioia, "First Music Streaming Service."

37   La redaction du Journal illustré, "Avis des Editeurs," in "L'art de vivre cent ans," f. 13.

38   Jenkins, *Images and Enterprise*, 96.

39   This film was somewhat difficult to work with. To make a print from the earliest Eastman roll film, the photograph emulsion had to be detached from the flexible paper surface and applied to a glass plate. Jenkins, *Images and Enterprise*, 101.

40   Jenkins, *Images and Enterprise*, 107. Due to the difficulty of printing from the roll film and the low-quality negative produced, studio photographers generally rejected the roll film, preferring to continue using the collodion glass plates. The Eastman Company moved to market this technology primarily to landscape photographers due to the relative ease of transporting roll film as opposed to the weighty glass plates.

41   Lacoste, "La photographie instantanée," 292. Chevreul's image would become an important part of Paul's advertising campaign, including on the illustrated catalog for the Office général de photographie published in 1889.

42   P. Nadar, "Sur les papiers positifs," 47; Nadar cited in Auer, *Paul Nadar: Le premier interview photographique*, 31–32.

43   The interview with Chevreul was not Paul's only experiment with instantaneous photography. He published a second photographic interview with the controversial General Boulanger on November 23, 1889. In this case, the rhetoric of novelty was even more pronounced. Introducing the interview with Boulanger, the editors write: "Everything is renewed. In this time of the Eiffel tower, luminous fountains, and telephones, journalism must follow the progress of science. He can no longer be satisfied with the vulgar procedures for which pen, ink, paper and memory are sufficient. . . . The public has become skeptical; he is wary of what he is told. Alongside each important declaration, he will need material proof." In addition to the interviews, Paul experimented with the different visual possibilities opened by the higher-speed film. He also contributed personally to the development of these technologies with the introduction of the Express Détective Nadar camera in 1888 and a type of high-speed photographic plate in 1893. A body of personal photographs demonstrates some of his early experiments with snapshot photographs. He also employed the camera on a journey to Turkestan (now part of the western provinces of the People's Republic of China) in 1890 and a second trip to Palestine in 1892. See Paul Nadar, "Entrevue photographique," *Le Figaro: Supplement littéraire*, November 23, 1889.

44   On the development of photographically illustrated newspapers in France, see Gervais, "L'illustration photographique." On this interview, see pages 108–30.

45   Gervais, "L'illustration photographique," 122.

46   P. Nadar, "Sur les papiers positifs," 49.

47   His early work focused on organic chemistry, examining animal fats and the saponification process and identifying several fatty acids. From early on in his career, his chemical research had important industrial applications, including the production of soap, the commercial manufacture of glycerin, and the production of a novel kind of candle, the stearin candle. He was appointed director of dyeing at the Manufacture royale des Gobelins in 1824. Here, he developed what might be his most well-known work, certainly to art historians, his book *De la loi du contraste simultané des couleurs*, which outlined a theory of contrasting colors. Lemay and Oesper, "Michel Eugène Chevreul,

1786–1889," 66–67; Chevreul, *De la loi du contraste simultané des couleurs*. On Chevreul and the standardization of color, see Kalba, *Color in the Age of Impressionism*. A bibliography of Chevreul's publications was commissioned on his 100th birthday in 1886; see Malloizel, *Oeuvres scientifiques de Michel-Eugène Chevreul*.

48   These were often embodied in the form of state funerals for the nation's "great men," including, for example, Victor Hugo's funeral the year before in 1885.

49   Ben-Amos, *Funerals, Politics, and Memory*, 137.

50   Citing Oliver Ihl's taxonomy of theoretical models for civic festivals as a means of reinvigorating the citizenry in service to a politically unified Republican nation, Ben-Amos notes that, during the Third Republic, "a secular and commemorative celebration that put the individual citizen at the centre" was favored over more overtly religious or revolutionary models. Ben-Amos, *Funerals, Politics, and Memory*, 141.

51   R. Chevreul, "La vie et l'oeuvre de Michel-Eugène Chevreul," 43. This phrase forms the subtitle for a text published on the centenary by the centenary's organizing committee. Malloizel, *Oeuvres scientifiques de Michel-Eugène Chevreul*.

52   Kalba, *Color in the Age of Impressionism*, 15.

53   Kalba, *Color in the Age of Impressionism*, 15.

54   This man is identified as Xu Jingcheng's translator in museum catalog entries, but I have not come across a reference to his name.

55   Qing dynasty China vigorously responded to the expansion of French imperialism in Annam and Tonkin (contemporary Vietnam), and fighting ensued for nine months. Though the war did not result in a clear military victory on either side, the diplomatic win was France's when China, in a one-sided colonial treaty, was forced to recognize France's protectorate over Tonkin, leading the way for the later establishment of the colony of French Indochina. See Harris, *Peking Gazette*, 238–52.

56   Valery, "Centenary of Photography," 163. Or, to put it otherwise, consider Alan Trachtenberg's description of a kind of "historicism-by-photography, [the] notion that historical knowledge proclaims its true value by its photographability." Trachtenberg, "Albums of War," 1.

57   Just several years earlier, Nadar *père* had in fact taken a deathbed image of Victor Hugo, another of the *grandes hommes* of the nineteenth century whose hagiographies were being solidified during the Third Republic.

58 Nadar, "L'art de vivre cent ans," f. 11.
59 The debate over the role of photographs as historical evidence also explicitly calls into question the nature of all historical evidence. See Tucker, "Entwined Practices."
60 Nelms, *Third Republic*, 198–99.
61 Nelms, *Third Republic*, 106–7.
62 Clark, *Paris and the Cliché of History*, 14.
63 Clark, *Paris and the Cliché of History*, 15.
64 Nadar, "L'art de vivre cent ans," f. 19.
65 Canales, *Tenth of a Second*, 119
66 Canales, *Tenth of a Second*, 121.
67 Canales, *Tenth of a Second*, 121–22.
68 See Gunthert, "La conquête de l'instantané."
69 Scientific experiments with high-speed photography were known to both the Nadars, who kept abreast of contemporary developments in photographic technology and applications of the medium in a variety of disciplines. Both Nadars had closely followed the physiologist and photographer Etienne-Jules Marey's experiments with chronophotography. Paul Nadar later published numerous articles by Marey in issues of *Paris-Photographe*, including in the inaugural 1891 issue. Both Marey and the Nadars had considered the role of the photographic document in historical narration. For the Nadars, this took the form of the photographic interview commemorating the life of Chevreul. Marey would later collaborate with the anarchist geographer Elisée Reclus on a proposal for a panoramic photographic chronology of human history, to be appended to the exterior of Reclus's planned but never built 1:100,000 scale replica of the earth for the 1900 Exposition universelle in Paris. See Marey, "L'analyse de mouvements par la photographie." Plans for the photographic installation are discussed in letters between Marey and Reclus; see "Lettre de Étienne-Jules Marey à Elisée Reclus," June 22, 1897, NAF22916, MF15058, ff. 137–240, Papiers Elisée Reclus, Département des manuscrits, Bibliothèque nationale de France, Paris.
70 Braun, *Picturing Time*, 61.
71 On the early trajectory of this formation, see Gervais, "Witness to War."
72 Bleichmar and Schwartz, "Visual History," 21.
73 Here, Chevreul is paraphrasing from one of his previous publications on the subject, "La verité sur l'invention de la photographie," from 1873. Reprising the title of Victor Fouque's 1867 book, Chevreul ar-

gues that daguerreotypy (attributed to Louis-Mandé Daguerre) and calotypy (attributed to Henry Fox Talbot) developed out of Niépce's work in heliography.

74   Chevreul, "La verité sur l'invention," 66.

75   Chevreul's passionate pronouncements in favor of Niépce were prompted by an 1871 paper given at the Institut de France by M. Legouvé titled "Daguerre proclamé l'inventeur de la photographie." Responding to Legouvé's arguments, Chevreul suggested that it was an essential part of the role of the Académie des sciences de l'Institut de France to accurately understand the development of inventions such as photography and to properly credit those "men whose work has pushed the limits of human knowledge." Chevreul, "Séance du lundi 30 octobre 1871," 1019.

76   Keller, "Photography, History, (Dis)Belief," 99–100.

77   Daston and Galison, *Objectivity*, 17.

78   Daston and Galison, *Objectivity*, 16.

79   Nadar, "L'art de vivre cent ans," f. 148. This is distinctly different from the relationship between imagination and invention argued for by Nadar and his collaborators discussed in chapter 1. Chevreul was an ardent proponent of the experimental method, while Nadar could rightly be categorized as one of those who "claimed invention when they had only imagination." Regarding Chevreul's approach to the experimental method, see Chevreul, *De la méthode a posteriori expérimentale*.

80   Tupicoff, *First Interview*. My thanks to Dennis Tupicoff for agreeing to speak with me about his film.

CHAPTER 4. ILLUMINATING INFRASTRUCTURES: THE FIRST PHOTOGRAPHS UNDERWATER AND UNDERGROUND

1   See, for example, Germaine Krull's photo book *Marseille* and László Moholy-Nagy's film *Impressionen vom alten Marseiller Hafen* (1929). Krull and Moholy-Nagy, alongside other photographers such as Herbert Bayer and Florence Henri, also photographed the Pont transbordeur at the entry of the Vieux port before its destruction in 1944.

2   While looking for a successor for the Marseille studio, Nadar met the Swiss photographer Frédéric Boissonnas at an exhibition in Lyon. Boissonnas later purchased the Marseille studio from Nadar in late 1901, eventually leaving it under the management of his assistant Fernand Detaille. The studio persevered under two more generations of the Detaille family. Gérard Detaille continues to be an advocate for

the history of the studio. For more on the history of this studio, see Detaille, *Marseille, un siècle d'images*.

3 The remarkable physical transformation of urban Paris throughout the nineteenth century is a canonical event within the historiography of Western European urban modernity and within the history of European Modernism. The scholarship on this topic is voluminous. See, for example, Clark, *Painting of Modern Life*; Harvey, *Paris, Capital of Modernity*.

4 That is, it was hoped that they would embody infrastructure's "political address," to borrow Brian Larkin's phrase, embodying "the way technologies come to represent the possibility of being modern, of having a future, or the foreclosing of that possibility." Larkin, "Politics and Poetics of Infrastructure," 333.

5 This was a subject that was also captured photographically; for example, see Édouard Baldus's album depicting the construction of the Paris–Lyon–Marseille railway line in Daniel, *Photographs of Édouard Baldus*, 78–90.

6 Montricher and Pascal, "Les docks, le bassin Napoléon et l'Avant-Port d'Arène," 1. A grisly infrastructural goal given that Liverpool had been one of the slave-trading capitals of Britain throughout the second half of the eighteenth century. On the principal ports in the development of France's maritime economy, including in the slave trade, see Forrest, "Port Cities of the French Atlantic," 22–32.

7 Forrest, "Port Cities of the French Atlantic," 24.

8 Roncayolo, *L'imaginaire du Marseille*, 83.

9 Borruey, *Le port moderne de Marseille*, 28.

10 Larouaire, "Un chantier au fond de la mer," *Le petit marseillais*, April 8, 1900, 1, Fonds Nadar, YB3-2340, Box 7, Département des estampes et de la photographie, Bibliothèque nationale de France, Paris.

11 On the history of underwater photography in France, see Adamowsky, *Mysterious Science of the Sea*, 151–55; Eigen, "Dark Space." See also Boutan, *La photographie sous-marine*.

12 In Verne's *20,000 Leagues Under the Sea*, first published in 1872, Professor Aronnax laments the tragedy of experiencing extraordinary underwater sights with no means to record them: "To sink into these depths where no man has ever ventured before! What a pity that we can keep nothing of them but the memory!" Ever conducive to impressing his captive guest, Captain Nemo offers the professor a camera in response. Aronnax notes that, somehow, even at such great depths, "the water was brilliantly lit through the windows, the sun could not have given a better light." A photograph was taken, and in the resulting

13 negative, "It was possible to see those rocks and caves and the whole extraordinary underwater landscape." Verne, *20,000 Leagues Under the Sea*, 114.

13 Most of the images in the album that can be firmly attributed to Nadar are those undertaken in the caisson "underwater." Of the nine photographs in the album that bear the stamp of the Nadar studio, two are photographs of above-water aspects of the construction project. The remaining images were reportedly taken by Boissonnas and Detaille in 1904. Aubenas, "Sous terre et sous la mer," 271.

14 Borruey, *Le port moderne de Marseille*, 25; cf. Poirel, *Mémoire sur les travaux à la mer*.

15 Aubenas, "Sous terre et sous la mer," 271.

16 See, for example, "Enlargement of the port of Marseille," *Scientific American Supplement*, no. 1147 (December 25, 1897): 18331. The coverage of the construction project in US publications like *Scientific American* was based on the translation and republication of articles from the French journal *La nature*. Representations of the specific construction techniques involved in the construction of the Pinède basin were also on display at the 1900 Exposition universelle in Paris. See "A huit mètres sous l'eau," *Le petit marseillais*, July 12, 1897, 1, Fonds Nadar, YB3–2340, Box 7; Jacomet, *Revue technique de l'Exposition universelle de 1900*, 259–63.

17 Rosalind Williams describes this as a particular characteristic of "underground" space, writing that "it is the combination of enclosure and verticality—a combination not found either in cities or in spaceships—that gives the image of an underworld its unique power as a model of technological environment." Williams, *Notes on the Underground*, 8. Likewise, Edward Eigen in his article on underwater photography argues that underwater photography always necessarily references the conditions of its own production. Eigen, "Dark Space," 93.

18 See, for example, Baldus's railway albums or Auguste Hippolyte Collard's work documenting Parisian bridges, among others. On Baldus, see Daniel, *Photographs of Édouard Baldus*, 78–90; on Collard, see McCauley, *Industrial Madness*, 196–224.

19 Born in Aix-en-Provence, Terris established a photography studio in Marseille in the 1860s and was commissioned to document the massive renovations of the city undertaken in the 1860s, as well as expansions made to the city's port.

20 Weiss, "Engineering, Photography, and the Construction of Modern Paris," 89.

21 Weiss, "Engineering, Photography, and the Construction of Modern Paris," 251.

22 Weiss situates photography as a kind of "paperwork" central to the bureaucratic operations of France's civil engineering corps and the renovation of Paris in the nineteenth century. Weiss, "Engineering, Photography, and the Construction of Modern Paris," 7–11. See also, for example, the discussion of how photography was used during the planning and construction of Parisian aqueducts in Weiss, "Making Engineering Visible." On photography and construction fundraising, see Baillargeon, "Construction Photography."

23 These images are also drawn into a longer tradition of representing construction sites. See, for example, Baridon et al., *L'art du chantier*.

24 Nadar, "L'emploi de l'air comprimé dans les travaux publics," unpublished manuscript, Fonds Nadar, YB3-2340, Box 7.

25 In "Problems Pertaining to Ears" in *Problemata*, Aristotle described divers "letting down a cauldron" to breathe underwater, likely a reference to what would later come to be known as a diving bell. Aristotle, *Collected Works of Aristotle*, 960.

26 John Smeaton was a British civil engineer known for his work on hydraulic lime concrete beginning in the eighteenth century. Nadar is probably referring to Smeaton's design for a diving bell for use during the construction of the Hexham Bridge Project.

27 As Brian Larkin has noted, it has become routine to repeat this assertion as articulated by Susan Leigh Star. However, as Larkin points out, this is only a "partial truth," as invisibility "is only one and at the extreme edge of a range of visibilities that move from unseen to grand spectacles and everything in between." Larkin, "Politics and Poetics of Infrastructure," 336; Star, "Ethnography of Infrastructure."

28 This line of thought rehearses much of what has been written about the aesthetic project of French modernity. See Harvey, *Paris, Capital of Modernity*.

29 As one can imagine, the smell is significant. There is, however, as this author can attest from personal experience, a certain grotesque sublimity in seeing the inner workings of a sewer network. Inquiring noses can visit the musée des égouts de Paris to smell for themselves.

30 Nadar, *When I Was a Photographer*, 93–94.

31 Press coverage of the operation also underlined the difficulty of the project, describing it as a kind of odyssey challenged not only because of the relative remoteness of the location but also in the logistical arrangement of batteries, wires, and lights that was necessitated by the system for illuminating the sewers. See Lacan, "Revue photographique," 1–2.

32   Aubenas, "Sous terre et sous la mer," 271. The projects resulted in seventy-three views of the catacombs and twenty-three pictures of the sewers.

33   Nadar, "Le dessus et le dessous de Paris." This text would be revised and republished in his son Paul's journal *Paris-Photographe* in 1893 and in *Quand j'étais photographe* in 1899.

34   Félix Nadar to Alphonse Davanne, cited in Aubenas, "Beyond the Portrait, Beyond the Artist," 106n52.

35   Belgrand penned a five-volume study reflecting on his contributions to Parisian infrastructure, the fifth volume of which explored the renovation of the Parisian sewers. Belgrand, *Les égouts de Paris*, 34.

36   Reid, *Paris Sewers and Sewermen*, 12.

37   Reid, *Paris Sewers and Sewermen*, 12.

38   Reid, *Paris Sewers and Sewermen*, 30.

39   Reid, *Paris Sewers and Sewermen*, 31–32.

40   Gandy, *Fabric of Space*, 41.

41   Williams, *Notes on the Underground*.

42   Nadar, *When I Was a Photographer*, 89. While this utopian vision of rebellion from below was brutally dashed, the circulation of waste enabled by the Parisian sewer network could literally fortify the masses. As Nadar notes in his account of the sewers, and as Victor Hugo recounts in a philosophical and historical interlude on the political possibility of waste management in *Les Misérables*, human feces had long been used as a potent agricultural fertilizer. This tradition had somewhat fallen out of fashion in urban life by the nineteenth century, necessitating the import of fertilizers like guano—the excrement of seabirds or bats—from countries like Peru. Nineteenth-century French socialist Pierre Leroux (to whom Victor Hugo was very sympathetic) proposed that the government collect excrement as a tax and use it to fertilize the nation's crops, a system he dubbed the "circulus." See Simmons, "Waste Not, Want Not."

43   We can only lament that the phrase "sewer socialism," coined in the 1930s United States to disparage Milwaukee socialists' reported bragging about the sewer system in their district, was not available to nineteenth-century commentators.

44   For the ensemble of prints with these annotations, see the prints in the collection of the musée des Arts et Métiers (Conservatoire national des arts et métiers) beginning with inventory number 10196.

45   Félix Nadar, "Photographie: "Les égoûts de Paris: Tunnel sous le quai de la mégisserie," print, 18.3 × 22.5 cm., ca. 1879–84, inventory

46   no. 10196–0018, collection of the Musée des art et métier, Paris, France.

46   This is the same donation that saw the "first" aerial photograph enter the museum's collections. In addition to images of the catacombs and sewers, there were also several "instantaneous" images taken of trains in motion by Nadar's son Paul. See the correspondence regarding the gift between Nadar *père* and Conservatoire national des arts et métiers director Aimé Laussedat: "Lettre de Aimé Laussdat à Félix Nadar," March 7, 1884, NAF24275, f. 364, Collection d'autographes formée par Félix et Paul Nadar, XVI La Fayette-Lemaire, Département des manuscrits, Bibliothèque nationale de France, Paris.

47   As Laussedat notes in his letter cited in the previous note, Nadar's photographs were to be in "good company," joining original works by "Nièpce, Daguerre, Talbot, Bayard, Legray, Poitevin. Woodbury, Becquerel, etc.," alongside examples of the "most interesting applications of this marvelous art."

48   Larkin, "Politics and Poetics of Infrastructure," 329.

49   Gestures toward doing so have more often been framed in terms of attempts to "represent capital" writ large. See Toscano and Kinkle, *Cartographies of the Absolute*.

50   David Harvey described this as the tension between alternative, and competing, theories of modernity. See Harvey, *Paris, Capital of Modernity*, 1–20.

51   Aubenas, "Sous terre et sous la mer," 271.

52   Ernest Lamé-Fleury cited in Aubenas, "Sous terre et sous la mer," 271. The text that Lamé-Fleury references is Paul Fassy's *Les catacombes, étude historique*. The text is also noted briefly by Nadar in his text on subterranean photography. See Nadar, *When I Was a Photographer*, 79.

53   As Shao-Chien Tseng has noted, the images also fit into a longer trajectory of topographic photography, conservation-oriented documentation of architecture, and images of contemporary construction projects. See Tseng, "Nadar's Photography of Subterranean Paris," 250.

54   Legacey, "Paris Catacombs," 511.

55   Philippe Muray cited in Pike, "Paris Souterrain," 177. The intellectual problem of siting Paris as "the capital of the nineteenth century," as Walter Benjamin put it, continues. The counter-chronological structure of this chapter, beginning in Marseille to rethink Paris, aims to point to the limitations of this approach to French photographic history.

56   Legacey, "Paris Catacombs," 511.

57 Legacey, "Paris Catacombs," 512.

58 Legacey, "Paris Catacombs," 513.

59 Legacey, "Paris Catacombs," 513.

60 Legacey, "Paris Catacombs," 514.

61 Legacey, "Paris Catacombs," 515.

62 Nadar, "Le dessus et le dessous de Paris".

63 Nadar, *When I Was a Photographer*, 78–79.

64 Nadar, *When I Was a Photographer*, 83.

65 Readers will recall from chapter 1 that the 1860s were also when Nadar was most invested in his aerial projects, including ballooning experiments and the production of his extant aerial photographs.

66 Nadar, *When I Was a Photographer*, 86.

67 Williams, *Notes on the Underground*, 17.

68 Williams, *Notes on the Underground*, 21.

69 Tseng also notes the emphasis on the electric lights. See Tseng, "Nadar's Photography of Subterranean Paris," 250.

70 On the vertical arm of the cross, a Latin Bible verse (Daniel 12:2) reads, "And many of them that sleep in the dust of the earth shall awake, some to everlasting life, and some to shame and everlasting contempt."

71 Nadar, *When I Was a Photographer*, 93.

72 *Le monde illustré* had also, earlier that year, published images based on Nadar's photographs of the Paris sewers. See G. B., "Un voyage dans les égouts de Paris," *Le monde illustré*, February 18, 1865 (vol.9, no. 410), 16.

73 Anonymous cited in Howes, *To Photograph Darkness*, note 18; see also "Les catacombes de Paris," *Le moniteur universel* 67, no. 8 (1862): 328.

74 Cited in Tseng, "Nadar's Photography of Subterranean Paris," 250. This reference is also noted in Howes, *To Photograph Darkness*, 15.

75 Liesegang, "Quelques observations sur la photographie," 42.

CHAPTER 5. WHEN I WAS A PHOTOGRAPHER, OR THE HISTORY OF PHOTOGRAPHY IN PHOTOGRAPHIC FIRSTS

1 Nadar, *When I Was a Photographer*, 1–2.

2 Most of the texts included in *Quand j'étais photographe* first appeared in *Paris-Photographe* under the series title "Souvenirs d'un atelier de

photographe," published between April 1891 and December 1894. Three texts not included in *Paris-Photographe* were later included in the book: "The Professional Secret," "The Bee Tamer," and "1830 and Thereabouts." "The Professional Secret" appears to have been published for the first time in *Quand j'étais photographe*, "The Bee Tamer" was first published in *Le magasin pittoresque* in 1897, and "1830 and Thereabouts" was first published in *Nouvelle revue international* in March 1899. See Robert, "Félix Nadar et la disparition de la photographie," 38. As Eduardo Cadava and Liana Theodoratou note in their 2015 English translation of *When I Was a Photographer*, there is some debate over the date of the original publication of the book version. There appear to have been two editions (one published in 1899 and another in 1900). The version referenced here is the 1900 edition (and the 2015 translation of this version), which includes a preface by Léon Daudet and a final chapter titled "1830 and Thereabouts." For quotations in English, I use the 2015 Cadava and Theodoratou translation. On the subject of the text in general, see Bann, "'When I Was a Photographer'"; Krauss, "Tracing Nadar"; Taws, "When I Was a Telegrapher."

3   This formulation borrows from the by now well-established tradition of thinking about "when old technologies were new." See Marvin, *When Old Technologies Were New*; Chun et al., *New Media, Old Media*.

4   Nadar, *When I Was a Photographer*, 1.

5   Alan Liu has also described this as the "narrative of the new media encounter," a series of rhetorical structures that distinguishes the new from the old via a series of established narrative tropes. Liu, "Imagining the New Media Encounter," 35.

6   The text was first published in the inaugural April 1891 issue of *Paris-Photographe*. See Nadar, "Balzac et le daguerréotype"; Nadar, *Quand j'étais photographe*, 1–8. The differences between the two texts are minimal; for example, a quotation is attributed to George Sand in the book version, and the final three paragraphs of the journal version are cut out of the book version. The journal essay also states that the text is an excerpt from an unpublished work, "Faces et profils, souvenirs du XIXe siècle."

7   Taussig, *Mimesis and Alterity*, 201.

8   Pinney, *Photography and Anthropology*, 70. Pinney's phrase "augurs and haruspices" references Walter Benjamin's 1931 essay "A Little History of Photography" in *One-Way Street*.

9   Nadar, *When I Was a Photographer*, 1.

10 Nadar, *When I Was a Photographer*, 1.

11 Nadar, *When I Was a Photographer*, 1.

12 Nadar, *When I Was a Photographer*, 2.

13 Nadar, *When I Was a Photographer*, 5.

14 Nadar, *When I Was a Photographer*, 4.

15 During Balzac's youth, Lucretius's *The Nature of Things* would have been a key text in the study of Latin in France. See Kavanagh, "Epicureanism Across the French Revolution," 99. Other scholars have discussed the relationship between Nadar's account of Balzac's theory of photography and Lucretius; see Eduardo Cadava, "Nadar's Photographopolis," in Nadar, *When I Was a Photographer*, xvi–xviii; Downing, "Lucretius at the Camera," 23–28.

16 Balzac, *Cousin Pons*, 128. This passage is also cited by Walter Benjamin in *The Arcades Project* in "Convolute Y" (on photography), surrounded by citations on the theme of the relationship between photography and painting. See Benjamin, *Arcades Project*, 687–88.

17 Balzac, *Cousin Pons*, 130. As Göran Blix has argued, Balzac was actively invested in exploring occult theories of material change, including Franz Friedrich Anton Mesmer's theory of animal magnetism. See Blix, "Occult Roots of Realism," 265.

18 Balzac, *Cousin Pons*, 130–32.

19 Liu, "Imagining the New Media Encounter."

20 Nadar, *When I Was a Photographer*, 2.

21 Nadar, *When I Was a Photographer*, 9.

22 Nadar, *When I Was a Photographer*, 9.

23 Nadar, *When I Was a Photographer*, 11.

24 Nadar notes that the young man claimed to have worked for Léopold Leclanché, "the son of an old friend of mine, the translator of Cellini's *Mémoires*." Nadar may have been collapsing his memories of Leopold Leclanché, who was indeed the translator of Benvenuto Cellini's texts into French, with the work of Leclanché's son, George Leclanché, an electrical engineer known primarily for his invention of the Leclanché cell, one of the first electrical batteries and the precursor to the dry cell battery. Gustave Trouvé was a prolific French electrical engineer who worked on electric motors for use in boats, bicycles, helicopters, and ornithopters, among other vehicles, and a portable electric headlamp known as the "photophore," among a plethora of other inventions. Paul-Gustave Froment was a French electrical engineer and instrument maker known for his early work on electric motors, one example of which is still held in the collection of the musée des Arts et Métiers in Paris. Froment also collaborated with Giovanni

Caselli on the construction of the pantelegraph. Marcel Deprez was a French electrical engineer known for his experiments with the long-distance transmission of electricity, most notably in an experiment to transmit electricity from Miesbach to Munich in 1882. Clément Ader was a French inventor and electrical engineer, known primarily for his work in aviation and on the telephone network in Paris. Giovanni Caselli was an Italian inventor and priest, known for his invention of the pantelegraph.

25 Nadar, *When I Was a Photographer*, 15.
26 Nadar, *When I Was a Photographer*, 14–15.
27 Nadar, *When I Was a Photographer*, 15.
28 Nadar, *When I Was a Photographer*, 16.
29 Nadar, *When I Was a Photographer*, 17.
30 Nadar, *When I Was a Photographer*, 17.
31 Nadar, *When I Was a Photographer*, 18.
32 Nadar, *When I Was a Photographer*, 18.
33 On the relationship between this story and the historical development of telegraphy, see Taws, "When I Was a Telegrapher."
34 Nadar, *When I Was a Photographer*, 21–22.
35 Nadar, *When I Was a Photographer*, 26.
36 Nadar, *When I Was a Photographer*, 26.
37 Nadar, *When I Was a Photographer*, 26.
38 Nadar, *When I Was a Photographer*, 119–29. On the development of chemistry during this period, see Ihde, *Development of Modern Chemistry*, 57–88.
39 Nadar, *When I Was a Photographer*, 119; Nadar, "Le docteur van Monckhoven," 301–8.
40 Nadar, *When I Was a Photographer*, 119. See "Suite du rapport sur les arts," 145–48.
41 For example, paraphrasing Fourcroy, Nadar described the invention of the modern pencil by Nicolas-Jacques Conté. Observing the impossibility of importing high-quality English graphite during wartime, Conté had invented a method for combining low-grade French-mined graphite with a clay mixture to create a suitable pencil lead. As Richard Taws has noted, Nadar also invoked Conté's contributions a second time, in his text on the use of balloons in the siege of Paris in 1870. Taws, "Conté's Machines," 247.
42 Nadar, *When I Was a Photographer*, 122.

43 Nadar, *When I Was a Photographer*, 122. Nicolas de Condorcet was a French philosopher and mathematician. Antoine Lavoisier was a French chemist, known for identifying the role of oxygen in combustion and compiling the first list of elements, among several other developments in modern chemistry. Gaspard Monge was a French mathematician, the inventor of descriptive geometry, and one of the founders of the École polytechnique in Paris. Jean-Antoine Chaptal was a French chemist and professor at the École polytechnique. Louis Nicolas Vauquelin was a French pharmacist and chemist. Jérôme Lalande was a French astronomer. Antoine François de Fourcroy was a French chemist who collaborated with Lavoisier, Louis-Bernard Guyton de Morveau, and Claude Louis Berthollet on the standardization of chemical nomenclature. Charles Bossut was a French mathematician and contributor to Denis Diderot and Jean le Rond d'Alembert's *Encyclopédie*. Jean Darcet was a French chemist and director of the porcelain works at Sèvres. Nicolas-Jacques Conté was a French painter and balloonist and the inventor of the modern pencil.

44 Pierre Jules César Janssen was a French astronomer and pioneer of astronomical photography. In the history of photography, he is perhaps best known for inventing the *"revolver photographique,"* a mechanism for sequential photography. The brothers Paul and Prosper Henry were also French astronomers who experimented with astronomical photography. Étienne-Jules Marey was a French physiologist who developed a system of chronophotography for studying human and animal locomotion.

45 Brewer, *Enlightenment Past*, 32.

46 Koselleck, *Practice of Conceptual History*, 228–29.

47 Koselleck, *Practice of Conceptual History*, 229. Frank Manuel has argued that we see one of the first iterations of this kind of ideology of progress in Anne Robert Jacques Turgot's 1750 lectures on world history at the Sorbonne. See Manuel, *Prophets of Paris*, 13–41, 61–101.

48 Nadar, *When I Was a Photographer*, 123.

49 Van Monckhoven published at least seven editions between around 1856 and 1884. For the earliest edition listed in the catalog of the Bibliothèque nationale, see Van Monckhoven, *Traité général de photographie*.

50 Nadar, *When I Was a Photographer*, 124.

51 See, for example, Alophe, *Le passé, le présent et l'avenir*; Belloc, *Les quatre branches*; Blanquart-Evrard, *La photographie*; Disdéri, *L'art de la photographie*; Figuier, *La photographie*; Ken, *Dissertations historiques*; Nègre, *De la gravure héliographique*; Thierry, *Daguerréotypie*;

Tissandier, *Les merveilles de la photographie*; Wey, "Comment le soleil est devenu peintre.

52 Nadar, *When I Was a Photographer*, 141.

53 Nadar, *When I Was a Photographer*, 142. The zeal for "small histories" in the absence of a "great history" is characteristic of the Romantic historiography of the nineteenth century. See White, *Metahistory*, 8.

54 Nadar, *When I Was a Photographer*, 141.

55 Bann, *Romanticism and the Rise of History*, 3.

56 Nadar, *When I Was a Photographer*, 141.

57 Nadar, *When I Was a Photographer*, 152.

58 Nadar, *When I Was a Photographer*, 155.

59 Nadar, *When I Was a Photographer*, 157.

60 Nadar, *When I Was a Photographer*, 158.

61 Victor Hugo cited in Nadar, *When I Was a Photographer*, 204. See Hugo, *Les Misérables*, 101.

62 Nadar, *When I Was a Photographer*, 204.

63 Running through the alternatives to Catholicism explored throughout his century, Nadar cites, among others, the Saint-Simonians, Barthélemy-Prosper Enfantin, Charles Fourier, and Étienne Cabet, describing Enfantin and the Saint-Simonians as those "who will soon achieve everything!" Nadar, *When I Was a Photographer*, 205. One of the lasting legacies of Henri de Saint-Simon and his disciples was the rhetorical project of constructing industry as a guiding mythology, and indeed cosmology, fully capable of superseding the reign of Christianity, offering nothing less than a re-enchantment of the world through growing knowledge of science and technology. See Picon, *Les Saints-Simoniens*.

64 Nadar, *When I Was a Photographer*, 204.

65 Nadar misspells Horeau's name as "Haureau" in the text. In Cadava and Theodorato's annotated translation of the text, they include a footnote reference to Jean-Barthélemy Hauréau, historian, writer, and director of manuscripts at the Bibliothèque nationale in Paris. Given Nadar's explicit reference to the architectural contributions of Horeau/Haureau, the reference is most likely instead to Hector Horeau, who submitted building plans for iron and glass structures to the competitions for both the Great Exhibition of 1851 in London and Les Halles in Paris. Nadar, *When I Was a Photographer*, 207.

66 Nadar, *When I Was a Photographer*, 207.

67 Think, for example, of the much more recent theorist Fredric Jameson's characterization of nineteenth-century European literature from the historical novel to scientific romances: "Capitalism demands in this sense a different experience of temporality from what was appropriate to a feudal or tribal system, to the *polis* or to the forbidden city of the sacred despot: it demands a *memory* of qualitative social change, a concrete vision of the past which we may expect to find completed by that far more abstract and empty conception of some future terminus which we sometimes call 'progress.'" Jameson, "Progress Versus Utopia," 284.

68 Sharma, *In the Meantime*, 9.

69 Sharma, *In the Meantime*, 7.

70 Nadar, *When I Was a Photographer*, 213. Louis Pasteur was a French biologist and chemist known for his discovery of vaccination and pasteurization processes. Jacques-Joseph Grancher was a French pediatrician known for his research on tuberculosis. Émile Roux was a French physician, bacteriologist, and immunologist. Grancher and Roux were early directors of departments at the Institut Pasteur, founded in 1887.

71 Nadar, *When I Was a Photographer*, 220.

72 Krauss, "Tracing Nadar," 29.

EPILOGUE: THE HISTORY OF PHOTOGRAPHY, AS TOLD TO ME BY NADAR

1 The print included in the exhibition was lent by Eastman Kodak Research Laboratories.

2 Newhall, *Photography 1839–1937*, 53. Newhall also noted that Nadar was responsible for the "first photographs taken by flashlight." The interview with Chevreul (in newspaper format) was included under Paul Nadar's name. Newhall's interest in aerial photography was personal, having worked as an aerial photograph interpreter in World War II. Newhall, *Photography 1839–1937*, 52; Newhall, *Focus*, 71–97.

3 Newhall, *Photography 1839–1937*, 105.

4 Allan Sekula would later comment on the mutability (and authorship) of aerial images in photographic theory and history in an article on American photographer Edward Steichen. In Newhall's narrative, the use of aerial photography in the twentieth century was a tactical realization of Nadar's early experiments: "Although Nadar took photographs in 1858 from a balloon, it was not until the Great War that the full possibilities of aerial photography were demonstrated." See Sekula, "Instrumental Image"; Newhall, *Photography 1839–1937*, 87.

5   On the nineteenth-century roots of celebrity and its imbrication with photography, see Marcus, *Drama of Celebrity*. The relevance of this aspect of Nadar's work was also acknowledged in a 1999 exhibition at the J. Paul Getty Museum of Nadar's portraits alongside those of Andy Warhol, curated by Gordon Baldwin and Judith Keller. See Baldwin and Keller, *Nadar–Warhol, Paris–New York*.

6   This approach follows what Craig Robertson has suggested as a move from "the archive as source" toward "the archive as subject." See Robertson, "Introduction," 1.

7   Newhall, *Photography 1839–1937*, 52. As the authors of the 2018 BNF exhibition catalog note, in this way, "Paul's personal heritage becomes emblematic of the French photographic heritage and draws its main axes." Aubenas, Lacoste, and Roubert, "La maison n'a pas de succursale," 21.

8   Newhall, *Photography 1839–1937*, 51.

9   *Photography 1839–1937, catalog of the exhibition* master checklist, Museum of Modern Art Library and Archives, accessed December 11, 2019, https://www.moma.org/documents/moma_master-checklist_387256.pdf.

10   Freund, *La photographie en France*, 49–70; Freund's *Photographie et société* was also published in English in 1980 as *Photography and Society*.

11   Freund, *Photography and Society*, 61.

12   Freund, *La photographie en France*, 57n1, 57n4.

13   See Benjamin, *Arcades Project*, 90–91, 309, 673–74, 681; Benjamin, "Paris, Capital of the Nineteenth Century," 6. Benjamin also described Nadar as a "master" of photography's early period. See Benjamin, "Small History of Photography," 240.

14   Aubenas, Lacoste, and Roubert, "La maison n'a pas de succursale," 10.

15   Viars, "Étude de douze plaques," 10.

16   Viars, "Étude de douze plaques," 10.

17   Viars, "Étude de douze plaques," 14.

18   Viars, "Étude de douze plaques," 14.

19   Aubenas, Lacoste, and Roubert, "La maison n'a pas de succursale," 10.

20   Among Nadar's earliest collectors were Gabriel Cromer and George Sirot, whom André Jammes lists as "first-generation" Nadar collectors. For an account of the collecting of the Nadar studio materials, see Jammes, "Nadar Studio"; Hellman, "Nadar dans les collections Nord-Américaines"; Falguière, "Le fonds d'atelier Nadar."

21 Jammes, "Nadar Studio," 116.
22 Chevallier, *Nadar*; Prinet and Dilasser, *Nadar*; Braive, *L'age de la photographie*.
23 Jammes, "Nadar Studio," 116. Jammes's centrality to the history of collecting nineteenth-century French photography cannot be underestimated.
24 Seniuta, "De la collection à la spéculation," 13.
25 Seniuta, "De la collection à la spéculation," 13.
26 The exhibition was accompanied by a catalog published by Aperture. See White, Jammes, and Sobieszek, *French Primitive Photography*.
27 Seniuta, "De la collection à la spéculation," 15.
28 Seniuta, "De la collection à la spéculation," 15.
29 Letter from Jammes to Wagstaff, 1976, quoted in Seniuta, "De la collection à la spéculation," 15–16.
30 Metropolitan Museum of Art, "First Major United States Exhibition of Photographs by Félix Nadar to Open at Metropolitan Museum," press release, January 1977, https://libmma.contentdm.oclc.org/digital/collection/p16028coll12/id/7948.
31 My goal here is not to create a comprehensive account of every exhibition featuring Nadar but rather to describe how many such exhibitions were entangled in the economics of collecting and the personas of several key figures in the North American historiography of photography.
32 Russell, "Art: Photographs by Nadar," 56.
33 Russell, "Art: Photographs by Nadar," 56.
34 *The Thrill of the Chase: The Wagstaff Collection of Photographs*, exhibition at the J. Paul Getty Museum, Los Angeles, March 15–July 31, 2016. See www.getty.edu/art/exhibitions/wagstaff.
35 Solomon-Godeau, *Photography at the Dock*, xxii.
36 Solomon-Godeau, *Photography at the Dock*, xxii.
37 Solomon-Godeau includes Victor Burgin, Douglas Crimp, Rosalind Krauss, Allan Sekula, Martha Rosler, and Christopher Phillips on this list. Interestingly, in the acknowledgments in *Photography at the Dock*, Solomon-Godeau notes that it is only because Krauss introduced her to the photography collections of the Bibliothèque nationale (no doubt including the Nadar collections) that she even began to study nineteenth-century French photography—an interesting material (and collegial) link between Nadar's archives and these

two scholars writing about photography in the twentieth century. Solomon-Godeau, *Photography at the Dock*, xxvii.

38 Solomon-Godeau, *Photography at the Dock*, xxiv.

39 Grundberg, "Photography View," 24. Interestingly, Grundberg speculated that the contextualists would end up losing the battle to the connoisseurs if only because of . . . new media: "The degree to which photography is no longer our culture's primary means of imaging the world is the degree to which it is an artifact. Thus, the adoption of the medium by museums, libraries and art history departments is not part of a late capitalist conspiracy aimed at blinding us to the real meaning of photographs, but merely a reflection of photography's devalued functional currency."

40 Solomon-Godeau, "Canon Fodder," 28–51.

41 Solomon-Godeau, "Calotypomania," 8.

42 Solomon-Godeau argues that Jammes's language—"master," "oeuvre," etc.—connects him to a legacy of European connoisseurship in the tradition of Bernard Berenson. Solomon-Godeau, "Calotypomania," 8–9.

43 Travis, "Introduction," 8.

44 During this same period, creativity was also being reconceptualized as something shared between artists and technologists, or even something engineers could learn from artists. See Turner, *From Counterculture to Cyberculture*.

45 See the summer 1978 issue of *October* and the September 1976 edition of *Artforum*. To this we might also add the Spring 1970 issue of *Aperture*, "French Primitive Photography," a title that partially borrows from the title of Nadar's essay "Les primitifs de photographie" in *Quand j'étais photographe*.

46 The *Artforum* issue also included a Nadar photograph on the cover.

47 Nadar, "My Life as a Photographer"; Crimp, "Positive/Negative"; Owens, "Photography 'En Abyme.'"

48 Kozloff, "Nadar and the Republic of Mind," 31.

49 Krauss, "Tracing Nadar."

50 Krauss, "Tracing Nadar," 30.

51 Krauss, "Tracing Nadar," 30.

52 Krauss, "Tracing Nadar," 30.

53 Krauss, "Tracing Nadar," 30.

54 And what an inventor! I'd be remiss not to acknowledge the likely perseverance of Nadar as a subject of cultural history due to his prowess as

a storyteller. His biography is undeniably compelling—possibly one reason he has been the subject of multiple biographies and few academic monographs. Of course, the problem of artistic originality, of copy versus original, was one of the defining problems of postmodern criticism. Nadar was thus likely appealing in more ways than one. While theorists like Douglas Crimp noted the ascendance of photographic authorship in the reorganization of collections—"portraits *of* Delacroix and Manet become portraits *by* Nadar and Carjat"—the histories I have outlined across this book make clear that "Nadar" as author was often interchangeable with Nadar as "subject." Crimp, "Museum's Old," 74.

55 Wagstaff, *Book of Photographs*.

56 Though Nadar's inclusion wasn't always a given. An exhibition organized by Berenice Abbott at Julien Levy's Madison Avenue gallery in the 1930s showed prints by Atget alongside portraits by Nadar that Abbott had purchased in Paris from Nadar's son Paul, and Levy reportedly "couldn't understand why in the world she wanted them." Van Haaften, *Berenice Abbott*, 168.

57 Montebello and Loyrette, "Director's Foreword," ix.

58 Montebello and Loyrette, "Director's Foreword," ix.

59 Aubenas, "Beyond the Portrait," 95.

60 The catalog for the 1994 exhibition also contained an essay by André Jammes (cited previously) that outlines a possible taxonomy of Nadar prints, including demonstration prints and those intended for display. See Jammes, "Nadar Studio," 115–17.

61 This is also characteristic of the most recent generation of Nadar scholarship, or, perhaps more accurately, scholarship that touches upon Nadar. Scholars have increasingly acknowledged Nadar's relevance to fields outside of photography, including history writing (in the case of Stephen Bann), optical telegraphy (for Richard Taws), or "graphic celebrity" (for Jillian Lerner). See Bann, "'When I Was a 'Photographer'"; Taws, "Conté's Machines"; Taws, "When I Was a Telegrapher"; Lerner, "Nadar's Signatures." Recent scholarship has, however, been much more attentive (and very productively so!) to the scientific dimension of Nadar's work. See, for example, Stiegler, *Nadar: Bilder der Moderne*.

BIBLIOGRAPHY

Adamowsky, Natascha. *The Mysterious Science of the Sea, 1775–1943*. Pickering and Chatto, 2015.

A. L. "La collection de Nadar." *Le radical*, June 16, 1897.

Alophe, Marie-Alexandre. *Le passé, le présent et l'avenir de la photographie: Manuel pratique de photographie* [The past, present, and future of photography: A practical manual of photography]. Paris: E. Dentu, 1861.

Andraud, Antoine. *Exposition universelle de 1855. Une dernière annexe au Palais de l'Industrie. Sciences industrielles. Beaux-arts. Philosophie* [1855 Universal Exhibition: A last annex to the Palace of Industry. Industrial sciences. Fine arts. Philosophy]. Paris: Guillaumin, 1855.

Angus, Siobhan. *Camera Geologica: An Elemental History of Photography*. Duke University Press, 2024.

Aristotle. *The Collected Works of Aristotle*. Edited by W. D. Ross and translated by E. S. Forster. Clarendon Press, 1927.

"Art. 3655." *Annales de la propriété industrielle, artistique et littéraire* 39 (1893): 172–77.

Aubenas, Sylvie. "Beyond the Portrait, Beyond the Artist." In *Nadar*, edited by Maria Morris Hambourg, Françoise Heilbrun, and Phillippe Néagu. Metropolitan Museum of Art, 1994.

Aubenas, Sylvie. "Sous terre et sous la mer" [Underground and under the sea]. In Aubenas and Lacoste, *Les Nadar: Une légende photographique*.

Aubenas, Sylvie, and Anne Lacoste, eds. *Les Nadar: Une légende photographique* [The Nadars: A photographic legend]. Bibliothèque nationale de France, 2018.

Aubenas, Sylvie, Anne Lacoste, and Paul-Louis Roubert. "La maison n'a pas de succursale" [The house has no branch]. In Aubenas and Lacoste, *Les Nadar: Une légende photographique*.

Auer, Michèle, ed. *Paul Nadar: Le premier interview photographique; Chevreul, Félix Nadar, Paul Nadar* [Paul Nadar: The first photographic interview]. Editions Ides et Calendes, 1999.

Ayrton, Hertha. *The Electric Arc*. D. van Nostrom, 1902.

Azoulay, Ariella Aïsha. *Potential History: Unlearning Imperialism*. Verso Books, 2019.

Babbage, Charles. *Ninth Bridgewater Treatise: A Fragment*. John Murray, 1838.

Baillargeon, Claude. "Construction Photography and the Rhetoric of Fundraising: The Maison Durandelle Sacré-Coeur Commission." *Visual Resources* 27, no. 2 (June 1, 2011): 113–28.

Baldwin, Gordon, and Judith Keller. *Nadar–Warhol, Paris–New York: Photography and Fame*. J. Paul Getty Museum, 1999.

Balzac, Honoré de. *Cousin Pons*. Translated by Ellen Marriage. J. M. Dent and Son and E. P. Dutton, 1910.

Bann, Stephen. *Romanticism and the Rise of History*. Twayne Publishers, 1995.

Bann, Stephen. *The Clothing of Clio: A Study of the Representation of History in Nineteenth-Century Britain and France*. Cambridge University Press, 1986.

Bann, Stephen. "'When I Was a Photographer': Nadar and History." *History and Theory* 48, no. 4 (2009): 95–111.

Bardon, Michèle. *Gustave de Ponton d'Amécourt: Un précurseur oublié* [Gustave de Ponton d'Amécourt: A forgotten precursor]. Self-published, 1983.

Baridon, Laurent, Valérie Nègre, Robert Carvais, Marie-Hélène Contal, Olivier Delarozière, and Christophe Feuillerat. *L'art du chantier: Construire et démolir du XVIe au XXIe siècle* [Art of the construction site: Building and demolishing from the sixteenth to twenty-first century]. Snoeck, 2018.

Barthes, Roland. *Camera Lucida: Reflections on Photography*. Translated by Richard Howard. Farrar, Straus and Giroux, 2010.

Batchen, Geoffrey. *Burning with Desire: The Conception of Photography*. MIT Press, 1999.

Baudry, Jérôme. "Examining Inventions, Shaping Property: The Savants and the French Patent System." *History of Science* 57, no. 1 (March 1, 2019): 62–80.

Begley, Adam. *The Great Nadar: The Man Behind the Camera*. Tim Duggan Books, 2017.

Béguet, Bruno, ed. *La science pour tous: Sur la vulgarisation scientifique en France de 1850 à 1914* [Science for all: On scientific popularization in

France from 1850 to 1914]. Bibliothèque du Conservatoire national des arts et métiers, 1990.

Belgrand, Eugène. *Les égouts de Paris* [The sewers of Paris]. Vol. 5 of *Les travaux souterrains de Paris* [The underground works of Paris]. Paris: Dunod, 1872.

Belloc, Auguste. *Les quatre branches de la photographie: Traité complet théorique et pratique des procédés de Daguerre, Talbot, Niépce de Saint-Victor et Archer, précédé des annales de la photographie et suivi d'éléments de chimie et d'optique appliqués à cet art* [The four branches of photography: Complete theoretical and practical treatise on the processes of Daguerre, Talbot, Niépce de Saint Victor et Archer, preceded by the annals of photography and followed by elements of chemistry and optics applied to this art]. Paris, 1855.

Ben-Amos, Avner. *Funerals, Politics, and Memory in Modern France, 1789–1996*. Oxford University Press, 2000.

Benjamin, Walter. "A Small History of Photography." In *One-Way Street, and Other Writings*, translated by Edmund Jephcott and Kingsley Shorter. NLB, 1979.

Benjamin, Walter. *One-Way Street, and Other Writings*. Translated by Edmund Jephcott and Kingsley Shorter. NLB, 1979.

Benjamin, Walter. "Paris, Capital of the Nineteenth Century <Exposé of 1935>." In *The Arcades Project*, translated by Howard Eiland and Kevin McLaughlin. Belknap Press of Harvard University Press, 2002.

Benjamin, Walter. *The Arcades Project*. Translated by Howard Eiland and Kevin McLaughlin. Belknap Press of Harvard University Press, 2002.

Biagioli, Mario. "Patent Republic: Representing Inventions, Constructing Rights and Authors." *Social Research* 73, no. 4 (2006): 1129–72.

Blanquart-Evrard, Louis-Désiré. *La photographie, ses origines, ses progrès, ses transformations* [Photography, its origins, progress, and transformations]. Lille: L. Danel, 1869.

Bleichmar, Daniela, and Vanessa R. Schwartz. "Visual History: The Past in Pictures." *Representations* 145, no. 1 (2019): 1–31.

Blix, Göran. "The Occult Roots of Realism: Balzac, Mesmer, and Second Sight." *Studies in Eighteenth-Century Culture* 36, no. 1 (2007): 261–80.

Borruey, René. *Le porte moderne de Marseille: Du dock au conteneur (1844–1974)*. Chambre de Commerce et d'Industrie Marseille-Provence, 1994.

Boutan, Louis. *La photographie sous-marine et les progrès de la photographie* [Underwater photography and the progress of photography]. Schleicher frères, 1900.

Braive, Michel-François. *L'age de la photographie, de Niépce à nos jours* [The age of photography, from Niépce to today]. Editions de la Connaissance, 1965.

Braive, Michel-François. "Nadar premier voyageur dans la lune" [Nadar, first voyager to the moon]. *France Aviation*, no. 61 (December 1959): 2–3.

Braun, Marta. *Picturing Time: The Work of Etienne-Jules Marey (1830–1904)*. University of Chicago Press, 1992.

Brewer, Daniel. *The Enlightenment Past: Reconstructing Eighteenth-Century French Thought*. Cambridge University Press, 2008.

Brunet, François. "Inventing the Literary Prehistory of Photography: From François Arago to Helmut Gernsheim." *History of Photography* 34, no. 4 (2010): 368–72.

Brunet, François. *La naissance de l'idée de photographie* [The birth of the idea of photography]. Presses Universitaires de France, 2012.

Cadava, Eduardo. *Words of Light: Theses on the Photography of History*. Princeton University Press, 1998.

Canales, Jimena. *A Tenth of a Second: A History*. University of Chicago Press, 2010.

Carey, James. "The Problem of Journalism History." In *James Carey: A Critical Reader*, edited by Eve Stryker Munson and Catherine A. Warren. University of Minnesota Press, 1997.

Challine, Éléonore, and Paul-Louis Roubert, eds. "Prix, coût, valeur: Pour une histoire économique de la photographie" [Price, cost, value: For an economic history of photography]. Special issue, *Photographica*, no. 8 (2024).

Chevallier, Alix, ed. *Nadar: Exposition 19 mars–16 mai 1965: Catalogue*. Bibliothèque nationale, 1965.

Chevreul, Michel-Eugène. *De la loi du contraste simultané des couleurs et de l'assortiment des objets colorés considérés d'après cette loi dans ses rapports avec la peinture, les tapisseries* [On the law of the simultaneous contrast of colors and the selection of colored objects according to this law as related to painting and tapestries]. Paris: Pitois-Levrault, 1839.

Chevreul, Michel-Eugène. *De la méthode a posteriori expérimentale et de la généralité de ses applications* [On the posteriori experimental method and the generality of its applications]. Paris: Dunod, 1870.

Chevreul, Michel-Eugène. "La verité sur l'invention de la photographie" [The truth about the invention of photography]. *Journal des savants*, February 1873, 65–82.

Chevreul, Michel-Eugène. "Séance du lundi 30 octobre 1871." *Compte rendu des séances de l'académie des sciences* 73 (July–December 1871): 1017–20.

Chevreul, Raoul. "La vie et l'oeuvre de Michel-Eugène Chevreul" [The life and work of Michel Eugène Chevreul]. In Roque, Bodo, and Viénot, *Michel-Eugène Chevreul, un savant, des couleurs!*

Chun, Wendy Hui Kyong. "Did Somebody Say New Media?" In Chun, Fisher, and Keenan, *New Media, Old Media*.

Chun, Wendy Hui Kyong, Anna Watkins Fisher, and Thomas Keenan, eds. *New Media, Old Media: A History and Theory Reader*. Routledge, 2016.

Clark, Catherine E. *Paris and the Cliché of History: The City and Photographs, 1860–1970*. Oxford University Press, 2018.

Clark, T. J. *The Painting of Modern Life: Paris in the Art of Manet and His Followers*. Rev. ed. Princeton University Press, 1999.

Clayson, Hollis. "Bright Lights, Brilliant Wit: Caricature and Electric Light in Later Nineteenth-Century Paris." In *Electric Worlds/Mondes Électriques: Creations, Circulations, Tensions, Transitions*, edited by Alain Beltran, Léonard Labarie, Pierre Lanthier, and Stéphanie Le Gallic. Peter Lang, 2014.

Collins, Paul. "Theatrophone—The 19th-Century iPod." *New Scientist*, January 9, 2008. https://www.newscientist.com/article/mg19726382-000-theatrophone-the-19th-century-ipod/.

Cosgrove, Denis, and William L. Fox. *Photography and Flight*. Reaktion Books, 2010.

"Cours de cassation, chambre de requêtes, 20 décembre, 1886" [Court of cassation, chamber of requests, December 20, 1886]. *Annales de la propriété industrielle et artistique* 33, no. 1 (1888): 42–44.

Crimp, Douglas. "Positive/Negative: A Note on Degas's Photographs." *October* 5 (1978): 89–100.

Crimp, Douglas. "The Museum's Old/The Library's New Subject." In *On the Museum's Ruins*. MIT Press, 1993.

Crosland, Maurice. "Popular Science and the Arts: Challenges to Cultural Authority in France Under the Second Empire." *British Journal for the History of Science* 34, no. 3 (2001): 301–22.

Daniel, Malcolm. *The Photographs of Édouard Baldus*. Metropolitan Museum of Art, 1994.

Daston, Lorraine. "The History of Science as European Self-Portraiture." *European Review* 14, no. 4 (2006): 523–36.

Daston, Lorraine, and Peter Galison. *Objectivity*. Princeton University Press, 2021.

Davies, Helen. *Emile and Isaac Pereire: Bankers, Socialists and Sephardic Jews in Nineteenth-Century France*. Manchester University Press, 2015.

de La Landelle, Gabriel. *Aviation, ou navigation aérienne* [Aviation, or air navigation]. Paris: E. Dentu, 1863.

de La Landelle, Gabriel. *Dans les airs, histoire elementaire de L'aeronautique* [In the air, an elementary history of aeronautics]. Paris: Haton, 1884.

de La Landelle, Gabriel. *Société d'encouragement pour la locomotion aérienne au moyen d'appareils plus lourds que l'air, Rapport du conseil d'administration sur le premier exercice 1864* [Society for the encouragement of aerial locomotion by machines heavier than air, Report from the administrative council on the first exercise]. Paris: J. Claye, 1865.

Delporte, Christian. *Les journalistes en France 1880–1950: Naissance et construction d'une profession* [Journalists in France 1880–1950: The birth and construction of a profession]. Editions du Seuil, 1999.

Detaille, Gérard. *Marseille, un siècle d'images* [Marseille, a century of images]. Éditions Parenthèses, 1998.

Dinkar, Niharika. "Pyrotechnics and Photography: Saltpeter and the Colonial History of Photographic Lighting." *Photographies* 14, no. 3 (2021): 395–420.

Disdéri, André-Adolphe-Eugène. *Application de la photographie à la reproduction des oeuvres d'art: Architecture, peinture, statuaire, orfèvrerie, émaux, ivoires, costumes, haute curiosité* [Application of photography to the reproduction of artworks: Architecture, painting, statuary, goldwork, enamels, ivories, costumes, high curiosities]. Paris, 1861.

Disdéri, André-Adolphe-Eugène. *L'art de la photographie*. Paris, 1862.

Dorrian, Mark, and Frédéric Pousin, eds. *Seeing from Above: The Aerial View in Visual Culture*. Bloomsbury Academic, 2019.

Doucet, Emily. "Excessive Luminosity: Félix Nadar's First Photographs by Electric Light." In *Le laboratoire/The Laboratory*, edited by Martha Langford and Zoë Tousignant. Artexte, 2021.

Doucet, Emily. "In History, the Future: Determinism in the Early History of Photography in France." *Communication +1* 7, no. 1 (2018). https://doi.org/10.7275/fvq1-t614.

Doucet, Emily. "The Idea Was in the Air: Nadar's Aerial Media." *Grey Room* 83 (June 2021): 112–37.

Downing, Eric. "Lucretius at the Camera: Ancient Atomism and Early Photographic Theory in Walter Benjamin's Berliner Chronik." *Germanic Review: Literature, Culture, Theory* 81, no. 1 (2006): 21–36.

Drie, Melissa van. "Hearing Through the Théâtrophone: Sonically Constructed Spaces and Embodied Listening in the Late Nineteenth-Century French Theatre." *SoundEffects—An Interdisciplinary Journal of Sound and Sound Experience* 5, no. 1 (2015): 73–90.

Edelman, Bernard. *Ownership of the Image: Elements for a Marxist Theory of Law*. Translated by Elizabeth Kingdom. Routledge and Kegan Paul, 1979.

Eder, Josef Maria. *History of Photography*. Translated by Edward Epstean. Dover Publications, 1945.

Edwards, Elizabeth. "Photography and the Material Performance of the Past." *History and Theory* 48, no. 4 (2009): 130–50.

Edwards, Elizabeth. *The Camera as Historian: Amateur Photographers and Historical Imagination, 1885–1918*. Duke University Press, 2012.

Edwards, Steve. "Why Pictures? From Art History to Business History and Back Again." *History of Photography* 44, no. 1 (2020): 3–15.

Eigen, Edward. "Dark Space and the Early Days of Photography as a Medium." *Grey Room*, no. 3 (March 1, 2001): 90–111.

Eigen, Edward. "On Purple and the Genesis of Photography, or the Natural History of an Exposure." In *Ocean Flowers: Impressions from Nature*, edited by Carol Armstrong and Catherine De Zegher. The Drawing Center and Princeton University Press, 2004

Emptoz, Gérard, Valérie Marchal, and Daniel Hangard, eds. *Aux sources de la propriété industrielle: Guide des archives de l'INPI* [The sources of industrial property: Guide to the INPI archives]. INPI, 2002.

"Exhibition of the French Photographic Society, Considered from an English Point of View." *British Journal of Photography* 10, no. 190 (1863): 212–13.

Exposition internationale. *Exposition universelle internationale de 1878. Section française. Deuxième groupe (classes 6 à 16), Éducation et enseignement, matériel et procédés des arts libéraux. Classe 16, Cartes et appareils de géographie et de cosmographie* [Universal exhibition of 1876. French section. Second group (class 6 to 16), education and teaching, material and processes of the liberal arts. Class 16, maps and geographic and cosmographic apparatuses]. Paris: Imprimerie Delalain, 1878.

Falguière, Mathilde. "Le fonds d'atelier Nadar: Sauvetage et traitement" [The Nadar studio archives: Rescue and treatment]. In Aubenas and Lacoste, *Les Nadar: Une légende photographique*.

Fassy, Paul. *Les catacombes, étude historique* [The catacombs, a historical study]. Paris: E. Dentu, 1861.

Feaster, Patrick. "Daguerreotyping the Voice: Léon Scott's Phonautographic Aspirations." In *Parole #1: The Body of the Voice/Stimmkörper*, edited by Annette Stahmer. Salon Verlag, 2009.

Figuier, Louis. *La photographie*. 1888. Reprint, Laffitte Reprints, 1983.

Flint, Kate. *Flash! Photography, Writing, and Surprising Illumination*. Oxford University Press, 2017.

Forrest, Alan. "The Port Cities of the French Atlantic." In *The Death of the French Atlantic: Trade, War, and Slavery in the Age of Revolution*. Oxford University Press, 2020.

Freund, Gisèle. *La photographie en France au dix-neuvième siècle: Essai de sociologie et d'esthétique* [Photography in France in the nineteenth century: An essay on sociology and aesthetics]. Adrienne Monnier, 1936.

Freund, Gisèle. *Photographie et société*. Éditions du Seuil, 1974.

Freund, Gisèle. *Photography and Society*. D. R. Godine, 1980.

Gaines, Jane M. *Contested Culture: The Image, the Voice, and the Law*. University of North Carolina Press, 2000.

Galvez-Behar, Gabriel. "The Patent System During the French Industrial Revolution: Institutional Change and Economic Effects." *Jahrbuch für Wirtschaftsgeschichte/Economic History Yearbook* 60, no. 1 (2019): 31–56.

Gandy, Matthew. *The Fabric of Space: Water, Modernity, and the Urban Imagination*. MIT Press, 2014.

Gasser, Martin. "Histories of Photography 1839–1939." *History of Photography* 16, no. 1 (1992): 50–60.

Gervais, Thierry. "Experimentations photographiques. La vision en plongée, de Nadar (1858) à Gaston Tissandier." In *Vues d'en haut*, edited by Angela Lampe. Centre Pompidou, 2013.

Gervais, Thierry. "Interview of Chevreul, France, 1886." In *Getting the Picture: The Visual Culture of the News*, edited by Jason E. Hill and Vanessa R. Schwartz. Bloomsbury, 2015.

Gervais, Thierry. "L'illustration photographique: Naissance du spectacle de l'information (1843–1914)" [Photographic illustration: Birth of the spectacle of information]. PhD diss., École des hautes études en sciences sociales, 2007.

Gervais, Thierry. "Un basculement du regard. Les débuts de la photographie aérienne 1855–1914" [A shift in perspective: The beginnings of aerial photography, 1855–1914]. *Études photographiques*, no. 9 (2001). https://journals.openedition.org/etudesphotographiques/916.

Gervais, Thierry. "Witness to War: The Uses of Photography in the Illustrated Press, 1855–1904." *Journal of Visual Culture* 9, no. 3 (2010): 370–84.

Gioia, Ted. "The First Music Streaming Service Was Invented in 1881—Discover the Thèâtrophone." *Open Culture*, October 2, 2019. https://www.openculture.com/2019/10/the-first-music-streaming-service-was-invented-in-1881-discover-the-theatrophone.html.

Gitelman, Lisa. *Always Already New: Media, History, and the Data of Culture*. MIT Press, 2008.

Gitelman, Lisa. *Scripts, Grooves, and Writing Machines: Representing Technology in the Edison Era*. Stanford University Press, 2000.

Gitelman, Lisa, and Geoffrey B. Pingree. *New Media, 1740–1915*. MIT Press, 2003.

Godfrey-Smith, Peter. *Theory and Reality: An Introduction to the Philosophy of Science*. University of Chicago Press, 2003.

Godin, Benoît. "Innovation: The History of a Category." Working paper, Institut national de la recherche scientifique, Centre Urbanisation Culture Société, 2008.

Gosling, Nigel. *Nadar*. Alfred A. Knopf, 1977.

Greaves, Roger. *Nadar: Ou, Le paradoxe vital [Nadar: Or, the vital paradox]*. Flammarion, 1980.

Grimm, Thomas. "L'art de vivre cent ans" [The art of living a hundred years]. *Le petit journal*, August 31, 1886.

Grundberg, Andy. "Photography View; Two Camps Battle over the Nature of the Medium." *New York Times*, August 14, 1983. https://www.nytimes.com/1983/08/14/arts/photography-view-two-camps-battle-over-the-nature-of-the-medium.html.

Gunthert, André. "La conquête de l'instantané: Archéologie de l'imaginaire photographique en France, 1841–1895." [The conquest of instantaneity: Archeology of the photographic imagination in France, 1841–1895]. PhD diss., Ecole des hautes etudes en sciences sociales, 1999.

Gunthert, André. "L'inventeur inconnu. Louis Figuier et la constitution de l'histoire de la photographie française" [The unknown inventor: Louis Figuier and the constitution of French photography]. *Études photographiques*, no. 16 (May 25, 2005): 6–18.

Haffner, Jeanne. *The View from Above: The Science of Social Space*. MIT Press, 2013.

Hambourg, Maria Morris, Françoise Heilbrun, and Philippe Néagu, eds. *Nadar*. Metropolitan Museum of Art, 1995.

Harris, Lane J., ed. *The Peking Gazette: A Reader in Nineteenth-Century Chinese History*. Brill, 2018.

Harvey, David. *Paris, Capital of Modernity*. Routledge, 2006.

Hayward, L. H. "A Review of Helicopter Patents: A Lecture Presented to the Members of the Helicopter Association of Great Britain on Friday, January 18, 1952." *Aircraft Engineering and Aerospace Technology* 24, no. 4 (1952): 92–105.

Heatherton, Christina. "Making the First International: Nineteenth-Century Regimes of Surveillance, Accumulation, Resistance, and Abolition." In *The Cambridge History of America and the World, vol. 2, 1820–1900*, edited by Kristin Hoganson and Jay Sexton. Cambridge University Press, 2021.

Hellman, Karen. "Nadar dans les collections Nord-Américains" [Nadar in North American collections]. In Aubenas and Lacoste, *Les Nadar: Une légende photographique*.

Henriot, Gabriel. "Catalogue des publications et des manuscrits composant la collection Nadar à la bibliothèque" [Catalog of publications and manuscripts in the Nadar collection at the library]. *Bulletin de la Bibliothèque et des travaux historiques, publié sous la direction de M. Marcel Poëte, inspecteur des travaux historiques, conservateur de la Bibliothèque de la ville de Paris* 6 (1913): 1–124.

Hill, Jason, and Vanessa R. Schwartz. *Getting the Picture: The Visual Culture of the News*. Bloomsbury Publishing, 2015.

Howes, Chris. *To Photograph Darkness: The History of Underground and Flash Photography*. Southern Illinois University Press, 1989.

Hugo, Victor. *Les Misérables*. Translated by Julie Rose. Modern Library, 2009.

Ihde, Aaron J. *The Development of Modern Chemistry*. Rev. ed. Dover Publications, 2012.

Iselin, J. F., and P. Le Neve Foster, eds. *Reports by the Juries on the Subjects in the Thirty-Six Classes into Which the Exhibition Was Divided*. London: Society for the Encouragement of Arts, Manufacture and Commerce, 1863.

Jacomet, Charles, ed. *Revue technique de l'Exposition universelle de 1900 par un comité d'ingénieurs, d'architectes, de professeurs et de constructeurs, quatrième partie, tome 2* [Technical review of the 1900 universal exhibition by a committee of engineers, architects, professors and builders, fourth part, volume 2]. E. Bernard, 1901.

Jameson, Fredric. "Progress Versus Utopia, or Can We Imagine the Future?" In *Archaeologies of the Future: The Desire Called Utopia and Other Science Fictions*. Verso, 2005.

Jammes, André. "The Nadar Studio, Dispersed and Recovered." Translated by Frederick Brown. In Hambourg, Heilbrun, and Néagu, *Nadar*.

Jenkins, Reese. *Images and Enterprise: Technology and the American Photographic Industry, 1839 to 1925*. Johns Hopkins University Press, 1975.

Jestaz, Juliette. "Soutenir l'aérostation pour mieux la tuer. La collection Nadar conservée à la Bibliothèque historique de la ville de Paris et au musée Carnavalet" [Supporting aerostation to better kill it. The Nadar collection preserved at the Historic Library of the City of Paris and the Carnavalet Museum]. *Revue des patrimoines* 35 (2018). https://journals.openedition.org/insitu/16768.

Kalba, Laura Anne. *Color in the Age of Impressionism: Commerce, Technology, and Art*. Pennsylvania State University Press, 2017.

Kaplan, Caren. *Aerial Aftermaths: Wartime from Above*. Duke University Press, 2018.

Kavanagh, Thomas M. "Epicureanism Across the French Revolution." In *Lucretius and Modernity: Epicurean Encounters Across Time and Disciplines*, edited by Jacques Lezra and Liza Blake. Palgrave Macmillan, 2016.

Keller, Ulrich. "Photography, History, (Dis)Belief." *Visual Resources* 26, no. 2 (2010): 95–111.

Ken, Alexandre. *Dissertations historiques, artistiques et scientifiques sur la photographie* [Historical, artistic, and scientific dissertations on photography]. Paris: Librairie nouvelle, 1864.

Keylor, William R. *Academy and Community: The Foundation of the French Historical Profession*. Harvard University Press, 1975.

Kluitenberg, Eric. *Book of Imaginary Media: Excavating the Ultimate Dream of Communication*. NAi Publishers, 2006.

Koselleck, Reinhart. *The Practice of Conceptual History: Timing History, Spacing Concepts*. Translated by Todd Presner, Kerstin Behnke, and Jobst Welge. Stanford University Press, 2002.

Kozloff, Max. "Nadar and the Republic of Mind." *Artforum* 15, no. 1 (1976): 29–39.

Krauss, Rosalind. "Tracing Nadar." *October* 5 (1978): 29–47.

Krull, Germaine. *Marseille*. Libraire Plon, 1935.

Lacan, Ernest. "Revue photographique." *Le moniteur de la photographie: Revue internationale des progrès du nouvel art* 4, no. 22 (1865): 1–2.

Lacoste, Anne. "La photographie instantanée" [Instantaneous photography]. In Aubenas and Lacoste, *Les Nadar: Une légende photographique*.

Lacoste, Anne. "Paul Nadar, l'ère de la modernité" [Paul Nadar, the era of modernity]. In Aubenas and Lacoste, *Les Nadar: Une légende photographique*.

Lampe, Angela, ed. *Vues d'en haut [Views from above]*. Centre Pompidou, 2013.

Larkin, Brian. "The Politics and Poetics of Infrastructure." *Annual Review of Anthropology* 42, no. 1 (2013): 327–43.

Larouaire, Léon. "Un chantier au fond de la mer" [A construction site at the bottom of the sea]. *Le petit marseillais*, April 8, 1900, 1.

"Le journal illustré à ses lecteurs ... L'art de vivre cent ans" [Le journal illustré to its readers ... the art of living a hundred years]. *Le journal illustré*, September 5, 1886, 282.

Legacey, Erin Marie. "The Paris Catacombs: Remains and Reunion Beneath the Postrevolutionary City." *French Historical Studies* 40, no. 3 (2017): 509–36.

Lemay, Pierre, and Ralph E. Oesper. "Michel Eugene Chevreul (1786–1889)." *Journal of Chemical Education* 25, no. 2 (1948): 62–67.

Leonardi, Nicoletta, and Simone Natale, eds. *Photography and Other Media in the Nineteenth Century*. Penn State University Press, 2018.

Lerner, Jillian. *Experimental Self-Portraits in Early French Photography*. Routledge, Taylor and Francis Group, 2021.

Lerner, Jillian. "Nadar's Signatures: Caricature, Self-Portrait, Publicity." *History of Photography* 41, no. 2 (2017): 108–25.

Liesegang, P. "Quelques observations sur la photographie à l'exposition internationale de Londres" [Some observations on the photography at the international exhibition of London]. *Le moniteur de la photographie*, June 1, 1862, 42–43.

Lissajous, Jules Antoine. "Mémoire sur l'étude optique des mouvements vibratoires" [Essay on the optical study of vibratory movements]. *Annales de chimie et de physique* 2, no. 51 (1847): 147–231.

Liu, Alan. "Imagining the New Media Encounter." In *Friending the Past: The Sense of History in the Digital Age*. University of Chicago Press, 2018.

Loh, Maria. *Titian Remade: Repetition and the Transformation of Early Modern Italian Art*. Getty Research Institute, 2007.

Loveday, Kiki. "The Pioneer Paradigm." *Feminist Media Histories* 8, no. 1 (2022): 165–80.

Malloizel, Godefroy. *Oeuvres scientifiques de Michel-Eugène Chevreul: Doyen des étudiants de France, 1806–1886* [Scientific Works of Michel-Eugène Chevreul: Dean of the students of France, 1806–1886]. Paris, 1886.

Manuel, Frank Edward. *The Prophets of Paris*. Harvard University Press, 1962.

Marcus, Sharon. *The Drama of Celebrity*. Princeton University Press, 2019.

Marey, Étienne-Jules. "L'analyse de mouvements par la photographie" [The analysis of movements by photography]. *Paris-Photographe*, April 1891, 5–12.

Marvin, Carolyn. *When Old Technologies Were New: Thinking About Electric Communication in the Late Nineteenth Century*. Oxford University Press, 1990.

McCauley, Elizabeth Anne. *Industrial Madness: Commerciale Photography in Paris, 1848–1871*. Yale University Press, 1994.

McCauley, Elizabeth Anne. "Writing Photography's History Before Newhall." *History of Photography* 21, no. 2 (1997): 87–101. https://doi.org/10.1080/03087298.1997.10443726.

Ministère des Postes et Télégraphes. *Exposition internationale d'électricité, Paris 1881: Catalogue général officiel*. Paris: A. Lahure, 1881.

Moholy-Nagy, László, dir. *Impressionen vom alten Marseiller Hafen* [Impressions of the old port of Marseille]. 1929.

Moitessier, M. A. "Procédé pour obtenir des épreuves positives à l'aide de la chambre noire; par M. A. Moitessier (extrait)" [Process for obtaining positive prints with a dark room; par M. A. Moitessier (extract)]. *Comptes rendus hebdomadaires des séances de l'Académies des sciences* 40 (January–June 1855): 120–21.

Montebello, Philippe de, and Henri Loyrette. "Director's Foreword." In *Nadar*, edited by Maria Morris Hambourg, Françoise Heilbrun, and Phillippe Néagu. Metropolitan Museum of Art, 1994.

Montricher, M. de, and M. Pascal. "Les docks, le bassin Napoléon et l'Avant-Port d'Arène" [The docks, the Napoleon basin, and the outer harbour of Arène]. *Nouvelles annales de la construction*, no. 9 (September 1855): 1–2.

Nadar, Félix. "Le dessus et le dessous de Paris" [Above and below Paris]. In *Paris-Guide, par le principaux écrivains et artistes de France; introduction par Victor Hugo* [Paris guide, by the principal writers and artists of France; introduction by Victor Hugo]. Paris: A. Lacroix and Verboeckhoven, 1867.

Nadar, Félix. *Le droit au vol* [The right to fly]. Paris: J. Hetzel, 1865.

Nadar, Félix. *Les ballons en 1870 C.E. qu'on aurait pu faire, ce qu'on a fait* [Balloons in 1870 CE, what could have been done, what was done]. E. Chatelain, 1870.

Nadar, Félix. "Les histoires du mois" [Stories of the month]. *Musée français-anglais* 24 (December 1856): 7.

Nadar, Félix. "Manifeste de l'autolocomotion aérienne" [Manifesto of aerial automotion]. *L'aéronaute*, numéro spécimen, [1863], 1–4.

Nadar, Félix. "Manifeste de l'autolocomotion aérienne" [Manifesto of aerial automotion]. *La presse*, August 7, 1863.

Nadar, Félix. *Mémoires du Géant: À terre & en l'air* Paris: E. Dentu, 1864.

Nadar, Félix. "My Life as a Photographer." Translated by Thomas Repensek. *October* 5 (1978): 7–28.

Nadar, Félix. "Obsidional Photography." In *When I Was a Photographer*, translated by Eduardo Cadava and Liana Theodoratou. MIT Press, 2015.

Nadar, Félix. *Quand j'étais photographe* [When I was a photographer]. Ernest Flammarion, 1900.

Nadar, Félix. *The Right to Fly*. Translated by James Spence Harry. London: Cassell, Petter, and Galpin, 1866.

Nadar, Félix. "Souvenirs d'un atelier photographe: Balzac et le daguerréotype" [Memories from a photographic studio: Balzac and the daguerreotype]. *Paris-Photographe* 1, no. 1 (1891): 13–17.

Nadar, Félix. "Souvenirs d'un atelier photographe: Le docteur van Monckhoven" [Memories from a photographic studio: Doctor van Monckhoven]. *Paris-Photographe* 2, no. 7 (1892): 301–8.

Nadar, Félix. "Souvenirs d'un atelier de photographe: La première épreuve de photographie aérostatique" [Memories from a photographic studio: The first test of aerostatic photography]. Pt. 1. *Paris-Photographe* 3, no. 4 (1893): 155–61

Nadar, Félix. "Souvenirs d'un atelier de photographe: La première épreuve de photographie aérostatique (suite)" [Memories from a photographic studio: The first test of aerostatic photography]. Pt. 2. *Paris-Photographe* 3, no. 5 (1893): 197–200.

Nadar, Félix. "Souvenirs d'un atelier de photographe: La première épreuve de photographie aérostatique (suite)" [Memories from a photographic studio: The first test of aerostatic photography]. Pt. 3. *Paris-Photographe* 3, no. 6 (1893): 245–48.

Nadar, Félix. *When I Was a Photographer*. Translated by Eduardo Cadava and Liana Theodoratou. MIT Press, 2015.

Nadar, Félix, and Paul Nadar. "L'art de vivre cent ans" [The art of living a hundred years]. *Le journal illustré*, September 3, 1886, 282–88.

Nadar, Paul. "Progrès et applications de la photographie (suite)." *Paris-Photographe* 2, no. 7 (1892): 279–91.

Nadar, Paul. "Sur les papiers positifs et négatifs Eastman: Chassis à rouleaux et porte-membrane pour leur emploi . . . Communication faite à la séance du 7 janvier 1887" [On Eastman's paper positives and negatives: Roll frames and film holders for their use . . . Communication made to the meeting of January 7, 1887]. *Bulletin de la société française de la photographie* 2, no. 3 (1887): 47.

Nègre, Charles. *De la gravure héliographique, son utilité, son origine, son application à l'étude de l'histoire, des arts et des sciences naturelles* [On heliographic engraving, its utility, origin, and its application to the study of history, arts, and natural sciences]. Nice: V.-E. Gauthier, 1866.

Nelms, Brenda. *The Third Republic and the Centennial of 1789.* Garland, 1987.

Nesbit, Molly. *Atget's Seven Albums.* Yale University Press, 1994.

Nesbit, Molly. "What Was an Author?" *Yale French Studies*, no. 73 (1987): 229–57.

Newhall, Beaumont. *Airborne Camera: The World from the Air and Outer Space.* Hastings House, 1969.

Newhall, Beaumont. *Focus: Memoirs of a Life in Photography.* Bullfinch Press, 1993.

Newhall, Beaumont. *Photography 1839–1937.* Museum of Modern Art, 1937.

Nieto-Galan, Agustí. *Science in the Public Sphere: A History of Lay Knowledge and Expertise.* Routledge, 2016.

O'Brien, Jean M. *Firsting and Lasting: Writing Indians Out of Existence in New England.* University of Minnesota Press, 2010.

Owens, Craig. "Photography 'En Abyme.'" *October* 5 (1978): 73–88.

Parks, Lisa, and Caren Kaplan, eds. *Life in the Age of Drone Warfare.* Duke University Press, 2017.

Picon, Antoine. *Les Saint-Simoniens: Raison, imaginaire, et utopie* [The Saint-Simonians: Reason, imagination, and utopia]. Belin, 2002.

Pierce, J. Mackenzie. "Writing at the Speed of Sound: Music Stenography and Recording Beyond the Phonograph." *19th-Century Music* 41, no. 2 (2017): 121–50.

Pike, David. "Paris Souterrain: Before and After the Revolution." *Dix-Neuf* 15, no. 2 (2011): 177–97.

Pinney, Christopher. *Photography and Anthropology.* Reaktion Books, 2011.

Pinney, Christopher. "Things Happen: Or, From Which Moment Does That Object Come?" In *Materiality*, edited by Daniel Miller. Duke University Press, 2005.

Poirel, V. *Mémoire sur les travaux à la mer ... comprenant l'historique des ouvrages exécutés au port d'Alger, et l'exposé complet et détaillé d'un système de foundation à la mer au moyen de blocs de béton* [Report on works at sea ... including the history of works carried out at the port of Algiers, and a complete and detailed presentation of a sea foundation system using concrete blocks]. Paris: Carilian-Goeury, 1841.

Ponton d'Amécourt, Gustave Vicomte de. *Collection de mémoires sur la locomotion aérienne sans ballons* [Collection of essays on aerial locomotion without balloons]. Paris: Gauthier-Villars, 1864.

Popper, Karl. *The Logic of Scientific Discovery*. Hutchinson, 1959. Reprint, Routledge, 2005.

Pottage, Alain, and Brad Sherman. *Figures of Invention: A History of Modern Patent Law*. Oxford University Press, 2010.

Prinet, Jean, and Antoinette Dilasser. *Nadar*. A. Colin, 1966.

Raichvarg, Daniel, and Jean Jacques. *Savants et ignorants: Une histoire de la vulgarisation des sciences* [Scholars and ignorants: A history of the popularization of the sciences]. Editions du Seuil, 1991.

Reid, Donald. *Paris Sewers and Sewermen: Realities and Representations*. Harvard University Press, 1991.

Reynes, Geneviève. "Chevreul interviewé par Nadar, premier document audiovisuel (1886)" [Chevreul interviewed by Nadar, the first audiovisual document]. *Gazette des beaux arts*, November 1981, 154–84.

Rice, Shelley. *Parisian Views*. MIT Press, 1997.

Robert, Paul-Louis. "Félix Nadar et la disparition de la photographie" [Félix Nadar and the disappearance of photography]. In Aubenas and Lacoste, *Les Nadar: Une légende photographique*.

Robertson, Craig. "Introduction." In *Media History and the Archive*, edited by Craig Robertson. Routledge, 2011.

Roncayolo, Marcel. *L'imaginaire de Marseille: Port, ville, pôle* [The imagination of Marseille: Port, city, hub]. ENS Éditions, 2014.

Roque, George, Bernard Bodo, and Françoise Viénot, eds. *Michel-Eugène Chevreul, Un savant, des couleurs!* [Michel Eugène Chevreul, a scholar, colors!]. Muséum national d'Histoire naturelle, 1997.

Russell, John. "Art: Photographs by Nadar." *New York Times*, February 4, 1977.

Schaffer, Simon. "Making up Discovery." In *Dimensions of Creativity*, edited by Margaret A. Boden. MIT Press, 1994.

Schaffer, Simon. "Scientific Discoveries and the End of Natural Philosophy." *Social Studies of Science* 16, no. 3 (August 1986): 387–420.

Schudson, Michael. "Question Authority: A History of the News Interview in American Journalism, 1860s–1930s." *Media, Culture and Society* 16, no. 4 (1994): 565–87.

Schwartz, Joan. "Archives, Records, and Power: The Making of Modern Memory." *Archival Science* 2 (2002): 1–19.

Schwartz, Vanessa R. *Spectacular Realities: Early Mass Culture in Fin-de-Siècle Paris*. University of California Press, 1998.

Seillan, Jean-Marie. "L'interview." In *La civilisation du journal: Histoire culturelle et littéraire de la presse française au XIXe siècle* [The civilization of the newspaper: A cultural and literary history of the French press in the nineteenth century], edited by Dominique Kalifa, Philippe Régnier, Marie-Ève Thérenty, and Alain Vaillant. Nouveau Monde Éditions, 2011.

Sekula, Allan. "Neither Metaphysics nor Positivism: Against the Bourgeois Histories of Photography, Part 1." Lecture, NSCAD Visual Resources Collection, 1980. Accessed September 18, 2023. https://nscad.cairnrepo.org/islandora/object/islandora%3A498 (no longer available).

Sekula, Allan. "The Instrumental Image: Steichen at War." *Artforum* 14, no. 4 (1975): 26–34.

Seniuta, Isabella. "De la collection à la spéculation: Portraits croisés des collectionneurs André Jammes et Sam Wagstaff" [From collection to speculation: Comparative portraits of the collectors André Jammes and Sam Wagstaff]. *Image and Narrative* 19, no. 4 (2018): 8–21.

Sharma, Sarah. *In the Meantime: Temporality and Cultural Politics*. Duke University Press, 2013.

Sheehan, Tanya, and Andres Zervigon, eds. *Photography and Its Origins*. Routledge, 2014.

Siegel, Steffen. "Afterword." In *First Exposures: Writings from the Beginning of Photography*, edited by Steffen Siegel. Getty Publications, 2017.

Siegel, Steffen, ed. *First Exposures: Writings from the Beginning of Photography*. Getty Publications, 2017.

Simmons, Dana. "Waste Not, Want Not: Excrement and Economy in Nineteenth-Century France." *Representations* 96, no. 1 (2006): 73–98.

Sinnema, Peter W. "Representing the Railway: Train Accidents and Trauma in the 'Illustrated London News.'" *Victorian Periodicals Review* 31, no. 2 (1998): 142–68.

Solomon-Godeau, Abigail. "Calotypomania: The Gourmet Guide to Nineteenth-Century Photography." In Solomon-Godeau, *Photography at the Dock*.

Solomon-Godeau, Abigail. "Canon Fodder: Authoring Eugène Atget." In Solomon-Godeau, *Photography at the Dock*.

Solomon-Godeau, Abigail. *Photography at the Dock: Essays on Photographic History, Institutions, and Practices*. University of Minnesota Press, 2009.

Star, Susan Leigh. "The Ethnography of Infrastructure." *American Behavioral Scientist* 42, no. 3 (1999): 377–91.

Sterne, Jonathan. *The Audible Past: Cultural Origins of Sound Reproduction*. Duke University Press, 2003.

Stiegler, Bernd. *Nadar: Bilder der Moderne*. König, 2019.

"Suite du rapport sur les arts qui ont servi à la défense de la république, et sur le nouveau procédé de tannage, découvre par le citoyen Armand Séguin, fait à la Convention nationale, le 14 nivôse, au nom du comité de salut public, par Fourcroy" [Continuation of a report on the arts which served in the defense of the republic, and on the tanning process discovered by

Citizen Armand Séguin, made to the National Convention, on 14 Nivôse, in the name of the Committee of Public Safety, by Fourcroy]. *Gazette nationale ou Le moniteur universel*, no. 109 (nivôse 19, l'an 3e; January 8, 1795). Facsimile edition, 3rd ser., 10 (1844): 145–48.

Stock, John T. "Bunsen's Batteries and the Electric Arc." *Journal of Chemical Education* 72, no. 2 (1995): 99–102.

Taussig, Michael. *Mimesis and Alterity: A Particular History of the Senses.* Routledge, 1993.

Taws, Richard. "Conté's Machines: Drawing, Atmosphere, Erasure." *Oxford Art Journal* 39, no. 2 (2016): 243–66.

Taws, Richard. "When I Was a Telegrapher." *nonsite.org*, no. 14, December 15, 2014. http://nonsite.org/article/when-i-was-a-telegrapher.

Thélot, Jérôme. *Les inventions littéraires de la photographie* [The literary inventions of photography]. Presses Universitaires de France, 2003.

Thierry, J. *Daguerréotypie: Franches explications sur l'emploi de sa liqueur invariable, sur les moyens qu'il met en usage pour en obtenir le maximum de sensibilité... Précédées d'une histoire abrégée de la photographie* [Daguerreotypy: Frank explanations on the use of its invariable liquor, on the means it uses to obtain maximum sensitivity... preceded by an abridged history of photography]. Paris: Lerebours et Secrétan, 1847.

Thorburn, David, and Henry Jenkins, eds. *Rethinking Media Change: The Aesthetics of Transition.* MIT Press, 2004.

Tissandier, Gaston. *La photographie.* 3rd ed. Paris: Hachette, 1882.

Tissandier, Gaston. *La photographie en ballon.* Paris: Gauthier-Villars, 1886.

Tissandier, Gaston. *Les merveilles de la photographie.* Paris: Hachette, 1874.

Toscano, Alberto, and Jeff Kinkle. *Cartographies of the Absolute.* Zero Books, 2015.

Trachtenberg, Alan. "Albums of War: On Reading Civil War Photographs." *Representations*, no. 9 (1985): 1–32.

Travis, David. "Introduction." In *Niépce to Atget: The First Century of Photography; From the Collection of André Jammes,* cataloged by Marie-Thérèse Jammes and André Jammes. Art Institute of Chicago, 1978.

Tresch, John. "The Prophet and the Pendulum: Sensational Science and Audiovisual Phantasmagoria Around 1848." *Grey Room* 43 (Spring 2011): 16–41.

Trouillot, Michel-Rolph. *Silencing the Past: Power and the Production of History.* Beacon Press, 1995.

Tseng, Shao-Chien. "Nadar's Photography of Subterranean Paris: Mapping the Urban Body." *History of Photography* 38, no. 3 (2014): 233–54.

Tucker, Jennifer. "Entwined Practices: Engagements with Photography in Historical Inquiry." *History and Theory* 48, no. 4 (2009): 1–8.

Tupicoff, Dennis, dir. *The First Interview*. Jungle Pictures, 2011. 15 min (27 min. extended version).

Turner, Fred. *From Counterculture to Cyberculture: Stewart Brand, the Whole Earth Network, and the Rise of Digital Utopianism*. University of Chicago Press, 2006.

Turnor, Christopher Hatton. *Astra Castra: Experiments and Adventures in the Atmosphere*. London: Chapman and Hall, 1865.

Valery, Paul. "The Centenary of Photography." In *Occasions*, translated by Jackson Matthews. Vol. 2 of *The Collected Works of Paul Valery*. Princeton University Press, 1970.

Van der Weyde, Henry. "Photography by Electric Light." *Journal of the Society of the Arts*, February 24, 1882, 370–71.

Van der Weyden, Rogier. "Henry van der Weyde." *History of Photography* 23, no. 1 (1999): 68–72.

Van Haaften, Julia. *Berenice Abbott: A Life in Photography*. W. W. Norton, 2018.

Van Monckhoven, Désiré. *Traité général de photographie . . . suivi des applications de cet art aux sciences, et de recherches sur l'action chimique de la lumière* [General treatise on photography . . . followed by the applications of this art to science, and research on the chemical action of light]. 2nd ed. Paris: A. Gaudin et frère, 1856.

Verne, Jules. *20,000 Leagues Under the Sea*. Translated by James Reeves. Chatto and Windus, 1956.

Viars, Dominique. "Étude de douze plaques de verre négatives au gélatinobromure d'argent vernies présentant des soulèvements d'émulsion, réalisées par l'atelier Nadar dans les années 1880–1886" [Study of twelve varnished gelatin silver bromide glass plates showing lifting emulsion, produced by the Nadar studio in the years 1880–1886], master's thesis, École national du patrimoine institut de formation des restaurateurs d'oeuvres d'art, 2001.

Wagstaff, Samuel. *A Book of Photographs from the Sam Wagstaff Collection*. Gray Press, 1978.

Wallace, Amy. "Studios of Nature: The Transformation of Artists' Studios, 1845–1900." PhD diss., University of Toronto, 2019.

Weiss, Sean. "Engineering, Photography, and the Construction of Modern Paris, 1857–1911." PhD diss., City University of New York, 2013.

Weiss, Sean. "Making Engineering Visible: Photography and the Politics of Drinking Water in Modern Paris." *Technology and Culture* 61, no. 3 (2020): 739–71.

Wentersdorf, Karl P. "The Underground Workshop of Oliver Wendell Holmes." *American Literature* 35, no. 1 (1963): 1–12.

Wey, Francis. "Comment le soleil est devenu peintre. Histoire du daguerréotype et de la photographie" [How the sun became a painter. History of the daguerreotype and of photography]. *Musée des familles* 20 (July 1853): 289–900.

White, Hayden. *Metahistory: The Historical Imagination in Nineteenth-Century Europe.* JHU Press, 1975.

White, Minor, André Jammes, and Robert Sobieszek. *French Primitive Photography.* Aperture Books, 1970.

Williams, Rosalind. *Notes on the Underground: An Essay on Technology, Society, and the Imagination.* MIT Press, 2008.

Wilson, E. L. "A Day Among the Paris Studios." *Photographic Times* and *American Photographer* 12, no. 142 (1882): 409–12.

INDEX

Page references in *italics* indicate figures.

Abbott, Berenice, 196n56
Académie des sciences, 74, 79, 124, 180n75
Ader, Clément, 74, 128, 176n35, 189n24
aerial photography, 16–43, 192n2; firstness, 12–13, 16–19, 20, 21, 24–26, 27–30, 34, 37, 39–41, 43, 44, 48, 59, 67, 87, 106, 118, 137, 139, 140, 185n46; idea of, 20–24, 43; interest in, 38; Paris, 16–18, *17*, 39; patents, 21; representations, 19; state and, 20; *1ères [Premières] épreuves en ballon: Trois vues aériennes de Paris* (Nadar), 41–42, *42*; *Ballonneau gonflé servant à Nadar de modèle pour le tableau "Le trainage du Géant"* (Nadar), 29–30, *30*; *Brevet d'invention de quinze ans déposé le 23 octobre 1858 par Félix Tournachon, dit Nadar, pour un Système de photographie aérostatique (n° 1BB38509)* (Nadar), *22*; "Les Nadaréostats," *Journal amusant*, no. 403 (September 19, 1863) (d'Arnoux), *29*; *Maquette d'hélicoptère à vapeur de Ponton d'Amécourt* (Nadar), *28*; *Nadar élevant la photographie à la hauteur de l'Art* (Daumier), *38*, 38–39; *Premier résultat de photographie aérostatique // Applications : Cadastre, Stratégie, etc // Cliché obtenu à l'altitude de 520 m par Nadar 1858* (Nadar), 16–18, *17*; *Vient de paraître: Le 1er n° de l'Aéronaute, moniteur de la Société générale d'aérostation et d'automation aérienne* (Doré), 27, *28*; *Vues aériennes du quartier de l'Étoile à Paris* (Nadar), 39, *40*. See also aviation

Algeria, 94
Algiers, 13–14, 99
Alophe (Adolphe Menut), 56
aluminum, 162n28
Andraud, Antoine: Nadar and, 20–21; *Exposition universelle de 1855: Une dernière annexe au Palais de l'industrie*, 18–19, 20, 24
annotations, 41, 50–51
Arago, François, 87, 170n24
Arago Laboratory, 99

archives, 83, 96, 160n34, 167n1, 193n6
Aristotle, 103, 183n25
Artforum (journal), 3, 147, 157n9, 195nn45–46
Art Institute of Chicago, 147
artistic practice, theories of, 36
astronomy, 86
Atget, Eugène, 146
Aubenas, Sylvie, 106, 150, 151
avant-garde, 4, 150
aviation, 31–33, 34, 162n29. *See also* aerial photography

Babbage, Charles, 161n15
Babinet, Jacques, 34–35, 36
Bagioli, Mario, 169n8
balloons: mechanisms, 23; photography, 16, 18–19, 21, 23, 27, 29–32, 33, 39, 40, 41–42, *42*, 96, 192n4; representations, 19, 24; *Ballonneau gonflé servant à Nadar de modèle pour le tableau "Le trainage du Géant"* (Nadar), 29–30, *30*. *See also Géant* (balloon)
Balzac, Honoré de, 129, 137; "Balzac and the Daguerreotype" (Nadar), 122–26, *127*, 131; *Honoré de Balzac* (Bisson), *127*
Bann, Stephen, 134
Barrault, Émile, 21, 50
Barthes, Roland, 157n1
Baudry, Jérôme, 168n6, 169n8
Bayer, Herbert, 180n1
Belgrand, Eugène, 106, 107, 184n35
Bell, Alexander Graham, 129
Ben-Amos, Avner, 79, 178n50
Benjamin, Walter, 3, 136, 142, 157n1, 157n6, 187n8, 188n16, 193n13
BHVP (Bibliothèque historique de la ville de Paris), 33, 161n7
Biblical literature, 114, 186n70
Bibliothèque historique de la ville de Paris (BHVP), 33, 161n7
Bibliothèque nationale de France (BNF), 3, 143, 167n1
Bisson, Louis-Auguste: *Honoré de Balzac*, *127*
Bleichmar, Daniela, 87
Blight, Henry, 166n69
Blix, Göran, 188n17
Blot, Jean-Baptiste, 130
BNF (Bibliothèque nationale de France), 3, 143, 167n1

Boissonnas, Frédéric, 91, 180n2; Pinède basin, 95; *Caisson ordinaire: Execution des déblais* (with Studio Nadar, Detaille), 99, *100*
Book of Photographs from the Sam Wagstaff Collection, A (exhibition), 149
Boutan, Louis, 99
Braive, Michel-François, 143–44
Bright, Henry, 37
British Museum, 164n50
Bunsen, Robert, 47, 55, 58, 105, 114
Butterfield, Herbert, 174n13

Cadava, Eduardo, 137, 157n6, 187n2
Cain, Julien, 143
calotypy, 87
Canales, Jimena, 86
capital, 11, 52, 79, 185n49
carbon arc lamp, 54–55, 57
Carey, James, 70–71, 174n13
Caselli, Giovanni, 128, 129–30, 189n24
catacombs, 91–93, 106, *111*, 111–20, *115*, *116*, *117*, *118*, 138, 142
Catholicism, 191n63
Cercle de la presse scientifique, 44, 45, 47, 49, 55, 59, 64, 65–66; *Le premier essai de photographie à la lumière électrique au Cercle de la Presse scientifique, avril 1859* (Nadar), 47, *48*
Chevreul, Michel-Eugène, 13; centenary, 69, 71, 78–81, 90, 135; color theory, 174n15; experimentation, 180n79; image of, 70; life, 78, 80, 83–84; Nadar and, 81; Niépce and, 180n75; photographic interview, 67, 70, 71, 72, 86–87, 89, 166n67, 179n69; photographs of, 71–72, 77, 78, 81–83, *82*, 177n41; photography's invention and, 87; reception, 78–81; *Cérémonie du centenaire de Chevreul*, 78, *79*; *Meeting of Chinese ambassador [Xu Jingcheng] and Michel-Eugène Chevreul at Atelier Nadar* (Nadar, Paul), *82*, 82–83
China, 82, 178n55
cholera, 106
Chun, Wendy Hui Kyong, 10, 159n22
civic engineering, 103, 183n22, 183n26
Clark, Catherine E., 84, 164n54
CNAM (Conservatoire national des arts et métiers), 33, 41, 110, 143
colonialism in France, 14, 24, 94, 120

color theories, 78
Columbus, Christopher, 25
Compagnie d'aérostiers, 24
conceptual art, 148
Conservatoire national des arts et métiers (CNAM), 33, 41, 110, 143
Conté, Nicolas-Jacques, 189n41
Corps des ponts et chaussées, 14, 96, 99
Couder, Auguste, 74
Cousin, Jules, 33
Crimp, Douglas, 147–48, 196n54
Cromer, Gabriel, 193n20

Daguerre, Louis-Mandé, 87, 180n73
daguerrotypy, 74, *127*; "Balzac and the Daguerreotype" (Nadar), 122, 126; France and, 124; materiality, 87; sound, 77, 175nn29–30
d'Arnoux, Charles Albert: "Les Nadaréostats," *Journal amusant*, no. 403 (September 19, 1863), *29*
Daston, Lorraine, 4, 88
Daumier, Honoré, *Nadar élevant la photographie à la hauteur de l'Art*, 38, *38*–39
David, Jacques-Louis, 85
Davy, Humphry, 55, 170n24
death, 37, 114–16; funerals, 178n48. See also catacombs
Delaage, Henry, 64–65; *Portrait d'Henry Delaage, h.e. [homme] de lettres // obtenu à la lumière électrique, avec renvois reflets et intermédiaires en glace dépolie // bd de Capucines 35, en 1859, ou 60 ou 61 par moi* (Nadar), 52, *53*, 59
de La Landelle, Gabriel, 166n69; on aluminum, 162n28; helicopter, 37; on invention, 36, 37; Nadar and, 26; on science, 35
Delporte, Christian, 175n24
Deprez, Marcel, 128, 129–30
Detaille, Fernand, 91, 95, 99, 180n2; Pinède basin, 95; *Caisson ordinaire: Execution des déblais* (with Studio Nadar, Boissonas), 99, *100*
Disdéri, André-Adolphe-Eugène, 134, 142
Doré, Gustave, *Vient de paraître: Le 1er n° de l'Aéronaute, moniteur de la Société générale d'aérostation et d'automation aérienne*, 27, *28*, 163n38
Dupuis-Delcourt, Jules-François, 32

Eastman, George, 76
Eastman film, 67, 72, 85, 176n39
Eastman Kodak, 9, 13, 76–77, 176n40; Paris, 69
École des ponts et chaussées, 103
economy, 101, 120, 168n7
Edelman, Bernard, 62, 168n4, 172n47
Edwards, Elizabeth, 173n9
Edwards, Steve, 11
Eigen, Edward, 182n17
electric light: carbon arc lamp, 54–55; construction, 101; firstness, 13, 65–66, 67, 110, 137; legal aspects, 60; patents, 55, 59, 169n17; photography, 87; portraits, 64–65; sewers and, 107; text and, 114; underwater photography, 104; Van der Weyde, 60–62; *Autoportrait de Pierre Petit dans son atelier, posant avec son matériel de lumière électrique* (Petit), *61*; *Dr. Trousseau* (Nadar), 60; "L'ossuaire (D'après la photographie faite à la lumière électrique par M. Nadar)" (Linton), 117, *117*; *Mrs. Langtry* (Van der Weyde), *63*
electric light photography, 57; Nadar's "first," 13, 44, 47–49, 50, 53, 56; patents, 13, 44–46, 50, 53; prints, 49–51; *Dr. Trousseau* (Nadar), 52, *53*; *La main de M. D*** banquier (étude chirographique), cliché obtenu à la lumière diurne, épreuve tirée en une heure à la lumière électrique*, 51, 51–52; *Le premier essai de photographie à la lumière électrique au Cercle de la Presse scientifique, avril 1859* (Nadar), 47, *48*; *Mario Uchard par la lumière électrique* (Nadar), 52–53, *54*; *Portrait d'Henry Delaage, h.e. [homme] de lettres // obtenu à la lumière électrique, avec renvois reflets et intermédiaires en glace dépolie //bd de Capucines 35, en 1859, ou 60 ou 61 par moi* (Nadar), 52, 53
engineering, 103
environment and infrastructure, 103
Exposition internationale d'électricité, 129, 130, 172n44
Exposition universelle (1855), 20
Exposition universelle (1867), 106

INDEX 219

Exposition universelle (1878), 39, 173n10
Exposition universelle (1889), 39, 70, 80, 84, 173n10

Fagel, Léon, 80
Faraday, Michael, 170n24
Fassy, Paul, 185n52
fidelity, 13, 83, 86
Figuier, Louis, 47
firstness, 9–10, 12; aerial photography, 12–13, 16, 18, 19, 20, 21, 24–26, 29, 34, 37, 39–41, 43, 44, 48, 59, 67, 87, 106, 118, 137, 139, 140, 185n46; canonization, 14; documentation, 48; electric light, 13, 65–66, 67, 110; excess and, 42; historical context, 2; materiality, 2; Nadar and, 1–2, 4, 9, 12–13, 15, 16, 19, 37–38, 39–41, 43, 45, 90, 93, 94, 99, 110, 111, 112, 121–23, 130; newness, 160n27; as object of study, 15; other firsts begotten by, 112; patents, 59; as process, 119; qualification, 97–99; as relation, 110; reproduction vs., 140; rhetoric of, 141; underwater photography, 13–14; as witness, 88
Flaherty, Robert, *Nanook of the North*, 123
Flint, Kate, 172n46
Fouque, Victor, 179n73
Fourcroy, Comte de, 131, 189n41
France: balloon experiments, 24–25; cartography, 20; colonialism, 14, 24, 94, 120; Corps des ponts et chaussées, 14; daguerrotypy and, 124; economy, 101, 120, 168n7; French Revolution, 24, 78, 131; heritage, 12; history, 70; imperialism, 94; industrialization, 10; journalism, 73; laws, 169n8; patents, 21–22; representations, 93; Republics, 70, 84, 174n10; Second Empire, 107, 144; self-image, 10. *See also* Marseille; Paris
French Primitive Photography (exhibition), 144, 147, 194n26
Fresnel, Augustin-Jean, 171n38
Freund, Gisèle, 3, 142, 157n6
Froment, Paul-Gustave, 128, 188n24
future, 125–26, 135

Gaiffe, Adolphe, 47, 55
Gaines, Jane M., 168n4
Galison, Peter, 88
Gazebon, 128–30
Géant (balloon), 12, 18, 29–32, 164n45; *Ballonneau gonflé servant à Nadar de modèle pour le tableau "Le trainage du Géant"* (Nadar), 29–30, *30*; *Mémoires du Géant: À terre & en l'air*, 20, 74–75, 161n6. *See also* balloons
Gervais, Thierry, 77
Giffard, Henri, 39
Girardin, Émile de, 136
Gitelman, Lisa, 10
Gobelins Manufactory, 78
Godard, Jules, 31, 164n42
Godard, Louis, 31, 164n42
Grignon, Gustave, 174n15
Grimm, Thomas, 67
Grove, William Robert, 171n38
Grundberg, Andy, 146, 195n39

Haiti, 24, 94
Hanover crash (*Géant*), 31
Harvey, David, 185n50
Haussmann, Baron, 106
helicopters: invention, 37; Nadar and, 12, 26–27, 30; representations, 19; *Maquette d'hélicoptère à vapeur de Ponton d'Amécourt* (Nadar), *28*
heliography, 87, 180n73
Henri, Florence, 180n1
Henry, Paul, 131
Henry, Prosper, 131
history: documentation, 84; historiography, 15, 157n3, 160n34, 191n53; images and, 84–85; invention and, 122; North American photography, 141; writing, 133–34
Holmes, Oliver Wendell, 161n7
Horeau, Hector, 135, 191n65
Hugo, Victor, *Les Misérables*, 71, 135, 184n42

Ihl, Oliver, 178n50
imperialism, 94
industrialization, 3, 10, 59, 131
infrastructure: civil, 94, 99, 105; environment and, 103; images, 96; Marseille, 94–95; ontology, 110; photography, 99, 104; politics, 181n4; scholarship on, 104; sewers, 110–11; technology and, 107

intellectual property, 45, 49
invention, 2, 36; creativity, 125; de La Landelle on, 36, 37; Europe, 25; fixing, 36; future and, 125–26; historicity and, 122; inventors, 25, 163n30; legal aspects, 21–22; listed, 128–30; materialization, 45; mediation, 18, 19; narrative, 18, 132; representation and, 23; science fiction, 162n21; theories, 36–37

Jameson, Fredric, 192n67
Jammes, André, 3, 143–45, 146, 147, 193n20, 194n23, 195n42, 196n60
Janssen, Jules, 86, 131
Jefferson, Thomas, 161n7
Jenkins, Reese, 176n40
Jestaz, Juliette, 165n59
Journal amusant, 27, *29*
journalism, 47, 49, 69, 70–78, 118–19, 174n19, 175n23
J. Paul Getty Museum, 145

Kaplan, Caren, 42
Keller, Ulrich, 88
Kluitenberg, Eric, 163n37
Kodak. *See* Eastman Kodak
Koselleck, Reinhart, 132
Kozloff, Max, 148
Krakow, Stanislas, 77
Krauss, Rosalind, 137, 148, 194n37

labor: death, 114–16; photography, 93, 96, *97*, 104, 108; representations, 116
Lacoste, Anne, 151
L'aéronaute (journal), 27, *28*, 34
Lamé-Fleury, Ernest, 111, 112
La nature (journal), 182n16
land surveying, 20
Larkin, Brian, 110, 181n4, 183n27
Larouaire, Léon, 96–97
Laussedat, Aimé, 41, 110, 185n47
Leclanché, George, 128
Leclanché, Leopold, 188n24
Le courier français (journal), 80; "Un siècle! . . ." (Willette), 80, *81*
Lecouturier, Henri, 47
legal disputes: electric light photography, 44–45; intellectual property, 45; as legal texts, 168n4; Liébert and Petit, 13, 44–45
Le Gray, Gustave, 56, 134

Le journal illustré (journal), 67, 72, 77
Lemmonyer, Émile, 64
Le monde illustré (newspaper), 116–17
Le petit journal (journal), 67
Le petit marseillais, 91, 96, *98*, *101*
Lerner, Jillian, 159n17
Les Inventions Nouvelles (journal), 37–38
Les Nadar: Une légende photographique (exhibition), 8, 151
Levy, Julien, 196n56
liberalism, 46–47
Library of Congress, 164n50
Liébert, Alphonse, 44, 58, 59, 60–62, 63, 64, 65
Liesegang, Raphael Eduard, 130
Lifton, Henry Duff, "L'ossuaire (D'après la photographie faite à la lumière électrique par M. Nadar)," 117, *117*
light: actinic, 54–55; carbon arc lamp, 54–55, 57; London, 57; Van der Weyde, 57–58, 59. *See also* electric light
L'illustration (journal), 83, 173n2
linography, 60
Lissajous, Jules Antoine, 74, 75, 176n32
lithography, 134
Liu, Alan, 187n5
London, 57
Loveday, Kiki, 158n11
Lucretius, 188n15
Lumière, Antoine, 172n43

machines heavier than air, 25
Manuel, Frank, 190n47
Mapplethorpe, Robert, 149–50
maps, 20, 21, 22; *Plan du port du Marseille indiquant les travaux projetés*, 95
Marconi, Guglielmo, 130
Marey, Étienne-Jules, 131, 190n44
Marseille, 91, 93–94, *100*; infrastructure, 94–95; Nadar and, 94–105; port, 13–14; *Embarcadère des blocs artificiels* (Terris), *102*; *Plan du port du Marseille indiquant les travaux projetés*, 95; *Travau sous la mer à Marseille (Cap Pinède)* (Nadar), 96, *97*
Mauclerc, M., 128
McCauley, Elizabeth Anne, 160n33, 171n31
media: consumption, 69; fidelity, 13; production, 69. *See also* new media

mediation: aerial photography, 16, 18, 25–26, 28, 29, 30–31, 34, 43; archives, 19; change (index of), 34; construction and, 103; electric light, 114; firstness, 2, 29, 45, 65, 97, 138, 139; immediacy, 83; interviews, 69, 73; invention, 18, 19, 46, 48, 49, 64; media history, 10, 11; Nadar family, 38, 65, 66, 73, 139; newness, 126; objectivity, 71; patents, 21, 49; priority, 71; remediation, 27; sites of, 12; storytelling, 37; underground and, 114
Menut, Adolphe, 56
Metropolitan Museum of Art (MoMA), 139, 145, 150
M. K. *See* Andraud, Antoine
Modernism, 3, 93, 94
modernity, 10, 103, 105, 135
Moitessier, Albert, 170n18
MoMA (Metropolitan Museum of Art), 139, 145, 150
Monnier, Adrienne, 157n6
movement, representation of, 86
Muray, Philippe, 112, 185n55
musée Carnavalet, 19, 143
Musée d'Orsay, 150
Muséum national d'Histoire naturelle, 80
music notation, 73

Nadar (exhibition), 150
Nadar, Adrien, 5–7, 8, 151, 159n14
Nadar, Atelier: *Caisson ordinaire: Execution des déblais* (with Boissonas, Detaille), 99, *100*; *Cérémonie du centenaire de Chevreul*, 78, *79*; *Meeting of Chinese ambassador [Xu Jingcheng] and Michel-Eugène Chevreul at Atelier Nadar*, *82*, 82–83
Nadar, Félix, 140–41; aerial interests, 18–19, 25, 27, 31, 32, 34, 36, 37–38, 42–43, 44, 48, 106, 110, 139, 161nn6–7, 164n45, 167n73, 186n65; aerial photography, 12–13, 16–19, 20, 27–30; Andraud and, 20–21; archive, 3, 19; cameras, 167n78; Chevreul and, 87, 89; collections, 32–34, 142–45, 151, 165nn56–57, 193n20; on creation, 125; daguerrotypy, 20, 31, 128, *129*; de La Landelle and, 26; descriptions of, 1; on dreaming, 130; electric light, 59, 65, 66, 93, 104, 107, 110, 111, 112, 114, 169n17; experiments, 56; family, 73, 128, 142–43, 151; finances, 106; firstness, 1–2, 4, 9, 12–13, 15, 16, 19, 37–38, 39–41, 43, 45, 90, 93, 94, 99, 110, 111, 112, 121–23, 130; Gazebon and, 128–29; helicopter photography, 12, 30; helicopters and, 26–27, 30; historiography, 4–5, 18; history and, 137, 140, 142; ideas for, 28; infrastructure, 14; inventing figure of, 4, 146; inventions, 11, 21, 128, 130, 132, 135–36; Lamé-Fleury and, 111; legacy of work, 93, 119; legal actions and patents, 5–8, 12, 19, 21, *22*, 23, 24, 45, 65, 66, 164n42; life, 128, 139, 142, 149; Marseille and, 93, 94–105, 180n2; mediation, 30–31; misdatings, 106; modernism, 12; museum(s), 12, 19; name, 5–7, 8, 16, 158n14; on newness, 124; newspaper, 12; objects, 137–38; Paris sewers, 91–93; photographic interview, 70, 71, 72, 80–81, 86–87, 89, 166n67; photographic work, 2–3; Pinède basin, 95; portraits, 2, 3–4, 27, 151, *151*; priority, 5, 7, 8, 9, 11, 12, 64; reception, 141, 145, 147–51; reception of work, 3, 40, 139; relationships, 31; representations, 4, *38*, 38–39; scholarship on, 196n61; self-fashioning, 5; self-images, 90; sense of significance, 7, 8, 19, 46, 145; sewers, 105–11, 114; sewers and catacombs, 14; signature, 5, 39, 47–48, 59, 91, 96, 110, 141, 159n17; on sound, 74; as storyteller, 195n54; studio(s), 5–7, *6*, 8–9, 11–12, 47, 56–57, 66, 78, 81, 89, 91, *92*, *93*, 128, 142–43, 151, 171n31, 182n13; subject position, 136, 196n54; temporality, 137; underground photography, 106, 130; *Panthéon Nadar*, 7, *7*, 8, 143
Nadar, Félix, photography and writings: "Balzac and the Daguerreotype" (Nadar), 122–26, *127*, 131; catacombs, 91–93, 106, *111*, 111–20, *115*, *116*, *117*, *118*, 142; "1830 and Thereabouts," 134–35; "The First Attempt at Aerostatic Photography," 40–41; "L'art de vivre cent ans" (with Paul Nadar), 67; "L'emploi de l'air comprimé dans les travaux publics," 103–4; "Manifesto of Aerial Locomotion," 33, 34; "The Primitives of

Photography," 133; writings in general, 50; 1ères [Premières] épreuves en ballon: Trois vues aériennes de Paris, 41–42, 42; Atelier de Nadar au 21, rue de Noailles à Marseille, 91, 92; Atelier de Nadar au 35, boulevard des Capucines à Paris, 6; Ballonneau gonflé servant à Nadar de modèle pour le tableau "Le trainage du Géant," 29–30, 30; Brevet d'invention de quinze ans déposé le 23 octobre 1858 par Félix Tournachon, dit Nadar, pour un Système de photographie aérostatique (n° 1BB38509), 22; Catacombes de Paris: Crypte n°9, 118; Catacombes de Paris: Façade n°11, 112; Catacombes de Paris: Façade n°12, 115; Catacombes de Paris: Mannequin n°12, 116; Dr. Trousseau, 52, 53, 60; Égouts de Paris: Chambre du Pont Notre-Dame, n°1, 107, 108; Égouts de Paris: Éclusée n°1, 107, 108; Égouts de Paris: Rue de Château d'Eau, n°2, 109; La main de M. D*** banquier (étude chirographique), cliché obtenu à la lumière diurne, épreuve tirée en une heure à la lumière électrique, 51, 51–52; Le droit au vol, 35, 163n40; Maquette d'hélicoptère à vapeur de Ponton d'Amécourt, 28; Le premier essai de photographie à la lumière électrique au Cercle de la Presse scientifique, avril 1859, 47, 48; Mario Uchard par la lumière électrique, 52–53, 54; Mémoires du Géant: À terre & en l'air, 20, 24, 32, 34–35, 40, 74–75, 161n6; Nadar élevant la photographie à la hauteur de l'Art (Daumier), 38, 38–39; Paris-Guide, 106, 113; Portrait d'Henry Delaage, h.e. [homme] de lettres // obtenu à la lumière électrique, avec renvois reflets et intermédiaires en glace dépolie // bd de Capucines 35, en 1859, ou 60 ou 61 par moi, 52, 53; Premier résultat de photographie aérostatique // Applications : Cadastre, Stratégie, etc // Cliché obtenu à l'altitude de 520 m par Nadar 1858, 16–18, 17; Quand j'étais photographe, 14, 41, 122, 123–24, 127, 128, 130, 131, 136, 137, 138, 141, 142, 161n6, 184n33, 186n2; Travau sous la mer à Marseille (Cap Pinède), 96, 97; Vues aériennes du quartier de l'Étoile à Paris, 39, 40

Nadar, Marthe, 9, 143, 144
Nadar, Paul: aerial photography, 39–40, 41; Chevreul (interview with), 13; history and, 141; "L'art de vivre cent ans" (with Félix Nadar), 67; life, 143; name, 8; Paris-Photographe (journal), 14, 41, 122, 126, 131, 133, 175n24, 184n33, 187n2, 187n6; photographic interview, 67; portraits, 81; on proofs, 76; studio, 8–9, 91, 141, 142; Meeting of Chinese ambassador [Xu Jingcheng] and Michel-Eugène Chevreul at Atelier Nadar, 82, 82–83; Premier résultat de photographie aérostatique // Applications : Cadastre, Stratégie, etc // Cliché obtenu à l'altitude de 520 m par Nadar 1858 (enlargement of Nadar's), 17
Nadar, Photographer (exhibition), 145
Nanook of the North (Flaherty), 123
Napoleon I, 20
natural light, photography, 44, 51, 52, 53, 57, 58
nature, 140; human and, 124
Nesbit, Molly, 160n33
Newhall, Beaumont, 139, 141–42, 192n2
new media, 10; commercial media, 159n22; discourse and ideology, 13; experience and, 137; historical experience, 14; narratives, 187n5; proliferation, 127; representation, 126
newness, 10, 111; already existing vs., 111; firstness, 160n27; mediation, 94; photographic interview, 72–73; power of, 123
Niépce, Nicéphore, 87, 132, 180n73
Niépce to Atget: Photography's First Century (exhibition), 147
nineteenth century: acceleration, 136; civics, 112–13; history of photography and, 132; United States, 147
North America, history and, 141
novelty, 21, 45, 46, 58, 62, 63, 64, 77, 81, 87, 88, 89, 90, 101, 103, 111, 122, 123, 126, 130, 168n5, 177n43; to acculturation, 126; patents, 66; technical, 122

objectivity, 70–71, 88
October (journal), 3, 147–48, 157n9, 195n45

originality, 4, 146
Owens, Craig, 147–48

panoramas, 80
Papin, Denis, 25
Paris: aerial photography, 16–18, *17*, 39; Arc de triomphe, 39; catacombs, 91–93, *111*, 111–20, *115*, *116*, *117*, *118*; Chevreul and, 80; Commune (1871), 110; economy, 94; Étoile (Quartier de), 39, *40*; infrastructure, 14, 181n3; light, 57, 58; Nadar's work held, 13; photography scene, 9; renovations, 107; sewers, 91–93; sewers and catacombs, 14, 105–11; siege (1870–71), 33, 161nn7; studios, 137; *Égouts de Paris: Chambre du Pont Notre-Dame, n°1* (Nadar), 107, *108*; *Égouts de Paris: Éclusée n°1* (Nadar), 107, *108*; *Égouts de Paris: Rue de Château d'Eau, n°2* (Nadar), *109*. See also France
Pasteur, Louis, 192n70
patents, 19, 21–22; applications, 50, 64; authorship, 147; electric light, 55, 59, 64, 169n17; electric light photography, 13, 44–46, 50, 53; firstness, 59; France, 63; law, 65, 162n17, 169n8; litigation, 46; Nadar, 50, 53–54; novelty, 45, 46, 66; photography, 49; private aspects of process, 46–47; revisions to process, 168n7; *Brevet d'invention de quinze ans déposé le 23 octobre 1858 par Félix Tournachon, dit Nadar, pour un Système de photographie aérostatique (n° 1BB38509)* (Nadar), *22*
Pereire brothers, 171n31
Petit, Pierre, 44, 172n52; electric light, 62; patents, 64; *Autoportrait de Pierre Petit dans son atelier, posant avec son matériel de lumière électrique*, *61*
Philadelphia Museum of Art, 147
phonoautograph, 75
phonography, 77, 130
photographic interview, 78; Chevreul and, 67, 70, 71, 72, 86–87, 89, 166n67; commemoration, 69; epistemology, 84; first, 67–69, 72, 83, 85; future and, 88–90; historical context, 69–70; "L'art de vivre cent ans" (Nadar and Nadar), 67; mediation, 83, 88; newness, 73, 78; novelty, 90; priority, 71; reception of, 72–73; representation, 75, 83, 85, 86; scholarship on, 69
photography: as allegory, 121–22; as art, 169n16; art history and, 146, 149; authorship, 149; balloons, 16, 18–19, 21, 23, 27, 29–32, 33, 39, 40, 41–42, *42*, 96, 192n4; canonization, 2; capitalism, 11, 52; color, 134; death and, 137; detail, 52; economics, 11, 45, 62, 144; efficiency of, 20; electric light, 57, 58, 59, 87; epistemology, 125; as evidence, 84, 179n59; experimentation and, 133–34; exposure time, 76, 108; firstness, 121–23, 127; future and, 137; historiography, 9, 13, 14; history of, 2, 4, 11, 71, 87, 124, 132, 140, 141, 142, 148, 150, 190n44; infrastructure, 93, 99, 102, 103, 104, 111, 119–20; instantaneity, 77, 85, 86, 177n43; invention, 87; journalism, 77, 118–19; labor, 93, 96, *97*, 104, 108; light, 44, 54–56, 58–59; manuals, 132, 133; maps, 21; materiality, 124–25; mediation, 28, 88, 117; as mediation, 122; natural light, 44, 51, 52, *53*; negatives, 49, 50, 52, 56, 119, 134, 143, 145; as new media, 122–27, 138, 148; newness, 10; newspapers and, 116–18; objectivity, 13, 85, 86; ontology, 148; origins, 9, 166n68; Paris and, 142; patents, 13, 62; Pinède basin, 95; portraits, 52; printing process, 49–50; representation, 88, 124, 195n39; reproductions, 60; roll system, 76, 176n40; science, 76, 86, 179n69; truth, 85; underwater, 96–99; United States, 147, 150
Photography 1839–1937 (exhibition), 139, 193n7
Pierce, J. Mackenzie, 74
Pinède docks, 99–103, *100*, 103, 107
Pinney, Christopher, 123, 157n4, 187n8
Plan du port du Marseille indiquant les travaux projetés, *95*
Ponton d'Amécourt, Gustave Vicomte de, 26, 27, 30, 37, 162n29, 163n30,

166n69; helicopter, 37; *Maquette d'hélicoptère à vapeur de Ponton d'Amécourt*, 28
positivism, 84
Pottage, Alain, 22–23, 162n20
prior art (patents), 45
progress, 35–36, 70
property relations, 62
public domain, 22
public health, 113
public spectacles, 79

religion, 135, 191n63
remediation, 27
Rencontre d'Arles, 157n2
representations: of accidents, 32; aerial photography, 19; balloons, 19, 24; France, 93; helicopters, 19; images, 85; invention and, 23; labor, 116; movement of, 86; Nadar, 4, 38–39; new media, 126; photographic interview, 75, 83, 85, 86; photography, 88, 124, 195n39; sound, 75; technology and, 122
reproductions: firstness vs., 140; mechanical, 41, 69, 71, 72, 77–78, 80, 126, 134; originality and reproduction (Sterne), 69; sound, 75, 176n33
Republicanism, 78–79, 85
revolutions: 1789, 80, 85; 1889, 84; French, 24, 78, 131; Haitian, 94
Reynes, Geneviève, 173n5
Robertson, Craig, 193n6
Roubaud, Félix, 47
Royal Society, 55
Russell, John, 145
Russia, 31

Saint-Simon, Henri de, 191n63
Sand, George, 35
Schaffer, Simon, 18, 36, 37
Schudson, Michael, 73
Schwartz, Vanessa R., 87
science: capital and, 79; discovery (theories of), 36; evidence, 87; facts, 84; objectivity, 76; progress, 35–36; proof, 76; scientists, 131–32
science fiction, 162n21
Scott de Martinville, Édouard-Léon, 75
Seillan, Jean-Marie, 174n19, 175n23

Sekula, Allan, 167n83, 192n4
self-canonization, 19
self-fashioning, 5
self-historicization, 10, 146
self-image, 10
self-images, 90
self-portraiture, 4, 5
self-positioning, 137, 146
self-promotion, 141
self-reflexivity, 148
Seniuta, Isabella, 144
Serrin, Victor, 47, 55, 170n25
sewers, 105–11, 137–38, 183n29, 184n43; electric light and, 107; infrastructure, 110–11; *Égouts de Paris: Chambre du Pont Notre-Dame, n°1* (Nadar), 107, *108*; *Égouts de Paris: Éclusée n°1* (Nadar), 107, *108*; *Égouts de Paris: Rue de Château d'Eau, n°2* (Nadar), *109*
SFP (Société française de la photographie), 51, 76–77, 134, 143, 157n2, 169n13
Sharma, Sarah, 136
Sherman, Brad, 22–23, 162n20
Siegel, Steffen, 9
Sino-French War, 82, 83
Sirot, George, 193n20
Smeaton, John, 103–4, 183n26
socialism, 135, 184n43
Société aérostatique et météorologique de France, 32, 34
Société française de la photographie (SFP), 51, 76–77, 134, 143, 157n2, 169n13
Solomon-Godeau, Abigail, 11, 145–46, 194n37, 195n42
sound: daguerrotypy, 77, 175nn29–30; recording, 74; representation, 75; reproduction, 75, 77, 88, 176n33; research, 75; transmission, 75; visualizing, 74, 176n32
spectators, 69, 173n6
speculation, 11, 18, 20, 21, 24, 33, 46, 88, 97, 128–31, 146
speed theory, 136
Star, Susan Leigh, 183n27
steam power, 135
Steichen, Edward, 192n4
stenography, 67, 72, 73–74, 77, 83, 84
Sterne, Jonathan, 69, 176n33

INDEX 225

Suez Canal, 94
Szathmari, Charles, 147

Tainter, Sumner, 129
Talbot, Henry Fox, 87, 132, 180n73
Taussig, Michael, 123
Taws, Richard, 189n41
technicity, 2, 116, 138
technology: acceleration, 136; change, 135; construction and, 99; infrastructure and, 107; modernity, 122; proliferation, 127; representation and, 122
telegraphy, 128, 130
temporality, 80
Tennis Court Oath (1789), 85
Terris, Adolphe, 103, 182n19; *Embarcadère des blocs artificiels*, 102, *102*
textiles, 78
théâtrophone, 74, 75
Theodoratou, Liana, 187n2
Third Republic, 13
Thompson, William, 99
Tissandier, Gaston, 32, 40, 58, 164n50
Tournachon, Adrien, 30
Tournachon, Ernestine, 11
Tournachon, Gaspard-Félix. *See* Nadar, Félix
Trachtenberg, Alan, 178n56
Travis, David, 147
Treaty of Tientsin, 83
Trouillot, Michel-Rolph, 157n3
Trousseau, Armand, *Dr. Trousseau* (Nadar), 52, *53*
Trouvé, Gustave, 128, 188n24
truth, 66, 70, 71
Tseng, Shao-Chien, 185n53, 186n69
Tupicoff, Dennis, 89, 180n80
Turnor, Christopher Hatton, 163n38

Uchard, Mario, *Mario Uchard par la lumière électrique* (Nadar), 52–53, *54*

underground photography, 93, 105–10, 111–19
underwater construction, 103–4
underwater photography, 13–14, 93–99, 104, 106, 130, 182n13
United States, 73
Universal Exhibition, 74
Un laboratoire des premières fois (exhibition), 157n2

Van der Weyde, Henry: economics, 58; electric light, 60–62; inventions, 171n38; legal disputes, 64; light, 57–58, 59; patents, 57–58, 171n40; studio, 62; *Mrs. Langtry*, *63*
Van Drie, Melissa, 176n35
Van Monckhoven, Désiré, 131–33, 190n49
Verne, Jules, 96, 99, 181n12
visual culture, 78, 144

Wagstaff, Samuel, 3, 143–45, 146, 149–51
Walker, William, 76
Wallace, Amy, 171n30
Warnerke, Leon, 76
waste, 105, 106–7
water, photography under, 13–14, 93–99, 104, 106, 130, 182n13
Weiss, Sean, 103, 183n22
Willette, Adolphe, "Un siècle! . . . ," 80, *81*
Williams, Rosalind, 114, 182n17
Wilson, E. L., 58
World War I, 139
World War II, 3, 167n83

Xu Jingcheng, 82–83, 178n54; *Meeting of Chinese ambassador [Xu Jingcheng] and Michel-Eugène Chevreul at Atelier Nadar* (Nadar, Paul), *82*, 82–83